H.L. Cycon R.G. Froese W. Kirsch
B. Simon

Schrödinger Operators

with Application to Quantum Mechanics and Global Geometry

With 2 Figures

Springer-Verlag
Berlin Heidelberg New York
London Paris Tokyo

Dr. Hans L. Cycon
Technische Universität Berlin
Fachbereich 3 – Mathematik
Straße des 17. Juni 135, D-1000 Berlin 12

Dr. Richard G. Froese
Department of Mathematics
University of British Columbia
Vancouver, B.C., Canada V6T 1W5

Professor Dr. Werner Kirsch
Institut für Mathematik
Universität Bochum
D-4630 Bochum, Fed. Rep. of Germany

Professor Dr. Barry Simon
California Institute of Technology
Department of Mathematics 253-37
Pasadena, CA 91125, USA

Editors

Wolf Beiglböck
Institut für Angewandte Mathematik
Universität Heidelberg
Im Neuenheimer Feld 294
D-6900 Heidelberg 1
Fed. Rep. of Germany

Joseph L. Birman
Department of Physics, The City College
of the City University of New York
New York, NY 10031, USA

Robert Geroch
University of Chicago
Enrico Fermi Institute
5640 Ellis Ave.
Chicago, IL 60637, USA

Elliott H. Lieb
Department of Physics
Joseph Henry Laboratories
Princeton University
Princeton, NJ 08540, USA

Tullio Regge
Istituto di Fisica Teorica
Università di Torino, C. so M. d'Azeglio, 46
I-10125 Torino, Italy

Walter Thirring
Institut für Theoretische Physik
der Universität Wien, Boltzmanngasse 5
A-1090 Wien, Austria

ISBN 3-540-16758-7 Springer-Verlag Berlin Heidelberg New York
ISBN 0-387-16758-7 Springer-Verlag New York Berlin Heidelberg

Library of Congress Cataloging-in-Publication Data. Schrödinger operators, with application to quantum mechanics and global geometry. (Texts and monographs in physics) Chapters 1–11 are revised notes taken from a summer course given in 1982 in Thurnau, West Germany by Barry Simon. "Springer Study Edition". Bibliography: p. Includes index. 1. Schrödinger operator. 2. Quantum theory. 3. Global differential geometry. I. Cycon, H.L. (Hans Ludwig), 1942-. II. Simon, Barry. III. Series. QC174.17.S6S37 1987 515.7'246 86-13953

Typesetting: ASCO Trade Typesetting Limited, Hongkong
Offset printing: Druckhaus Beltz, 6944 Hemsbach
Bookbinding: J. Schäffer GmbH & Co. KG, 6718 Grünstadt
2153/3150-543210

Preface

In the summer of 1982, I gave a course of lectures in a castle in the small town of Thurnau outside of Bayreuth, West Germany, whose university hosted the lecture series. The Summer School was supported by the Volkswagen foundation and organized by Professor C. Simader, assisted by Dr. H. Leinfelder. I am grateful to these institutions and individuals for making the school, and thus this monograph, possible.

About 40 students took part in a grueling schedule involving about 45 hours of lectures spread over eight days! My goal was to survey the theory of Schrödinger operators emphasizing recent results. While I would emphasize that one was not supposed to know all of Volumes 1 – 4 of Reed and Simon (as some of the students feared!), a strong grounding in basic functional analysis and some previous exposure to Schrödinger operators was useful to the students, and will be useful to the reader of this monograph.

Loosely speaking, Chaps. 1 – 11 of this monograph represent "notes" of those lectures taken by three of the "students" who were there. While the general organization does follow mine, I would emphasize that what follows is far from a transcription of my lectures. Even with 45 hours, many details had to be skipped, and quite often Cycon, Froese and Kirsch have had to flesh out some rather dry bones. Moreover, they have occasionally rearranged my arguments, replaced them with better ones and even corrected some mistakes!

Some results such as Lieb's theorem (Theorem 3.17) that were relevant to the material of the lectures but appeared during the preparation of the monograph have been included.

Chapter 11 of the lectures concerns some beautiful ideas of Witten reducing the Morse inequalities to the calculation of the asymptotics of eigenvalues of cleverly chosen Schrödinger operators (on manifolds) in the semiclassical limit. When I understood the supersymmetric proof of the Gauss-Bonnet-Chern theorem (essentially due to Patodi) in the summer of 1984, and, in particular, using Schrödinger operator ideas found a transparent approach to its analytic part, it seemed natural to combine it with Chap. 11, and so I wrote a twelfth chapter. Since I was aware that Chaps. 11 and 12 would likely be of interest to a wider class of readers with less of an analytic background, I have included in Chap. 12 some elementary material (mainly on Sobolev estimates) that have been freely used in earlier chapters.

Los Angeles, Fall 1986 *Barry Simon*

Contents

1. Self-Adjointness

Self-adjointness of Schrödinger operators has been a fundamental mathematical problem since the beginning of quantum mechanics. It is equivalent to the unique solvability of the time-dependent Schrödinger equation, and it plays a basic role in the foundations of quantum mechanics, since only self-adjoint operators can ben understood as quantum mechanical observables (in the sense of *von Neumann* [361]).

It is an extensive subject with a large literature (see e.g. [293, 107, 196]) and the references given there), and it has been considerably overworked. There are only a few open problems, the most famous being Jörgens' conjecture (see [293, p. 339; 71, 317]).

We will not go into an exhaustive overview, but rather pick out some subjects which seen to us to be worth emphasizing. We will begin with a short review of the basic perturbation theorems and then discuss two typical classes of perturbations. Then we will discuss Kato's inequality. Finally, using an idea of Kato, we give some details of the proof of the theorem of Leinfelder and Simader on singular magnetic fields.

1.1 Basic Perturbation Theorems

First, we give some definitions (see [293, p. 162] for a more detailed discussion). We denote by A and B, densely-defined linear operators in a Hilbert space H, and by $D(A)$ and $Q(A)$, the operator domain and form domain of A respectively.

Definition 1.1. Let A be self-adjoint. Then B is said to be *A-bounded* if and only if

(i) $D(A) \subseteq D(B)$
(ii) there are constants $a, b > 0$ such that

$$\|B\varphi\| \leq a\|A\varphi\| + b\|\varphi\| \quad \text{for} \quad \varphi \in D(A) . \tag{1.1}$$

The infimum of all such a is called the *A-bound* (or relative-bound) of B. There is an analogous notion for quadratic forms:

Definition 1.2. Let A be self-adjoint and bounded from below. Then a symmetric operator B is said to be *A-form bounded* if and only if

(i) $Q(A) \subseteq Q(B)$
(ii) there are constants a, b > 0 such that

$$|\langle\varphi, B\varphi\rangle| \leq a\langle\varphi, A\varphi\rangle + b\langle\varphi,\varphi\rangle \quad \text{for} \quad \varphi \in Q(A) \ .$$

The infimum of all such a is called the *A-form-bound* (relative form-bound) of B.

Note that the operators in the above definitions do not need to be self-adjoint or symmetric [196, p. 190, p. 319]. We require it here because later propositions will be easier to state or prove for the self-adjoint case.

A subspace in H is called a *core* for A if it is dense in $D(A)$ in the graph norm. It is called a *form core* if it is dense in $Q(A)$ in the form norm.

There is an elementary criterion for relative boundedness.

Proposition 1.3. (i) Assume A to be self-adjoint and $D(A) \subseteq D(B)$. Then B is A-bounded if and only if $B(A + \mathrm{i})^{-1}$ is bounded. The A-bound of B is equal to

$$\lim_{|\gamma|\to\infty} \|B(A + \mathrm{i}\gamma)^{-1}\| \ .$$

(ii) (form version). Assume A to be self-adjoint, bounded from below and $Q(A) \subseteq Q(B)$. Then B is A-form-bounded if and only if $(A + \mathrm{i})^{-1/2} B (A + \mathrm{i})^{-1/2}$ is bounded. The A-form-bound of B is equal to

$$\lim_{|\gamma|\to\infty} \|(A + \mathrm{i}\gamma)^{-1/2} B (A + \mathrm{i}\gamma)^{-1/2}\| \ .$$

The assertion (i) can easily be seen by replacing φ by $(A + \mathrm{i}\gamma)^{-1}\psi$ in (1.1) and observing that $\|B(A + \mathrm{i}\gamma)^{-1}\| \leq [a + (b/|\gamma|)]$. (ii) follows analogously. Note that there is an extension of this notion which we use occasionally: We say that B is *A-compact* if and only if $B(A + \mathrm{i})^{-1}$ is compact. Here i can be replaced by any point of the resolvent set.

Now we will state the basic perturbation theorem which was proven by Kato over 30 years ago, and which works for most perturbations of practical interest.

Theorem 1.4 (Kato-Rellich). Suppose that A is self-adjoint, B is symmetric and A-bounded with A-bound $a < 1$. Then $A + B$ [which is defined on $D(A)$] is self adjoint, and any core for A is also a core for $A + B$.

We give a sketch of the proof. Note that self-adjointness of A is equivalent to $\mathrm{Ran}(A \pm \mathrm{i}\mu) = \mathrm{H}$ for some $\mu > 0$ [292, Theorem VIII.3]. Then, as above, we conclude from (1.1) that

$$\|B(A \pm \mathrm{i}\mu)^{-1}\| \leq a + \frac{b}{\mu} \ .$$

Thus, for μ large enough $C := B(A \pm \mathrm{i}\mu)^{-1}$ has norm less than 1, and this implies that $\mathrm{Ran}(1 + C) = \mathrm{H}$. This, together with the equation

$$(1 + C)(A \pm i\mu)\varphi = (A + B \pm i\mu)\varphi \quad \varphi \in D(A)$$

and the self-adjointness of A, implies that $\operatorname{Ran}(A + B \pm i\mu) = \mathrm{H}$. The second part of the theorem is a simple consequence of (1.1).

There are various improvements due to *Kato* [196] and *Wüst* [371] for the case $a = 1$, but in fact all the perturbations one usually deals with in the theory of Schrödinger operators have relative bound 0.

There is also a form version of Theorem 1.4 (due to Kato, Lax, Lions, Milgram and Nelson):

Theorem 1.5 (KLMN). Suppose that A is self-adjoint and bounded from below and that B is symmetric and A-bounded with form-bound $a < 1$. Then

(i) the sum of the quadratic forms of A and B is a closed symmetric form on $Q(A)$ which is bounded from below.
(ii) There exists a unique self-adjoint operator associated with this form which we call the form sum of A and B.
(iii) Any form core for A is also a form core for $A + B$.

For a proof, see [293, Theorem X.17]. We will denote the form sum by $A \dotplus B$ when we want to emphasize the form character of the sum, otherwise we will write $A + B$.

Note that in spite of the parallelism between operators and forms, there is a fundamental asymmetry. There are symmetric operators which are closed but not self-adjoint. But a closed form which is bounded from below is automatically the form of a unique, self-adjoint operator [196, Theorem VI.2.1]. The form analog of essential self-adjointness, however, does exist: a suitable set being a form core. If one defines something to be a closed quadratic form, it is automatic that the associated operator is self-adjoint—one knows nothing, however, about the operator domain or the form domain. It is therefore a nontrivial fact that a convenient set (e.g. C_0^∞) is a form core.

1.2 The Classes S_ν and K_ν

In this book, we will study the sum $-\Delta + V$ in virtually all cases. But occasionally we will also study $(-i\nabla + a)^2 + V$ as operators or forms in the Hilbert space $L^2(\mathbb{R}^\nu)$. Here V is a real-valued function on $\mathbb{R}^\nu$ describing the electrostatic potential, and a is a vector-valued function which describes the magnetic potential. We denote by H_0 the self-adjoint representation of $-\Delta$ in $L^2(\mathbb{R}^\nu)$. In reasonable cases, one can think of V as a perturbation of H_0. Physically, this is motivated by the uncertainty principle which allows the kinetic energy to control some singularities of V if they are not too severe. This phenomenon has no classical analog. This is also practical since the Laplacian has an explicit eigenfunction expansion and integral kernel, and one knows everything about operator cores, etc.

There are two classes of perturbations we will discuss here. The class S_ν, which is an (almost maximal) class of operator perturbations of H_0 and the class K_ν which is the form analog of S_ν. S_ν was introduced originally by *Stummel* [352], and has been discussed by several authors (see e.g. [308]).

Definition 1.6. Let V be a real-valued, measurable function on $\mathbb{R}^\nu$. We say that $V \in S_\nu$ if and only if

a) $$\lim_{\alpha \downarrow 0}\left[\sup_x \int_{|x-y|\le\alpha} |x-y|^{4-\nu}|V(y)|^2 d^\nu y\right] = 0 \quad \text{if} \quad \nu > 4$$

b) $$\lim_{\alpha \downarrow 0}\left[\sup_x \int_{|x-y|\le\alpha} \ln(|x-y|)^{-1}|V(y)|^2 d^\nu y\right] = 0 \quad \text{if} \quad \nu = 4$$

c) $$\sup_x \int_{|x-y|\le 1} |V(y)|^2 d^\nu y < \infty \quad \text{if} \quad \nu \le 3 \; .$$

For the reader who is disturbed by the lack of symmetry in the above definition, we remark that for $\nu \le 3$,

$$\sup_x \int_{|x-y|\le 1} |V(y)|^2 d^\nu y < \infty$$

is equivalent to

$$\lim_{\alpha \downarrow 0}\left[\sup_x \int_{|x-y|\le\alpha} |x-y|^{4-\nu}|V(y)|^2 d^\nu y\right] = 0 \; .$$

We define a S_ν-norm on S_ν by

$$\|V\|_{S_\nu} := \sup_x \int_{|x-y|\le 1} K(x, y; \nu)|V(y)|^2 d^\nu y \; ,$$

where K is the kernel in the above definition of S_ν. We now state (and prove) a theorem which shows how these quantities arise naturally. We denote, by $\|\cdot\|_{p,q}$, the operator norm for operators from $L^p(\mathbb{R}^\nu)$ to $L^q(\mathbb{R}^\nu)$, and by $\|\cdot\|_p$ the norm in $L^p(\mathbb{R}^\nu)$.

Theorem 1.7. $V \in S_\nu$ if and only if

$$\lim_{E\to\infty} \|(H_0 + E)^{-2}|V|^2\|_{\infty,\infty} = 0 \; . \tag{1.2}$$

Proof. As with all functions of H_0, $(H_0 + E)^{-2}$ is a convolution operator with an explicit kernel $Q(x-y, E)$ [293, Theorem IX.29]. It has the following properties (see [308, Theorem 3.1, Chap. 6]).

1. $Q(x-y,E) \geq 0$,

2. $$Q(x-y,E) = \begin{cases} 0(|x-y|^{4-\nu}) & \text{if} \quad \nu > 4 \\ 0(\ln|x-y|^{-1}) & \text{if} \quad \nu = 4 \\ C & \text{if} \quad \nu \leq 3 \end{cases} \quad \text{as} \quad |x-y| \to 0 ,$$

3. $\sup_{|x-y|>\delta} e^{|x-y|} Q(x-y,E) \to 0$ as $E \to \infty$, for any $\delta > 0$.

Using the elementary fact that

$$\sup_x \int_{|x-y| \leq 1} |V(y)|^2 \, dy < \infty$$

for any $V \in S_\nu$, it is not hard to see that $V \in S_\nu$ if and only if $\sup_x \int Q(x-y,E)|V(y)|^2 d^\nu y \to 0$ as $E \to \infty$. This gives the result, since $Q(\cdot - y, E)|V(y)|^2$ is a positive integral kernel and $\|A\|_{\infty,\infty} = \|A1\|_\infty$ holds for any A with positive integral kernel. □

The above result has an L^2 consequence by a standard "duality and interpolation" argument:

Corollary 1.8. If $V \in S_\nu$, then

$$\|(H_0 + E)^{-1} V\|_{2,2} \to 0 \quad \text{as} \quad E \to \infty . \tag{1.3}$$

Proof. Let $V \in S_\nu$. Then it is enough to show that

$$\|(H_0 + E)^{-1}|V|\|_{2,2}^2 \leq \|(H_0 + E)^{-2}|V|^2\|_{\infty,\infty} , \tag{1.4}$$

since (1.3) follows then by Theorem 1.7. Assume for a moment that V is bounded, and consider the function

$$F(z) := |V|^{2z}(H_0 + E)^{-2}|V|^{2-2z} \quad z \in \mathbb{C} .$$

$F(z)$ is an operator-valued function which is L^1 and L^∞-bounded and analytic in the interior of the strip $\{z \in \mathbb{C} | \operatorname{Re} z \in [0,1]\}$. Thus, by the Stein interpolation theorem [293, Theorem IX.21] and, using that (by duality)

$$\|(H_0 + E)^{-2}|V|^2\|_{\infty,\infty} = \||V|^2(H_0 + E)^{-2}\|_{1,1} ,$$

we get

$$\||V|(H_0 + E)^{-2}|V|\|_{2,2} \leq \|(H_0 + E)^{-2}|V|\|_{\infty,\infty} .$$

Since

$$\||V|(H_0 + E)^{-2}|V|\|_{2,2} = \|(H_0 + E)^{-1}|V|\|_{2,2}^2 ,$$

(1.4) follows for bounded V's, and by an approximation argument, also for all $V \in S_\nu$. □

Remark. Note that Corollary 1.8 implies that if $V \in S_\nu$, then it is H_0-bounded with H_0-bound 0 by Proposition 1.3 (Proposition 1.3 has to be slightly modified for the semibounded case we are considering here).

One might think that since S_ν is telling us something about L^∞-bounds and L^∞ is "stronger" than L^2, there would be no way going from L^2-bounds to S_ν. So the following theorem is interesting.

Theorem 1.9. Suppose there are $a, b > 0$ and a δ with $0 < \delta < 1$ such that, for all $0 < \varepsilon < 1$ and all $\varphi \in D(H_0)$

$$\|V\varphi\|_2^2 \le \varepsilon \|H_0\varphi\|_2^2 + a\exp(b\varepsilon^{-\delta})\|\varphi\|_2^2 .$$

Then $V \in S_\nu$.

Proof. We just have to pick the right φ's. Fix $y \in \mathbb{R}^\nu$, $t \in \mathbb{R}^+$, and consider the integral kernel

$$\varphi(x) := \sqrt{\exp(-tH_0)(x, y)} .$$

Then, noting that $\|\varphi\|_2 = 1$ and (by scaling)

$$\|H_0\varphi\|_2 = ct^{-2} \quad \text{for suitable} \quad c > 0$$

we have

$$[\exp(-tH_0)|V|^2](y) \le c\varepsilon t^{-2} + a\exp(b\varepsilon^{-\delta}) . \tag{1.5}$$

Now, take $\varepsilon := (1 + |\ln t|)^{-\gamma}$, where $\gamma := 2/(1 + \delta)$, and multiply (1.5) by $t\exp(-tE)$ for $E > 0$. Then the R.H.S. of (1.5) is integrable in t and its integral goes to zero as $E \to \infty$. Now if we use the identity

$$(H_0 + E)^{-2} = \int_0^\infty t\mathrm{e}^{-tH_0}\mathrm{e}^{-tE}\,dt$$

we get (1.2), and therefore $V \in S_\nu$ by Theorem 1.7. □

The second class of potentials we are considering here is K_ν, which is the form analog of S_ν. This type of potentials was first introduced by *Kato* [193]. See also *Schechter* [308] for related classes. K_ν was studied in some detail by *Aizenman* and *Simon* [7], and *Simon* [334].

Definition 1.10. Let V be a real-valued measurable function on $\mathbb{R}^\nu$. We say that $V \in K_\nu$ if and only if

a) $\lim\limits_{\alpha \downarrow 0} \left[\sup\limits_x \int\limits_{|x-y| \le \alpha} |x - y|^{2-\nu}|V(y)|\,d^\nu y\right] = 0, \quad \text{if} \quad \nu > 2$

b) $\lim_{\alpha\downarrow 0}\left[\sup_x \int_{|x-y|\leq\alpha} \ln|(x-y)|^{-1}|V(y)|d^\nu y\right] = 0, \quad \text{if} \quad \nu = 2$

c) $\sup_x \int_{|x-y|\leq 1} |V(y)|d^\nu y < \infty, \quad \text{if} \quad \nu = 1 .$

We also define a K_ν-norm by

$$\|V\|_{K_\nu} := \sup_x \int_{|x-y|\leq 1} \tilde{K}(x, y; \nu)|V(y)|d^\nu y$$

where $\tilde{K}$ is the kernel in the above definition of K_ν. Then virtually everything goes through as before.

Theorem 1.11 [7]. $V \in K_\nu$ if and only if

$$\lim_{E\to\infty} \|(H_0 + E)^{-1}|V|\|_{\infty,\infty} = 0 .$$

The proof is the same as in Theorem 1.7.

Theorem 1.12 [7]. Suppose there are $a, b > 0$ and a δ with $0 < \delta < 1$ such that, for all $0 < \varepsilon < 1$ and all $\varphi \in Q(H_0)$

$$\langle\varphi, |V|\varphi\rangle \leq \varepsilon\langle\varphi, H_0\varphi\rangle + a\exp(b\varepsilon^{-\delta})\|\varphi\|_2^2 .$$

Then $V \in K_\nu$.

The proof is again like that in Theorem 1.9 above (see also [7, Theorem 4.9]).

Remarks. (1) Both of the classes S_ν and K_ν have some nice properties:

a) If $\mu \leq \nu$, then $K_\mu \subseteq K_\nu$ and $S_\mu \subseteq S_\nu$. By these inclusions we mean the following. Suppose $W \in K_\mu$ (resp. S_μ), and there is a linear surjective map $T: \mathbb{R}^\nu \to \mathbb{R}^\mu$ and $V(x) := W(T(x))$. Then $V \in K_\nu$ (resp. S_ν). The canonical example to think of here is an N-body system with $\nu = N\mu$, where a point $x \in \mathbb{R}^\nu$ is thought of as an N-tuple of μ-dimensional vectors $x = \langle x_1, \ldots, x_N\rangle$ and $Tx := x_i - x_j$ for some $i, j \in \{1, \ldots, N\}$, $i \neq j$.

b) There are some L_p-estimates which tell you when a potential is in K_ν (resp. S_ν), i.e.

$$L^p_{\text{unif}} \subseteq S_\nu \quad \text{if} \quad \begin{cases} p > \frac{\nu}{2} & \text{for} \quad \nu \geq 4 \\ p = 2 & \text{for} \quad \nu < 4 \end{cases}$$

and

$$L^p_{\text{unif}} \subseteq K_\nu \quad \text{if} \quad \begin{cases} p > \frac{\nu}{2} & \text{for} \quad \nu \geq 2 \\ p = 2 & \text{for} \quad \nu < 2 \end{cases}$$

where

$$L^p_{\text{unif}} := \{V | \sup_x \int_{|x-y|\leq 1} |V(y)|^p \, dy < \infty\} .$$

The proof is a straightforward application of Hölder's inequality (see [7, Proposition 4.3]).

(2) If $V \in K_\nu$, then V is H_0-form bounded with relative bound 0. This follows again analogously from Proposition 1.3(ii), Theorem 1.11 and a corollary analogous to Corollary 1.8.

The classes K_ν and S_ν, however, are not the "maximal" classes with respect to the perturbation theorems, that is, one just misses the "borderline cases." This can be seen in the following:

Example. (a) Let $\nu \geq 3$ and

$$V(x) := |x|^{-2} |\ln|x||^{-\delta} .$$

Then $V \in K_\nu$ if and only if $\delta > 1$, but V is H_0-form bounded with bound 0 if and only if $\delta > 0$.

(b) Let $\nu \geq 5$ and V as in (a). Then $V \in S_\nu$ if and only if $\delta > 1/2$ but it is H_0-bounded with bound 0 if and only if $\delta > 0$. (a) is a consequence of [7, Theorem 4.11] and general perturbation properties (see [293, Chap. X.2]). (b) has a similar proof.

Remark. The above example shows that it is false that S_ν is contained in K_ν.

1.3 Kato's Inequality and All That

We will now sketch a set of ideas which go back to *Kato* [193], and which were subsequently studied by *Simon* [322, 327] (see also *Hess*, *Schrader* and *Uhlenbrock* [163]).

Let us first consider a vector potential a (magnetic potential), and a scalar V (electric potential) satisfying

$$\begin{aligned} &a \in L^2_{\text{loc}}(\mathbb{R}^\nu)^\nu \\ &V \in L^1_{\text{loc}}(\mathbb{R}^\nu), \quad V \geq 0 , \end{aligned} \tag{1.6}$$

Then the formal expression

$$\tau := (-i\nabla - a)^2 + V$$

is associated with a quadratic form $h_{\max}$ (called the maximal form) defined by

$$Q(h_{\max}) := \{\varphi \in L^2(\mathbb{R}^\nu) | (\nabla - ia)\varphi \in L^2(\mathbb{R}^\nu)^\nu, \, V^{1/2}\varphi \in L^2(\mathbb{R}^\nu)\}$$

and

$$h_{\max}(\varphi,\psi) := \sum_{j=1}^{\nu} \langle(\partial_j - ia_j)\varphi, (\partial_j - ia_j)\psi\rangle + \langle V^{1/2}\varphi, V^{1/2}\psi\rangle$$

for $\varphi, \psi \in Q(h_{\max})$; ($\partial_j := \partial/\partial x_j$). Note that $h_{\max}$ is a closed, positive form (since it is the sum of $(\nu + 1)$ positive closed forms), and therefore there exists a self-adjoint, positive operator H associated with $h_{\max}$, with

$$Q(H) = Q(h_{\max}) \quad \text{and}$$

$$\langle H\varphi, \psi\rangle = h_{\max}(\varphi,\psi) \quad \text{for} \quad \varphi, \psi \in D(H) \quad [196] \ .$$

Note also that (1.6) are the weakest possible conditions for defining a (closable positive) quadratic form associated with τ on $C_0^\infty(\mathbb{R}^\nu)$. The closure of this form [which is the restriction of $h_{\max}$ to $C_0^\infty(\mathbb{R}^\nu)$] is called $h_{\min}$. Our first theorem now says that these two forms coincide. Thus, the self-adjoint operator associated with the formal expression τ is, in a sense, unique.

Theorem 1.13 [329, 195]. $C_0^\infty(\mathbb{R}^\nu)$ is a form core for H.

We give only a sketch of the proof (see [329]).

Step 1.

$$e^{-tH}: L^2(\mathbb{R}^\nu) \to L^\infty(\mathbb{R}^\nu), \quad t \in \mathbb{R}^+ \ . \tag{1.7}$$

We only need to show that

$$|e^{-tH}\varphi| \le e^{-tH_0}|\varphi|, \quad \varphi \in L^2(\mathbb{R}^\nu) \tag{1.8}$$

(which is the semigroup version of Kato's inequality, sometimes also called Kato-Simon inequality or diamagnetic inequality; see [327]), since (1.7) follows from (1.8) by using Young's inequality and the fact that $\exp(-tH_0)$ is a convolution with an L^2-integral kernel.

We know that H is a form sum of $\nu + 1$ operators. Therefore, we can use a generalized version of Trotter's product formula (shown by *Kato* and *Masuda* [198]) and get

$$\exp(-tH) = s - \lim_{n\to\infty} \left[\exp\left(\frac{t}{n}D_1^2\right)\exp\left(\frac{t}{n}D_2^2\right)\ldots\exp\left(\frac{t}{n}D_\nu^2\right)\exp\left(-\frac{t}{n}V\right)\right]^n , \tag{1.9}$$

where

$$D_j := \partial_j - ia_j, \quad j \in \{1, \ldots, \nu\} \ .$$

Now, let

$$\lambda_j(x) := \int_0^{x_j} a(x_1, \ldots, x_{j-1}, y, x_{j+1}, \ldots, x_\nu)\, dy \ .$$

Then [329]

$$-\mathrm{i}D_j = \mathrm{e}^{\mathrm{i}\lambda_j}(-\mathrm{i}\partial_j)\mathrm{e}^{-\mathrm{i}\lambda_j} .$$

(Note that, in a "physicist's language", this means that in one dimension, magnetic vector potentials can always be removed by a gauge transformation.) Therefore

$$\exp\left(\frac{t}{n}D_j^2\right) = \exp(i\lambda)_j \exp\left(\frac{t}{n}\partial_j^2\right)\exp(-\mathrm{i}\lambda_j) , \quad \text{so that}$$

$$|\exp(tD_j^2)\varphi| \leq \exp(t\partial_j^2)|\varphi|, \quad \varphi \in L^2(\mathbb{R}^\nu) . \tag{1.10}$$

Now (1.8) follows from (1.10), (1.9) and $|\exp(-tV/n)| \leq 1$.

Step 2. $L^\infty(\mathbb{R}^\nu) \cap Q(H)$ is a form core for H.

This follows from (1.7) and the fact that $\operatorname{Ran}[\exp(-tH)]$ is a form core for H by the sepectral theorem.

Step 3. $L^\infty_{\text{comp}}(\mathbb{R}^\nu) \cap Q(H)$ is a form core for H [where $L^\infty_{\text{comp}}(\mathbb{R}^\nu) := \{\varphi \in L^2(\mathbb{R}^\nu) | \varphi \in L^\infty(\mathbb{R}^\nu), \operatorname{supp}\varphi \text{ is compact}\}$].

This follows by a usual cut-off approximation argument, i.e. choose $\eta \in C_0^\infty(\mathbb{R}^\nu)$ with $\eta = 1$ near 0, then consider, for any $\varphi \in L^\infty \cap Q(H)$

$$\varphi_n(x) := \eta\left(\frac{x}{n}\right)\varphi(x) \quad (n \in \mathbb{N})$$

then $\varphi_n \to \varphi$, $(n \to \infty)$ in the form sense. Now the proof will be finished by

Step 4. $C_0^\infty(\mathbb{R}^\nu)$ is a form core for H.

This follows by a standard mollifier argument, i.e. choose $j \in C_0^\infty(\mathbb{R}^\nu)$ such that $\int j(x)d^\nu x = 1$; set $j_\varepsilon := \varepsilon^{-\nu} j(x/\varepsilon)$, then for $\varphi \in L^\infty_{\text{comp}} \cap Q(H)$ $\varphi_\varepsilon := j_\varepsilon * \varphi \in C_0^\infty$ and $\varphi_\varepsilon \to \varphi$, $(\varepsilon \to 0)$ in the form sense. □

Note that in the last two steps, it is crucial that the approximated function is in L^∞.

The next theorem is also a well-known result [193].

Theorem 1.14. Let $V \geq 0$, $V \in L^2_{\text{loc}}(\mathbb{R}^\nu)$ and $a = 0$. Then $H := H_0 + V$ is essentially self-adjoint on $C_0^\infty(\mathbb{R}^\nu)$, i.e. $C_0^\infty(\mathbb{R}^\nu)$ is an operator core for H, and its closure is the form sum.

The proof is exactly the same as in Theorem 1.13 (replacing form cores by operator cores and form domains by operator domains) with one additional step. Once one notices that $L^\infty(\mathbb{R}^\nu) \cap D(H)$ is an operator core for H one uses the formula

$$H(\eta\varphi) = \eta H\varphi + 2\nabla\eta \cdot \nabla\varphi - \varphi\Delta\eta \tag{1.11}$$

for $\varphi \in L^\infty(\mathbb{R}^\nu) \cap D(H)$ and $\eta \in C_0^\infty(\mathbb{R}^\nu)$. The right-hand side of (1.11) makes sense since we know from Theorem 1.13 that $\varphi \in Q(H)$ and therefore $\nabla\varphi \in L^2(\mathbb{R}^\nu)^\nu$. Equation (1.11) can then be used to show the analogous steps of Step 3 and Step 4 in Theorem 1.13. □

1.4 The Leinfelder-Simader Theorem

Our last theorem in this chapter is a result due to *Leinfelder* and *Simader* [229]. It finishes the problem of self-adjointness of Schrödinger operators with singular potentials and $V \geq 0$ by giving a definitive result.

Theorem 1.15 (Leinfelder, Simader [229]). Let $V \geq 0$, $V \in L^2_{\text{loc}}(\mathbb{R}^\nu)$ and $a \in L^4_{\text{loc}}(\mathbb{R}^\nu)^\nu$ and $\nabla \cdot a \in L^2_{\text{loc}}(\mathbb{R}^\nu)$. Then H [the operator associated with the maximal form of $(-i\nabla - a)^2 + V$] is essentially self-adjoint on $C_0^\infty(\mathbb{R}^\nu)$.

Though not explicitly mentioned in [229], the key lemma in the proof of Leinfelder and Simader is

Lemma 1.16 (Kato's Version [197]). Let $\varphi \in L^\infty_{\text{comp}}(\mathbb{R}^\nu)$, $a \in L^4_{\text{loc}}(\mathbb{R}^\nu)^\nu$. If $\nabla\varphi \in L^2(\mathbb{R}^\nu)^\nu$ and $-\Delta\varphi + 2ia\cdot\nabla\varphi \in L^2(\mathbb{R}^\nu)$, then $\Delta\varphi \in L^2(\mathbb{R}^\nu)$ and $\nabla\varphi \in L^4(\mathbb{R}^\nu)^\nu$.

Proof (of Lemma 1.16) [227]. By a scaling argument, it is clear that without loss one can choose $\operatorname{supp}\varphi$ to be contained in the unit ball B_1. One needs, as a basic step, the following inequality which goes back to *Gagliardo* [127] and *Nirenberg* [264]

$$\|\nabla\varphi\|_{2p} \leq d(p)\|\varphi\|_\infty \|\Delta\varphi\|_p \tag{1.12}$$

for any $p \in (1, \infty)$, $\varphi \in L^\infty_{\text{comp}}$ and a suitable constant $d(p)$ depending on p. (Note $\|\nabla\varphi\| := \| |\nabla\varphi| \|$). Equation (1.12) can be shown by using

$$\|\partial_j\varphi\|_{2p}^{2p} = \lim_{\varepsilon\to 0} \int \{[(\partial_j\varphi)^2 + \varepsilon]^{p-1}\partial_j\varphi\}\partial_j\varphi$$

partial integration and controlling all second derivatives by the formula $\|D^2\varphi\|_p \leq d'(p)\|\Delta\varphi\|_p$ (see [350, p. 59]). If we choose $1 < p_0 \leq p_1 < \infty$ and only concern ourselves with $p \in [p_0, p_1]$, then $d := \max d(p)$ can be chosen independently of p.

From (1.12) we get, for $\varepsilon > 0$ and $p \in [p_0, p_1]$

$$\|\nabla\varphi\|_{2p} \leq \tfrac{1}{2}d\varepsilon^{-1}\|\varphi\|_\infty + \tfrac{1}{2}d\varepsilon\|\Delta\varphi\|_p .$$

Thus, with $g := -\Delta\varphi + 2ia\cdot\nabla\varphi$

$$\|\nabla\varphi\|_{2p} \leq \tfrac{1}{2}d\varepsilon^{-1}\|\varphi\|_\infty + \tfrac{1}{2}d\varepsilon\|g\|_p + \tfrac{1}{2}d\varepsilon\|2ia\cdot\nabla\varphi\|_p .$$

Now, since $\operatorname{supp}\varphi$ is in the unit ball, if we choose $p \leq p_1 \leq 2$, we can always estimate $\|g\|_p$ by $\|g\|_2$, and by using $a \in L^4(\mathbb{R}^\nu)^\nu$ and Hölder's inequality, we get

$$\|\nabla\varphi\|_{2p} \leq c(\varepsilon) + \tfrac{1}{2}\tilde{c}\varepsilon\|\nabla\varphi\|_r$$

for some $\tilde{c}$, $c(\varepsilon) > 0$, where $1/r = 1/p - 1/4$. Now take

$$r_n := \left(\frac{1}{4} + \frac{1}{2^{n+1}}\right)^{-1}, \quad n \in \mathbb{N} .$$

Then $r_n \leq 4$ and $r_n \nearrow 4$ and

$$\frac{1}{2p_n} = \frac{1}{2}\left(\frac{1}{r_n} + \frac{1}{4}\right) = \frac{1}{r_{n+1}} .$$

If we choose $\varepsilon > 0$ suitably, we get inductively that $|\nabla\varphi| \in L^{r_n}$ and $\|\nabla\varphi\|_{r_n} \leq D + 1/2\|\nabla\varphi\|_{r_n}$, for some constant D and all r_n. Here we used the fact that $r_n \leq 2p_n$ and that supp φ is contained in the unit ball. This implies $\|\nabla\varphi\|_{r_n} \leq 2D$, $(n \in \mathbb{N})$ and therefore $\|\nabla\varphi\|_4 < \infty$, and this proves Lemma 1.16. □

Having this result, the proof of Theorem 1.15 is as elementary as the above theorems.

Proof (of Theorem 1.15). The only problem in following the proof of Theorem 1.13 is Step 4, since the mollifier j_ε does not commute with $(\nabla - ia)$. All other steps work as in Theorems 1.13 and 1.14, i.e. we can prove as above that $L^\infty_{\text{comp}} \cap D(H)$ is an operator core for H. So, for $\varphi \in L^\infty_{\text{comp}} \cap D(H)$

$$H\varphi = -\Delta\varphi + 2ia\cdot(\nabla\varphi) + (-i\nabla\cdot a + a^2 + V)\varphi . \tag{1.13}$$

By the assumptions of Theorem 1.15 and Lemma 1.16, each individual term in (1.13) is in $L^2(\mathbb{R}^\nu)$ and $\nabla\varphi \in L^4(\mathbb{R}^\nu)^\nu$. This suffices to show that the "mollified" sequence $\varphi_\varepsilon := j_\varepsilon * \varphi$ converges to φ in the operator norm as $\varepsilon \to 0$. □

2. L^p-Properties of Eigenfunctions, and All That

In this chapter, we study properties of eigenfunctions and some consequences for the spectrum of H.

We begin with some semigroup properties which turn out to be useful for showing essential self-adjointness of $H_0 + V$ when the negative part of V is in K_ν (Section 2.1). In Sects. 2.2 and 3, we give some estimates for eigenfunctions, which we use in Sect. 2.4, to give a characterization of the spectrum of H.

In Sect. 2.5, we make some assertions about positive solutions, and in Sect. 2.6 we give an alternative proof of the result of Zelditch, that the time evolution $\exp(-itH)$ has a weak integral kernel under suitable hypotheses on V.

We will only prove a few things, and refer the reader to the review article of *Simon* [334] which has fairly complete references and results. Some of the results are also contained in the Brownian motion paper of *Aizenman* and *Simon* [7].

2.1 Semigroup Properties

The first theorem states a basic "smoothing" property of the semigroup associated with $H = H_0 + V$ where H_0 is the self-adjoint realization of $(-\Delta)$. We will give a complete proof of it. The following Corollary 2.2 is an immediate consequence of the $L^2 \to L^\infty$-boundedness of the semigroup. It is an extension of Theorem 1.14, i.e. it gives essential self-adjointness of H if V_- (the negative part of V) is in K_ν. In the last proposition, we give (without proof) a semigroup criterion for V being in K_ν if V is negative. This illustrates the "naturalness" of the class K_ν for these L^p-properties.

Theorem 2.1 [7]. If $V \in K_\nu$ and $t > 0$, then $\exp(-tH)$ is a bounded operator from L^p to L^q for all $1 \leq p \leq q \leq \infty$.

Remark. Note that $V \in K_\nu$ implies that V is H_0-form bounded with relative bound 0. So $H := H_0 + V$ is well defined and self-adjoint as a form sum (see Theorem 1.5).

Proof (of Theorem 2.1). We divide it into six steps.

Step 1. $\exp(-tH)$: $L^\infty(\mathbb{R}^\nu) \to L^\infty(\mathbb{R}^\nu)$ is bounded for small t.

We have, for $V \in K_\nu$

$$\lim_{t \searrow 0} \left\| \int_0^t e^{-sH_0} |V| \, ds \right\|_{\infty,\infty} = 0 \; . \tag{2.1}$$

To verify (2.1), we note that $\exp(-tH_0)$ has an explicit integral kernel

$$P(x,y,t) := (4\pi t)^{-\nu/2} \exp\left(-\frac{|x-y|^2}{4t}\right).$$

Moreover, explicit integration shows that

$$Q(x,y,t) := \int_0^t P(x,y,s)\,ds$$

behaves outside the region $A := \{(x,y) \in \mathbb{R}^{2\nu} \mid |x-y| \le 4\sqrt{t}\}$ like

$$|x-y|^{\gamma} \exp\left(-\frac{|x-y|^2}{4t}\right) \tag{2.2}$$

for suitable γ real and inside the region A like

$$\begin{array}{ll} |x-y|^{-(\nu-2)} & \text{if } \nu \ge 3 \\ \ln|x-y|^{-1} & \text{if } \nu = 2 \\ \mathrm{c} & \text{if } \nu = 1\,, \end{array}$$

i.e. like the kernel in Definition 1.10 of K_ν. Thus,

$$\lim_{t \searrow 0} \left\| \int_0^t \mathrm{e}^{-sH_0} |V|\,ds \right\|_{\infty,\infty} \le \lim_{t \searrow 0} \left\| \int_{|x-y| \le 4\sqrt{t}} Q(x,y,t)|V(y)|d^\nu y \right\|_{\infty}$$

$$+ \lim_{t \searrow 0} \left\| \int_{|x-y| > 4\sqrt{t}} Q(x,y,t)|V(y)|d^\nu y \right\|_{\infty}.$$

The first term on the R.H.S. vanishes since $V \in K_\nu$, and the second because of (2.2), if we use the fact $V \in L^1_{\text{unif}}$.

Therefore, we can choose a $t_0 > 0$ such that

$$\alpha := \left\| \int_0^{t_0} \mathrm{e}^{-tH_0} |V|\,ds \right\|_{\infty,\infty} < 1\,.$$

Now assume for a moment that $V \in C_0^\infty$. Then we can expand the semigroup by the Dyson-Phillips expansion:

$$\mathrm{e}^{-tH} = \sum_{j=0}^{\infty} T_j \qquad \text{where}$$

$$T_0 := \mathrm{e}^{-tH_0} \qquad \text{and}$$

$$T_j := \int_{0 \le \sum_{i=1}^{j} s_i \le t} \prod_{i=1}^{j} ds_i \exp(-s_1 H_0) V \exp(-s_2 H_0) V \ldots$$

$$\exp(-s_j H_0) V \exp\left[-\left(t - \sum_{i=1}^{j} s_i\right) H_0\right].$$

Now we calculate the norm of T_j as an operator from L^∞ to L^∞. Let $t \leq t_0$ and denote by $|T_j|$ the operator which has replaced V by $|V|$ in T_j. Then

$$\begin{aligned}
\|T_j\|_{\infty,\infty} &\leq \| |T_j| 1 \|_\infty \\
&\leq \left\| \int_{\substack{0 \leq s_i \leq t \\ 1 \leq i \leq j}} \prod_{i=1}^{j} ds_i \exp(-s_1 H_0)|V| \exp(-s_2 H_0) \ldots 1 \right\|_\infty \\
&\leq \prod_{i=1}^{j} \left(\left\| \int_0^t ds_i \exp(-s_i H_0)|V| \right\|_{\infty,\infty} \right) \left\| \exp\left[-\left(t - \sum_{i=1}^{j} s_i \right) H_0 \right] 1 \right\|_\infty \\
&\leq \alpha^j .
\end{aligned}$$

The last inequality holds since the last factor on the L.H.S. is equal to 1. Therefore, we have

$$\|e^{-tH}\|_{\infty,\infty} \leq \sum_{j=0}^{\infty} \alpha^j < \infty$$

since $\alpha < 1$.

Now let $V \in K_\nu$. Using the continuity of $\int \tilde{K}(x, y, \nu) V(y) d^\nu y$ in x [7, Theorem 4.15] where $\tilde{K}$ is the kernel in Definition 1.10 of K_ν, one can easily find a sequence $\{V_n\}_{n \in \mathbb{N}} \subseteq C_0^\infty$ such that $V_n \to V$, $(n \to \infty)$ in the K_ν-norm. Then we have $\|\exp(-sH_0)(V_n - V_m)\|_{\infty,\infty} \to 0$ when $n, m \to \infty$, by arguments similar to those we used to verify (2.1).

Now, if we denote by T_j^n the operator which has replaced V by V_n, we can conclude, by a suitable "telescoping" argument, that $\{T_j^n\}_{n \in \mathbb{N}}$ is a Cauchy sequence in $B(L^\infty, L^\infty)$, and therefore also $\{\exp(-tH_n)\}_{n \in \mathbb{N}}$, where $H_n := H_0 + V_n$. Thus, a limiting argument yields

$$\|e^{-tH}\|_{\infty,\infty} < \infty$$

for general $V \in K_\nu$.

Step 2. By the semigroup property, we know that $\exp(-tH)$ is bounded from L^∞ to L^∞ for all $t \in \mathbb{R}^+$.

Step 3. We claim that

$$|[\exp(-tH)f](x)|^2 \leq [\exp(-tH_0)|f|^2](x)\{\exp[-t(H_0 + 2V)]1\}(x) \quad (2.3)$$

holds for $f \in L^2$ and a.e. $x \in \mathbb{R}^\nu$. Assume for a moment that $V \in C_0^\infty$. Consider for $\varphi \in C_0^\infty$, $\varphi \geq 0$

$$F(z) := \langle \exp[-t(H_0 + zV)]|f|^{2-z}, \varphi \rangle, \quad 0 \leq \operatorname{Re} z \leq 2 .$$

Then this is an analytic function in z. When $\operatorname{Re} z = 0$, then $|F(z)| \leq \langle \exp(-tH_0)|f|^2, \varphi \rangle$, since the imaginary part of z just gives a phase factor (use the Trotter product formula). When $\operatorname{Re} z = 2$, then $|f|^{2-z}$ gives a phase factor. So

$$|F(z)| \leq \langle \exp[-t(H_0 + 2V)]1, \varphi\rangle, \quad \text{if } \operatorname{Re} z = 2 .$$

Therefore, by Hadamard's three line theorem [293, p. 33] we conclude, for $\operatorname{Re} z = 1$

$$\begin{aligned} |F(z)| &\leq \langle \exp(-tH)|f|, \varphi\rangle \\ &\leq [\langle \exp(-tH_0)|f|^2, \varphi\rangle]^{1/2} \{\langle \exp[-t(H_0 + 2V)]1, \varphi\rangle\}^{1/2} . \end{aligned}$$

Furthermore, since $\exp(-tH)$ is positivity preserving [293, Theorem X.55], we have

$$|\mathrm{e}^{-tH} f| \leq \mathrm{e}^{-tH}|f| , \qquad \text{and hence}$$

$$|\langle \exp(-tH)f, \varphi\rangle|^2 \leq \langle \exp(-tH_0)|f|^2, \varphi\rangle \langle \exp[-t(H_0 + 2V)]1, \varphi\rangle . \quad (2.4)$$

Now let $x \in \mathbb{R}^\nu$, and choose $\varphi(x') := \varphi_\varepsilon(x - x')$, $x' \in \mathbb{R}^\nu$, $\varepsilon > 0$ where $\varphi_\varepsilon(\cdot) := \varepsilon^{-\nu}\psi(\cdot/\varepsilon)$, with $\psi \in C_0^\infty$, $\psi \geq 0$, $\|\psi\|_1 = 1$. Then (2.4) reads as

$$\begin{aligned} &|\exp(-tH)f * \varphi_\varepsilon|^2(x) \\ &\quad \leq [\exp(-tH_0)|f|^2 * \varphi_\varepsilon](x)\{\exp(-t(H_0 + 2V)]1 * \varphi_\varepsilon\}(x) , \end{aligned}$$

and this gives (2.3) for $V \in C_0^\infty$ when $\varepsilon \to 0$ for a.e. x.

If $V \in K_\nu$, we can approximate V again by C_0^∞-functions in the K_ν-norm and get (2.4), and therefore (2.3) in this case.

Step 4. $\exp(-tH)$ is bounded from L^2 to L^∞. Because $2V \in K_\nu$, we know that $\|\exp(-t(H_0 + 2V)1\|_\infty \leq c$ by Step 1. Since $\exp(-tH_0)$ is a convolution with a smooth decaying function, $\exp(-tH_0)$ is bounded from L^1 to L^∞. Thus, we can estimate, by (2.3),

$$\|\mathrm{e}^{-tH} f\|_\infty^2 \leq c\| |f|^2 \|_1 \leq c\|f\|_2^2 ,$$

Step 5. $\exp(-tH)$ is bounded from L^1 to L^2 by Step 4 and duality.

Step 6. $\exp(-tH)$ is bounded from L^p to L^q for $1 \leq p \leq q \leq \infty$.

We know, by Step 4 and Step 5 and the semigroup property, that $\exp(-tH)$ is bounded from L^1 to L^∞. Furthermore, since $\exp(-tH)\varphi \in L^2$ for φ in a suitable dense set (say C_0^∞) of L^1, we can conclude by duality and Step 1 that $\exp(-tH)\varphi \in L^1$, and that $\exp(-tH)$ is bounded from L^∞ to L^∞. So we have boundedness of $\exp(-tH)$ if

$$(p, q) = \begin{cases} (\infty, \infty) & \text{by Step 1} \\ (1, 1) & \text{by duality} \\ (1, \infty) & \text{just proven} . \end{cases}$$

Now, by the Riesz-Thorin interpolation theorem [293, Theorem IX.17], $\exp(-tH)$ is bounded from L^p to L^q for all (p^{-1}, q^{-1}) in the convex set $\{(p^{-1}, q^{-1}) \in \mathbb{R}^2 | 0 \leq q^{-1} \leq p^{-1} \leq 1\}$, and this proves the theorem. □

The above proof is very similar to that in [334], where Simon used Brownian motion techniques. In fact, our proof is an analyst's translation of the Brownian motion proof. Equation (2.3) is more transparent in the Brownian motion language since it is just the Schwarz inequality in path space. The interpolation argument we used is borrowed from *Guerra*, *Rosen* and *Simon* [145].

Theorem 2.1 was originally proven, using the above ideas, by *Carmona* [58] and *Simon* [331] independently.

Somewhat earlier, it was proven using semigroup analytical methods by *Kovalenko* and *Semenov* [219]. There were also slightly different results by *Herbst* and *Sloan* [161] which motivated some of this work.

One obvious consequence of Theorem 2.1 is the following:

Corollary 2.2. If $V \in L^2_{\text{loc}}(\mathbb{R}^\nu)$ and $V_- := \max(0, -V) \in K_\nu$, then $H_0 + V$ is essentially self-adjoint on $C_0^\infty(\mathbb{R}^\nu)$.

This is just an extension of Theorem 1.14. The proof is identical, except that we replace Step 1 of that proof by Theorem 2.1 with $p = 2$ and $q = \infty$, i.e. by the fact that $\exp(-tH)$ is bounded from L^2 to L^∞.

We should also remark that $V \in K_\nu$ is almost necessary for L^∞-semigroup boundedness, for one has the following proposition which we will not prove (see [334, Theorem A2.1]).

Proposition 2.3. Let $V \leq 0$. Then $V \in K_\nu$ if and only if $\exp(-tH)$ is bounded from L^∞ to L^∞ with $\lim_{t \searrow 0} \|\exp(-tH)\|_{\infty,\infty} = 1$.

This proposition says that K_ν is the "natural" class for this L^p-property.

Remark. If one keeps track of how the $L_p \to L_q$ norm of $\exp(-tH)$ behaves in t, and if one sees for which (p, q) this norm is integrable, one obtains some information about (powers) of the resolvent of H mapping from L^p to L^q. This leads to (analogs of) Sobolev estimates where H_0 is replaced by H; (see [334], Sect. B2).

2.2 Estimates on Eigenfunctions

In this section, we state without proof two basic results concerning eigenfunctions, and we give some interesting applications. We denote $V_- := \max(0, -V)$ and

$$K^\nu_{\text{loc}} := \{V | V\varphi \in K_\nu \text{ for any } \varphi \in C_0^\infty(\mathbb{R}^\nu)\} \ .$$

Note that $K^\nu_{\text{loc}} \subseteq L^1_{\text{loc}}(\mathbb{R}^\nu)$.

The first main result is the following:

Theorem 2.4 (subsolution estimate) [7, 334, 131]. Suppose $V \in K_{\text{loc}}^{\nu}$ and let $Hu = Eu$ in an open set $\Omega \subseteq \mathbb{R}^{\nu}$ (in the distributional sense), i.e. in the sense that $\Delta u \in L_{\text{loc}}^1$, $Vu \in L_{\text{loc}}^1$ and

$$\langle -\Delta\varphi, u\rangle + \langle (V - E)\varphi, u\rangle = 0 \quad \text{for all } \varphi \in C_0^{\infty}.$$

Then

(i) u is (a.e. equal to) a continuous function in Ω.
(ii) for $x \in \Omega$ and any $r > 0$ with $B := \{y \,|\, |x - y| \leq r\} \subseteq \Omega$ we have

$$|u(x)| \leq c \int_{|x-y| \leq r} |u(y)| d^{\nu} y \tag{2.5}$$

where c depends on r and the K_{ν}-norm of $V_{-}\chi_B$ (χ_B = characteristic function of B). In particular, if $V_{-} \in K_{\nu}$ and $\Omega = \mathbb{R}^{\nu}$, then c can be chosen independently of x.

This estimate is very useful, for example, if u is an eigenfunction (i.e. $u \in L^2$), then (2.5) implies that it goes pointwise to 0 at ∞. Also, if one has exponential decay in some average sense, one gets pointwise exponential decay.

Note that V can always be replaced by $V - E$, therefore the assertions hold also for solutions of $Hu = Eu$ (the constants, however, will be E-dependent). Equation (2.5) is called a *subsolution estimate* because if $u \geq 0$, then one has only to require the distributional inequality $Hu \leq 0$ (i.e. u has only to be a subsolution) for (2.5) to hold. This is proven in [7, Theorem 6.1]. Note also that it generalizes the well-known estimates on (sub-) harmonic functions when $V = 0$.

The other result is

Theorem 2.5 (Harnack Inequality) [7, 334, 131]. Suppose $V \in K_{\text{loc}}^{\nu}$, let $\Omega \subseteq \mathbb{R}^{\nu}$ be an open set, $Hu = Eu$ (in the distributional sense), $u \not\equiv 0$ and $u \geq 0$ on Ω. Let K be a compact set $K \subseteq \Omega$. Then the following estimate holds

$$c^{-1} \leq \frac{u(x)}{u(y)} \leq c \quad \text{for } x, y \in K \tag{2.6}$$

where c depends only on K, Ω and on local K^{ν}-norms of V.

The proof of this theorem (in [7]) consists essentially in estimating a probabilistic representation of the Poisson kernel of H. *Brossard* [55] and *Zhao* [381] have further studied this Poisson kernel using probabilistic methods. They establish that the singularities of this Poisson kernel at the boundary are the same as those for harmonic functions.

We should note that K_{ν} is almost the optimal class for which these estimates hold. There is a theorem which says that if $V \leq 0$ and one has a "strong" Harnack inequality, then $V \in K_{\text{loc}}^{\nu}$ [7, Theorem 1.1]. This is illustrated by the following

Example. Let $\nu = 3$, and consider potentials that behave at the origin like

$$V(x) = |x|^{-2}(\ln|x|)^{-1}, \quad |x| < \tfrac{1}{2}, x \in \mathbb{R}^3 .$$

By the example after Theorem 1.12, we know that this is just the border line case where $V \notin K_\nu$. Furthermore, a straightforward calculation shows that the eigensolution u of $Hu = 0$ behaves at 0 like

$$u(x) = -\ln|x|, \quad |x| < 1 \ .$$

Thus, because of the logarithmic singularity of u at 0, the Harnack inequality and the subsolution estimates fail.

This example also shows that K_ν is "exactly" the class where the eigensolutions are (locally) bounded.

Note, however, that for potentials of any sign the (strong) Harnack inequality is not a sufficient condition for V to be in K_ν. There are examples of heavily oscillating potentials which are not in K_ν, but where $\exp(-tH)$ is a bounded semigroup in L^∞. So it is quite likely that a Harnack inequality still would hold (see [7, Example 3, Appendix 1]).

We should mention that there is a result of *Brezis* and *Kato* [54] where it is shown that if V is H_0-form bounded with bound 0, then the eigenfunctions are in all L^p, $p < \infty$ (but they are not necessarily bounded).

2.3 Local Estimates on Gradients

In this section, we show some simple L^2_{loc} bounds on ∇u for eigenfunctions, and that these bounds depend only on local norms on V. These estimates will be useful in the next section where we give a characterization of the spectrum of H.

We start with the following key lemma.

Lemma 2.6. If $u \in L^\infty_{\text{loc}}$ and $\Delta u \in L^1_{\text{loc}}$, then $\nabla u \in L^2_{\text{loc}}$ and for $\varphi \in C_0^\infty$

$$\int \varphi |\nabla u|^2 d^\nu x = \tfrac{1}{2} \int \Delta\varphi |u|^2 d^\nu x - \int \varphi u \Delta \bar{u}\, d^\nu x \ . \tag{2.7}$$

In particular, if u is an eigensolution, i.e. $Hu = 0$, then for $\varphi \in C_0^\infty$, $\varphi \geq 0$

$$\int \varphi |\nabla u|^2 d^\nu x \leq \tfrac{1}{2} \int \Delta\varphi |u|^2 d^\nu x + \int V_- \varphi |u|^2 d^\nu x \ . \tag{2.8}$$

Proof. Equation (2.7) follows just by integration by parts twice for smooth functions and then by a mollifier argument. Equation (2.8) follows from (2.7). □

Together with the subsolution estimate, this leads to an L^1-estimate for the gradient.

Theorem 2.7. Suppose $V_+ := \max\{0, V\} \in K^\nu_{\text{loc}}$, $V_- \in K_\nu$ and $Hu = 0$. Then $\nabla u \in L^2_{\text{loc}}$ and for K compact in the open bounded set $\Omega \subseteq \mathbb{R}^\nu$

$$\int_K |\nabla u|^2 d^\nu x \leq c \left[\int_\Omega |u|\, d^\nu x \right]^2 , \tag{2.9}$$

where the constant c depends only on local norms on V and on Ω and K.

Proof. Since $V \in K_{\text{loc}}^\nu$, we have $u \in L_{\text{loc}}^\infty$ by Theorem 2.4. This, together with $V \in L_{\text{loc}}^1$ implies $\Delta u \in L_{\text{loc}}^1$. Thus, if we choose $\varphi \in C_0^\infty$ such that $\varphi(x) = 1$ for $x \in K$ and $\operatorname{supp} \varphi =: W' \subseteq \Omega$, (2.8) implies

$$\int_K |\nabla u|^2 d^\nu x \le c_1 \sup_{x \in W'} |u(x)|^2 \tag{2.10}$$

for suitable c_1 depending on Ω and the L_{loc}^1 norm of V_-. Now choosing W'' open such that $W' \subseteq W'' \subseteq \Omega$, we conclude from (2.5) that

$$\sup_{x \in W''} |u(x)| \le c_2 \int_\Omega |u(x)| \, d^\nu x$$

for suitable c_2 depending on Ω and local K_ν-norms of V_-. This, together with (2.10), implies (2.9). □

The following L^2-estimate on "rings" of hypercubes will be useful in the next section.

Corollary 2.8. Let $V_+ \in K_{\text{loc}}^\nu$, $V_- \in K_\nu$ and C_r be the hypercube

$$C_r := \{x \in \mathbb{R}^\nu | \max |x_i| \le r\} \quad \text{for } r \in \mathbb{N} \ .$$

Then for any eigensolution u of $Hu = 0$ and any $r \in \mathbb{N}$

$$\int_{C_{r+1} \setminus C_r} |\nabla u|^2 d^\nu x \le c \int_{C_{r+2} \setminus C_{r-1}} |u|^2 \ , \tag{2.11}$$

where c is independent of r (it depends only on local norms of V_-).

Proof. Let K be a unit cube, and K' a cube of side 3 centered at the center of K. Then, by (2.9) and the Schwarz inequality

$$\begin{aligned} \int_K |\nabla u|^2 d^\nu x &\le \tilde{c} \left[\int_{K'} |u| \, d^\nu x \right]^2 \\ &\le c \int_{K'} |u|^2 d^\nu x \end{aligned}$$

for suitable constants $\tilde{c}$, c. Now (2.11) follows by adding up these estimates for a partition of $C_{r+1} \setminus C_r$ into unit cubes. □

2.4 Eigenfunctions and Spectrum (Sch'nol's Theorem)

For Schrödinger operators with K_ν-potentials there is an interesting characterization of the spectrum [which we denote by $\sigma(H)$]. It consists essentially (i.e. up to a closure) of "eigenvalues" with polynomially bounded eigensolutions. This has an important application in Chap. 9, where we discuss the spectrum of random Jacobi matrices.

The first theorem is a result of *Sch'nol* [307] (1957), who assumed that V is bounded from below. It was rediscovered by *Simon* [332], who proved it for more general V's, with a very different proof. Our proof here is essentially Sch'nol's proof, which we extend to $V_- \in K_\nu$ by using Harnack's inequality.

Theorem 2.9 (Sch'nol, Simon). Let $V_+ \in K^\nu_{\text{loc}}$. $V_- \in K_\nu$ and $Hu = Eu$, where u is polynomially bounded. Then $E \in \sigma(H)$.

Proof. Note that it suffices to prove the assertion only for $E = 0$, since E can always be absorbed in V.

If $u \in L^2$, then obviously $0 \in \sigma(H)$, since it is an eigenfunction. So assume $u \notin L^2$. Let $C_r, r = 1, 2, \ldots$ be the hypercubes defined in Corollary 2.8. Choose $\eta_r \in C_0^\infty(\mathbb{R}^\nu)$ with $\operatorname{supp} \eta_r \subseteq C_{r+1}$ and $\eta_r(x) = 1$ for $x \in C_r$ such that $\|\Delta\eta_r\|_\infty \le D$ and $\|\nabla\eta_r\|_\infty \le D$ with a suitable $D > 0$ independent of r. Let

$$w_r := \eta_r u / \|\eta_r u\|_2 \ .$$

We will show that

$$\|Hw_{r_n}\| \to 0 \quad (r_n \to \infty)$$

for a suitable subsequence $\{r_n\} \subseteq \mathbb{N}$. This implies $0 \in \sigma(H)$, since we have a Weyl sequence (see [292, Theorem VII.12]).

Since $Hu = 0$ and

$$\Delta(\eta_r u) = (\Delta\eta_r)u + 2\nabla\eta_r \cdot \nabla u + \eta_r \Delta u \ ,$$

we get

$$H(\eta_r u) = -\Delta\eta_r u - 2\nabla\eta_r \nabla u \ .$$

Thus, since $\|\Delta\eta_r\|_\infty$ and $\|\nabla\eta_r\|_\infty$ are uniformly bounded,

$$\begin{aligned} \|H\eta_r u\|^2 &\le c' \int_{C_{r+1}\setminus C_r} (|u|^2 + |\nabla u|^2) d^\nu x \\ &\le c'' \int_{C_{r+2}\setminus C_{r-1}} |u|^2 d^\nu x \end{aligned} \tag{2.12}$$

by Corollary 2.8.

Let $M(r) := \int_{C_r} |u|^2 d^\nu x$. Then (2.12) implies

$$\|Hw_r\|^2 \le c \frac{M(r+2) - M(r-1)}{\|\eta_r u\|^2} \le \frac{M(r+2) - M(r-1)}{M(r-1)} \ .$$

Now assume there is no subsequence $\{r_n\}$ such that

$$\frac{M(r_n+2) - M(r_n-1)}{M(r_n-1)} \to 0, \quad (r_n \to \infty) \ .$$

Then there exists a $R \in \mathbb{N}$ and an $\alpha > 0$ such that

$$\frac{M(r+2) - M(r-1)}{M(r-1)} \geq \alpha > 0 \quad \text{if } r > R \ .$$

This implies that

$$M(r+2) \geq (1+\alpha)M(r-1) \quad \text{and}$$

$$M(r+3) \geq (1+\alpha)M(r) \quad \text{for } r \geq R \ .$$

Thus, by induction we get

$$M(R+3k) \geq (1+\alpha)^k M(R) \quad \text{for any } k \in \mathbb{N} \ .$$

But this means that $M(R)$ has an exponential growth, which is a contradiction of the hypothesis that u grows polynomially. □

Remark. A direct consequence of our proof of this theorem is that if $Hu = Eu$, u is polynomially bounded, and $u \notin L^2$, then $E \in \sigma_{\text{ess}}(H)$.

A kind of converse of Theorem 2.9 can also be proven, using trace class-valued measures and eigenfunction expansions. It can be found in *Simon*'s review article [334, Theorem C5.4] or in [219]. See also [46]. We state it without proof.

We say that an assertion $A(E)$ holds *H-spectrally almost everywhere* if and only if $E_\Delta(H) = 0$, where $\Delta := \{E \in \mathbb{R} | A(E) \text{ does not hold}\}$ and $E_\Delta(H)$ is the spectral projection of H on Δ. Then we have

Theorem 2.10 [334]. If $V \in K_\nu$, then H-spectrally a.e. there exists a polynomially-bounded solution of $Hu = Eu$.

Note that this does not imply that for any $E \in \sigma(H)$, $Hu = Eu$ has a polynomially-bounded solution! Combining these two theorems one gets

Corollary 2.11. If $V \in K_\nu$, then $\sigma(H)$ is the closure of the set of all E for which $Hu = Eu$ has a polynomially-bounded solution.

Proof. If $\Delta \subseteq \mathbb{R}$ is the set where $Hu = Eu$ has polynomially-bounded solutions, then by Theorem 2.9, $\Delta \subseteq \sigma(H)$. Suppose that Δ is not dense in $\sigma(H)$. Then $\sigma(H) \backslash \bar{\Delta}$ contains a open set $S \subseteq \sigma(H)$ with $E_S(H) > 0$. But this contradicts Theorem 2.10. □

2.5 The Allegretto-Piepenbrink Theorem

Here we will discuss a theorem which was originally shown by *Allegretto* [9, 10, 11] and *Piepenbrink* [283, 284] and *Moss* and *Piepenbrink* [254]. It states that "eigenvalues" below the spectrum have some positive eigensolutions. We will prove it under very weak regularity hypotheses.

Theorem 2.12. Let $V_- \in K_\nu$ and $V_+ \in K_{\log}^\nu$. Then $Hu = Eu$ has a nonzero distributional solution which is everywhere nonnegative if and only if $\inf \sigma(H) \geq E$.

Proof. Suppose $\inf \sigma(H) \geq E$. Let $\{f_n\}_{n \in \mathbb{N}}$ be a sequence of C_0^∞-functions which are nonzero and positive with

$$\operatorname{supp} f_n \subseteq \{x \in \mathbb{R}^\nu \mid n \leq |x| \leq 2n\} .$$

Let $u_n := c_n(H - E + n^{-1})^{-1} f_n$, where c_n is chosen such that $u_n(0) = 1$. u_n is everywhere nonnegative since $(H - E + n^{-1})$ has a positivity preserving resolvent [295, p. 204]. Clearly, u_n obeys $Hu_n = (E - n^{-1})u_n$ for the region $|x| < n$. Thus, by Harnack's inequality (2.6), for any $R > 0$ we know that $u_n(x) > 0$ if $|x| < R$, and we can indeed normalize u_n such that $u_n(0) = 1$. Moreover, by Harnack's inequality, we find $C_R > 0$ such that

$$C_R^{-1} \leq u_n(x) \leq C_R \quad \text{if } |x| < R .$$

By passing to a subsequence, we can be sure that u_n has a limit point u in the weak-star L_{loc}^∞-sense, so that $\langle u_n, \varphi \rangle \to \langle u, \varphi \rangle$, $(n \to \infty)$ for all $\varphi \in L^1$ with $\operatorname{supp} \varphi$ compact. It is easy to see that u is a distributional solution of $Hu = Eu$, and that

$$C_R^{-1} \leq u(x) \leq C_R$$

so that u is a nonnegative and not identically zero.

Conversely, suppose $Hu = Eu$ has a nonzero nonnegative solution. By Harnack's inequality, u is strictly positive, and by Theorem 2.7

$$g := u^{-1} \nabla u \quad \text{is in } L_{\text{loc}}^2 .$$

We will prove that, for $\varphi \in C_0^\infty$

$$\langle \varphi, (H - E)\varphi \rangle = \tfrac{1}{2} \|\nabla \varphi - g\varphi\|_2^2 , \tag{2.13}$$

which implies that $H - E \geq 0$.

We first prove, (2.13), assuming $u \in C^\infty$. Then, by direct calculation (as operators)

$$u^{-1} \nabla u^2 \nabla u^{-1} = [\Delta - u^{-1}(\Delta u)] = -(H - E) ,$$

so

$$\langle \varphi, (H - E)\varphi \rangle = \|u \nabla u^{-1} \varphi\|^2 = \|\nabla \varphi - (u^{-1} \nabla u)\varphi\|^2$$

proving (2.13) in that case.

Given general u and V as in the assumption, we know that u is continuous and locally bounded away from zero (Theorems 2.4, 2.5). Let $u_\delta \in C^\infty$ be u-convoluted with an approximative identity j_δ. Let $V_\delta := u_\delta^{-1}(\Delta u_\delta) + E$ and $g_\delta :=$

$u_\delta^{-1}\nabla u_\delta$. Then $(H_0 + V_\delta)u_\delta = Eu_\delta$, so by the above

$$\langle \varphi, (H_0 + V_\delta - E)\varphi \rangle = \|\nabla\varphi - g_\delta\varphi\|^2 \ .$$

But since $u_\delta \to u$ local uniformly (u is continuous!) and $\nabla u_\delta \to \nabla u$ in L^2_{loc}, we have that $g_\delta \to g$ in L^2_{loc} and $V_\delta \to V$ in L^1_{loc} as $\delta \to 0$. This proves (2.13) in general. □

Agmon [4] has made a deep and complete analysis of all positive solutions of $Hu = Eu$ if V is periodic.

2.6 Integral Kernels for $\exp(-tH)$

It is a comforting fact to learn that some operators have integral kernels. There is a very general theorem which implies the existence of an integral kernel: the theorem of Dunford and Pettis (see *Treves* [358]).

Theorem 2.13. Let (M, μ) be a separable measure space, and $\mathscr{L}$ a separable Banach space. Let A be a bounded operator from $\mathscr{L}$ to $L^\infty(M, d\mu)$. Then there exists a unique (up to sets of μ-measure zero) weakly measurable function K from M to $\mathscr{L}^*$ such that, for each $f \in \mathscr{L}$ and a.e. $x \in M$

$$Af(x) = \langle K(x), f \rangle \ .$$

Moreover, $\|K\|_\infty = \|A\|$.

In particular, choosing $\mathscr{L} = L^p(M, d\mu)$, $1 \le p < \infty$, so that $\mathscr{L}^* = L^q(M, d\mu)$ with $q^{-1} + p^{-1} = 1$, and noting the trivial converse of Theorem 2.13, we have

Corollary 2.14. If A is bounded from L^p to L^∞, then there is a measurable function, K, on $M \times M$ obeying

$$\sup_{x \in M} \left[\int_M |K(x,y)|^q d^\nu y \right]^{1/q} = \|A\|_{p,\infty} < \infty \ , \tag{2.14}$$

so that, for any $f \in L^p$

$$(Af)(x) = \int K(x,y) f(y) d^\nu y \ . \tag{2.15}$$

Conversely, if A: $L^p \to L^p$ has an integral kernel K in the sense of (2.15) obeying (2.14), then A is a bounded map from L^p to L^∞.

There are some results which state that the semigroup $\exp(-tH)$ has a uniformly-bounded, jointly continuous integral kernel [334, Theorem B7.1]. But the results for $\exp(-tH)$ are much weaker. Consideration of the "free" case (i.e. $V = 0$) shows we cannot hope that there are integral operators in the sense of (2.14) and (2.15) for $p = 2$, since this kernel has no decay. Thus, we need a weaker notion of integral kernel.

We say that an operator A has a *weak integral kernel* $K(x, y)$ if and only if

$K \in L^1_{\text{loc}}(\mathbb{R}^\nu \times \mathbb{R}^\nu)$, and for all L^∞-functions with compact support f, g we have

$$\langle f, Ag \rangle = \int_{\mathbb{R}^\nu \times \mathbb{R}^\nu} f(x)K(x,y)g(y)d^\nu x\, d^\nu y \;.$$

Then we have the following result.

Theorem 2.15. Suppose V is a C^∞-function obeying

$$|(D^\alpha V)(x)| \le C_\alpha(1 + |x|)^{k(\alpha)}$$

for all multi-indices α where either $k(\alpha) = k_0 - |\alpha|$ with $k_0 < 1$ or $k(\alpha) = 0\,(C_\alpha > 0$ suitable). Then $\exp(-\mathrm{i}tH)$ has a weak integral kernel $P(x,y,t)$ for all $t \neq 0$, and it is jointly C^∞ on $\mathbb{R}^\nu \times \mathbb{R}^\nu \times (\mathbb{R}\backslash\{0\})$.

This was proven by *Fujiwara* [120, 121] in a series of papers; see also *Fujiwara* [122] and *Kitada* [209], *Kitada* and *Kumanogo* [211]. See also *Zelditch* [380] for an alternative proof. The restrictions on V are undoubtedly too strong. *Zelditch* [380] has eased the conditions, e.g. he has the following

Theorem 2.16. Let $V(x) = \sum_{\alpha=1}^m V_\alpha(T_\alpha x)$ where $x \in \mathbb{R}^\nu$, V_α is a function on $\mathbb{R}^{u_\alpha}$ with $\hat{V}_\alpha(k) \in L^2(\mathbb{R}^{u_\alpha})$, and T_α is a linear map of R^ν onto R^{u_α}. Then $\exp(-\mathrm{i}tH)$ has a weak integral kernel.

We give here a proof different from that of Zelditch, due to *Cycon, Leinfelder* and *Simon* [73] (see also [266]).

Proof. For simplicity, we assume for a moment $m = 1$ and $V = V_\alpha$. We have

$$\mathrm{e}^{-\mathrm{i}tH} = \mathrm{e}^{-\mathrm{i}tH}\mathrm{e}^{\mathrm{i}tH_0}\mathrm{e}^{-\mathrm{i}tH_0} \;.$$

We know that $\exp(-\mathrm{i}tH_0)$ is bounded from L^1 to L^∞. So, by Corollary 2.14, it is enough to show that

$$U(t) := \mathrm{e}^{-\mathrm{i}tH}\mathrm{e}^{\mathrm{i}tH_0}$$

is bounded from L^∞ to L^∞. Assume first that $V \in C_0^\infty$. Then

$$U(t) = 1 + \int_0^t U(s)V(s)\,ds \;,$$

where $V(s) := \exp(\mathrm{i}sH_0)V\exp(-\mathrm{i}sH_0)$. Note that $V(s)$ has the integral kernel

$$V(x,y;s) := \frac{1}{(2\pi)^\nu s^\nu}\exp\left(\frac{x^2 - y^2}{2\mathrm{i}s}\right)\hat{V}\left(\frac{x-y}{2s}\right) \;,$$

Now we expand $U(t)$ by the Dyson-Phillips expansion

$$U(t) = \sum_{n=0}^\infty Q_n(t) \;,$$

where $Q_n(t)$ are integral operators with kernels $Q_n(x, y, t)$ defined by $Q_0(x, y, t) := 1$ and

$$Q_n(x, y, t) := \int_{0 \le s_1 \le s_2 \ldots s_n \le t} \prod_{j=1}^{n-1} ds_j \int_{\mathbb{R}^\nu} \prod_{j=1}^{n} dx_j V(x_{j-1}, x_j; s_j) \ ,$$

And we estimate the operators

$$\begin{aligned} \|Q_n\|_{\infty,\infty} &\le \int_{0 \le s_1 \ldots \le t} \prod_{j=1}^{n-1} ds_j \int_{\mathbb{R}^\nu} \prod_{j=1}^{n} dx_j (2\pi s_j)^{-\nu} \left| \hat{V}\left(\frac{x_{j-1} - x_j}{2s_j}\right)\right| \\ &\le c^n \frac{t^n}{n!} \|\hat{V}\|_1^n \ . \end{aligned}$$

Therefore, we have

$$\|U(t)\|_{\infty,\infty} \le \sum_{n=1}^{\infty} \|Q_n\|_{\infty,\infty} \le \exp(ct\|\hat{V}\|_1)$$

for $V \in C_0^\infty$. Now an approximation argument gives the result for general V.

To handle the general case (including some N-body situations), we can always modify V_α so that T_α is actually an orthogonal projection onto a subspace X_α of R^ν. Let $Y_\alpha = X_\alpha^\perp$ and δ_{y_α}, the $\nu - \mu_\alpha$ dimensional δ-function in the Y_α variables. If $W_\alpha = V_\alpha \cdot T_\alpha$, then

$$\hat{W}_\alpha = (2\pi)^{(\mu_\alpha - \nu)/2} \delta_{Y_\alpha} \hat{V}_\alpha \cdot T_\alpha \ ,$$

so $\hat{W}_\alpha$ is a measure of total bounded variation. The above proof is easily seen to extend to such measures. □

3. Geometric Methods for Bound States

In this chapter, we develop methods that make explicit use of the geometry of phase space to investigate bound states (that is, discrete spectrum). We apply these methods to determine the essential spectrum and to distinguish the cases of infinitely many, finitely many and no bound states below the essential spectrum.

Among the theorems we will prove by geometric methods are the celebrated HVZ-Theorem on the essential spectrum of N-body Schrödinger operators, a theorem due to *Klaus* [214] on the essential spectrum of a one-dimensional Schrödinger operator with infinitely many wells further and further apart, and a theorem on the nonexistence of very negative ions due to *Ruskai* [302, 303] and *Sigal* [310, 313].

Geometric methods were already used in the works of *Zhislin* [382] and *Jörgens* and *Weidmann* [187], as well as in a different context in Lax-Phillips theory and in quantum field theory by Haag.

A systematic use of geometric ideas in Schrödinger operator theory started with the works by *Enss* [94], *Deift* and *Simon* [78] and *Simon* [323]. Further developments appear in *Morgan* [250], *Morgan* and *Simon* [251] and *Sigal* [310, 312, 313]. In this chapter we develop most closely the approach of Sigal. Geometric ideas will play a major role in later chapters.

3.1 Partitions of Unity and the IMS Localization Formula

The main tools we are going to work with are appropriately chosen partitions of unity in the following sense:

Definition 3.1 A family of functions $\{J_a\}_{a \in A}$ indexed by a set A is called a *partition of unity* if

(i) $0 \le J_a(x) \le 1$ for all $x \in \mathbb{R}^\nu$,
(ii) $\sum J_a^2(x) = 1$ for all $x \in \mathbb{R}^\nu$,
(iii) $\{J_a\}$ is locally finite, i.e. on any compact set K we have $J_a = 0$ for all but finitely many $a \in A$.
(iv) $J_a \in C^\infty$.
(v) $\sup_{x \in \mathbb{R}^\nu} \sum_{a \in A} |\nabla J_a(x)|^2 < \infty$.

Note that a definition of partition of unity that is more common in mathematics requires $\sum J_a(x) = 1$ instead of (ii). Nevertheless, for us the square will be very convenient.

The key to the geometric approach presented here is the following localization formula:

Theorem 3.2 (IMS Localization Formula). Let $\{J_a\}_{a\in A}$ be a partition of unity, and let $H = H_0 + V$ for a potential $V \in K_\nu$. Then

$$H = \sum_{a\in A} J_a H J_a - \sum_{a\in A} |\nabla J_a|^2 . \tag{3.1}$$

We call the term $\sum_{a\in A} |\nabla J_a|^2$ the *localization error*. The above formula appeared, at least implicitly, in *Ismagilov* [177], and was rediscovered by *Morgan* [250] and used in *Morgan* and *Simon* [251]. It was *I.M. Sigal* [310] who discovered its importance in the present context.

Remark. Since $V \in K_\nu$, $\varphi \in D(H)$ implies $J_a \varphi \in D(H)$ (and the same for the form domains). Thus, (3.1) makes sense.

Proof. Straightforward computations show

$$[J_a, [J_a, H]] = -2(\nabla J_a)^2 \quad \text{and}$$
$$[J_a, [J_a, H]] = J_a^2 H + H J_a^2 - 2J_a H J_a .$$

Summing over all $a \in A$, we end up with (3.1). □

We give a first application of the IMS-localization formula due to *Morgan* [250]:

Proposition 3.3. Let $V \in L^1_{\text{loc}}(\mathbb{R}^\nu)$, and assume that for a partition of unity $\{J_i\}$ we have

$$VJ_i^2 \le aJ_i H_0 J_i + bJ_i^2 \quad \text{with } a, b \text{ independent of } i \in I$$

then

$$V \le aH_0 + \tilde{b} \quad \text{with } \tilde{b} = b + \sup_{x\in\mathbb{R}^d} \sum |\nabla J_i(x)|^2 .$$

Proof.

$$V = \sum_{i\in I} VJ_i^2 \le a\sum J_i H_0 J_i + b = aH_0 + \sum |\nabla J_i|^2 + b . \quad \square$$

It is well known that $V = -\alpha/|x|^2$ in $\mathbb{R}^3$ is relatively form bounded with relative bound $a < 1$ (resp. $a \le 1$) if $\alpha < 1/4$ (resp. $\alpha \le 1/4$). By Proposition 3.3 we can conclude that

$$W = -\sum_{i\in I} \frac{\alpha_i \exp(-\delta|x - x_i|)}{|x - x_i|^2}$$

is form bounded with $a < 1$ (resp. $a \leq 1$) if, for all $i \in I$, $\alpha_i \leq C < 1/4$ (resp. $\alpha_i \leq 1/4$) and the distance between the points x_i is bounded away from zero.

3.2 Multiparticle Schrödinger Operators

Before we continue our discussion of the geometric methods, we first fix some notations for N-body operators. The reader may consult *Reed* and *Simon* III and IV [294, 295] for further information.

The positions of N particles each moving in μ-dimensional space is represented by a vector $x = (x_1, \ldots, x_N) \in \mathbb{R}^{N\mu}$, where each x_i is a vector in $\mathbb{R}^\mu$, giving the position of the ith particle. The corresponding free Hamiltonian (kinetic energy) is given by

$$\tilde{H}_0 = -\sum_{i=1}^{N} \frac{1}{2m_i} \Delta_i \ . \tag{3.2}$$

Here m_i is the mass of the ith particle and Δ_i is a μ-dimensional Laplacian in the x_i variables.

We will consider a potential V that comes from pair potentials, i.e.

$$V(x) = \sum_{i<j} V_{ij}(x) \ , \qquad \text{where} \tag{3.3}$$

$$V_{ij}(x) = f_{ij}(x_i - x_j) \tag{3.4}$$

for functions $f_{ij}\colon \mathbb{R}^\mu \to \mathbb{R}$.

In the following, we will assume that the functions f_{ij} are relatively compact with respect to the (μ-dimensional) Laplacian, i.e.

$$f_{ij}(-\Delta + 1)^{-1} \quad \text{is compact} \ . \tag{3.5}$$

We refer to *Reed* and *Simon* IV [295, Sect. XIII.4] for this notion and further details.

To investigate Schrödinger operators with pair potentials, it is convenient to remove the center of mass motion, a procedure that is well known in classical mechanics. For this, let

$$R = R(x) := \frac{1}{M} \sum_{i=1}^{N} m_i x_i$$

with $M := \sum m_i$. $R(x)$ is the center of mass of x. Define $X := \{x \in \mathbb{R}^{N\mu} | \sum m_i x_i = 0\}$. In the $(N-1)\mu$-dimensional vector space X, we choose suitable coordinates $y_1, \ldots, y_{N-1} \in \mathbb{R}^\mu$. By "coordinates" we mean linear mappings $y_1, \ldots, y_{N-1}$ from $\mathbb{R}^{N\mu}$ to $\mathbb{R}^\mu$ such that the linear mapping $y(x) = [y_1(x), \ldots, y_{N-1}(x)]$ gives a linear isomorphism of $X \subset \mathbb{R}^{N\mu}$ and $\mathbb{R}^{(N-1)\mu}$ and $y_i(x) = 0$ if $x_1 = \cdots = x_N$.

We may choose the coordinate $y_1, \ldots, y_{N-1}$ in a way that is convenient for the problem under consideration. One possible choice are the so-called *atomic coordinates* $y_i := x_i - x_N$, $i = 1, \ldots, N-1$. As the name suggests, these coordinates are particularly useful in atomic physics where usually one particle is distinguished (the nucleus with coordinates x_N).

Once we have chosen our new coordinate system $y_1, \ldots, y_{N-1}$ of X, we compute the Laplacian on $\mathbb{R}^{N\mu}$ in terms of the coordinates $y_1, y_2, \ldots, y_{N-1}, R$ of $\mathbb{R}^{N\mu}$. Doing this we obtain in general cross terms of the form $\nabla_{y_i}\nabla_{y_j}$ $(i \neq j)$. Those expressions are called *Hughes-Eckart terms.* However, we obtain no cross terms of the form $\nabla_{y_i}\nabla_R$.

Therefore $\tilde{H}_0$ splits into a tensor product:

$$\tilde{H}_0 = \left(-\frac{1}{2M}\Delta_R\right) \otimes \mathbb{1}_{L^2(X)} + \mathbb{1}_{L^2(\mathbb{R}^\mu)} \otimes H_0 \ , \tag{3.6}$$

where Δ_R is the Laplacian with respect to the R-variable acting on $L^2(\mathbb{R}^\mu)$ and H_0 acts on $L^2(X) \cong L^2(\mathbb{R}^{(N-1)\mu})$.

The exact form of H_0 depends, of course, on our choice of $y_1, \ldots, y_{N-1}$. For atomic coordinates, for example, H_0 is given by

$$H_0 = -\sum_{i=1}^{N-1} \frac{1}{2\mu_i}\Delta_{y_i} + \sum_{i<j} \frac{1}{m_N} \nabla_{y_i} \cdot \nabla_{y_j} \tag{3.7}$$

with $1/\mu_i := 1/m_i + 1/m_N$. A pair potential V does not depend on the coordinate R, thus the Hamiltonian $\tilde{H} = \tilde{H}_0 + V$ splits into

$$\tilde{H} = -\left(\frac{1}{2M}\Delta_R\right) \otimes \mathbb{1} + \mathbb{1} \otimes H \tag{3.8}$$

with $H = H_0 + V$. Equation (3.8) expresses that the center of mass of our system will move like a free particle, whereas the relative motion of the particles is governed by the Hamiltonian H.

There is an interesting, more systematic way to look at the separation of the center of mass motion, due to *Sigalov* and *Sigal* [315]. Let us introduce a scalar product $\tilde{g}(\cdot, \cdot)$ in $\mathbb{R}^{\mu N}$ by $\tilde{g}(x, y) = \sum_{i=1}^{N} 2m_i x_i \cdot y_i$. Then $\tilde{H}_0$ is the Laplace-Beltrami operator with respect to the scalar product $\tilde{g}$.

We note that the Laplace-Beltrami operator can be defined on any Riemannian manifold with Riemannian structure g (see e.g. *Spivak* [348] or virtually any book on Riemannian geometry). We will restrict ourselves to the case of linear spaces with scalar product.

Let $x_1, \ldots, x_n$ be the coordinates in $\mathbb{R}^n$ and $\tilde{g}(\cdot, \cdot)$ be a scalar product in $\mathbb{R}^n$. Then $\tilde{g}(x, y) = x^t \tilde{G} y$ for a suitable matrix $\tilde{G}$. The *Laplace-Beltrami* $\Delta_{\tilde{g}}$ operator with respect to g is defined by

$$\Delta_{\tilde{g}} := \sum_{i,j=1}^{n} \frac{\partial}{\partial x_i} (\tilde{G}^{-1})_{ij} \frac{\partial}{\partial x_j} \ . \tag{3.9}$$

Specializing to $\mathbb{R}^{\mu N}$ with the above introduced scalar product, we see that (3.9) is just a complicated way to introduce $\tilde{H}_0$.

The operator π defined by

$$\pi(x) = (x_1 - R(x), x_2 - R(x), \dots, x_N - R(x))$$

is the orthogonal (w.r.t. $\tilde{g}$) projection onto X. Then $\mathbb{R}^{\mu N}$ is the direct sum of the subspace X and the (μ-dimensional) space $X^\perp = \{x \in \mathbb{R}^{\mu N} | x_i = x_j \text{ all } i, j\}$. Let us denote the scalar product $\tilde{g}$ restricted to X by g. Then H_0 is the Laplace-Beltrami operator on X with respect to g.

For a two-body system condition (3.5) forces the potential $V(x)$ to decay, at least in some weak sense, as $|x|$ goes to infinity. In sharp contrast to this, in an N-body system (with $N > 2$), V will not decay at infinity even in the case that all f_{ij} have compact support. This is due to the fact that the vector $(y_1, \dots, y_{N-1})$ may go to infinity while, e.g., y_1 may remain constant. This fact that V will not decay in certain "tubes" around the direction $x_i - x_j$ makes the general N-body theory so complicated (and so rich!).

A common approximation that is used in atomic physics is to take the nuclear mass to be infinite, that is, one looks at the operator that results after removing the center of mass, using atomic coordinates and then taking the mass of one particle to infinity. This operator looks much like an $N - 1$ body Hamiltonian before its center of mass term is removed, but with additional potentials added that only depend on the location of the particles relative to the origin. Certain arguments in the theory of N-body systems must be slightly modified to handle this situation. Since these modifications are always simple, we will settle for placing the reader on notice that one should look for these places and make the appropriate modifications if one is interested in this infinite mass situation. Even though we have not explicitly given the proofs in the infinite mass case, we will occasionally discuss this case and use the results corresponding to those we have proven in the case where all masses are finite.

To take into account that some of the particles may remain close together while others will move away, we introduce the notion of clusters. As a rule, our partitions of unity introduced later will reflect the cluster structure of phase space.

By a *cluster decomposition* a, we mean a set $a = \{A_1, \dots, A_K\}$ such that

$$\bigcup_{i=1}^{K} A_i = \{1, \dots, N\}$$

and $A_i \cap A_j = \phi$ for $i \neq j$. The elements A_i of a are called *clusters*, and $\# a$ will denote K, the number of clusters of a. We use the notation $(ij) \subset a$ to express that i and j belong to the same cluster of a.

For a given cluster decomposition a, we define the *intercluster interaction* by

$$I_a := \sum_{(ij) \not\subset a} V_{ij} \tag{3.10}$$

and the *internal Hamiltonian*

$$H(a) = H - I_a \ . \tag{3.11}$$

Let $a = \{A_1, \dots, A_k\}$ be a cluster decomposition. We define the center of mass of the cluster A_i by

$$R_i^a := R_i^a(x) := \frac{1}{M_i} \sum_{j \in A_i} m_j x_j$$

with $M_i = \sum_{j \in A_i} m_j$. Let us set

$$X^a := \{x \in X \,|\, R_i^a(x) = 0 \quad i = 1, \dots, j\} \ . \tag{3.12}$$

Then X splits into a direct sum

$$X = X^a \oplus X_a \ ,$$

where $X_a := \{x \in X \,|\, x_i = x_j \text{ if } (ij) \subset a\}$. X^a and X_a are orthogonal with respect to the above introduced scalar product g. The Hilbert space $L^2(X)$ splits into a tensor product: $L^2(X) \cong L^2(X^a) \otimes L^2(X_a)$ and

$$H_0 = (h_0^a \otimes \mathbb{1}) + (\mathbb{1} \otimes T_a) \tag{3.13}$$

similar to the removal of the center of mass [see (3.6)]. [In (3.13) h_0^a is the Laplace-Beltrami operator corresponding to g restricted to X^a.] Since $H(a)$ does not depend on any interaction term between different clusters in a, we also have

$$H(a) = (h^a \otimes \mathbb{1}) + (\mathbb{1} \otimes T_a) \ . \tag{3.14}$$

The operator h^a describes the internal dynamics in the clusters of a. Let ε^a denote the set of eigenvalues of h^a. We define the set T of *thresholds* of H by

$$T = \bigcup_{\# a > 1} \varepsilon^a \tag{3.15}$$

with the convention that $\varepsilon^a = \{0\}$ if $\# a = N$; thus $0 \in T$.

3.3 The HVZ-Theorem

Now we are ready to define a partition of unity which will allow us to determine the essential spectrum of N-body operators. This partition of unity was introduced by *Simon* [323]. A related partition into sets was used by *Ruelle* [299] in quantum field theory.

Definition 3.4. *A Ruelle-Simon partition of unity* is a partition of unity $\{J_a\}$

indexed by all two cluster decompositions a (i.e. all a with $\# a = 2$) with the following properties:

(i) J_a is homogeneous of degree zero outside the unit sphere, i.e. $J_a(\lambda r) = J_a(r)$ for all $\lambda > 1$, $|r| = 1$.
(ii) There exists a constant $C > 0$ such that

$$\operatorname{supp} J_a \cap \{x \| x| > 1\} \subset \{x \| x_i - x_j| \geq C|x| \text{ for all } (ij) \not\subset a\} .$$

Remarks. (1) We require homogeneity (i) only outside the unit sphere to avoid a singularity at the origin. For the moment, we are not interested in the region around the origin, where all particles are close together.

(2) Condition (ii) says that J_a lives where the particles in different clusters of a are far away from each other. Note that on supp J_a, two particles belonging to the same cluster need not be close to each other.

Of course, we have to prove the existence of a Ruelle-Simon partition of unity.

Proposition 3.5. There exists a Ruelle-Simon partition of unity.

Proof. Once we have a (locally) finite cover $\{U_n\}$ of the whole space, it is a standard procedure to construct a partition of unity, $\{j_n\}$, with $\operatorname{supp} j_n \subset U_n$. Moreover, in the present case it is enough to construct the partition of unity on the unit sphere since we may extend it to the exterior by homogeneity, and to the interior in an (almost) arbitrary way. Thus, it suffices to find a constant $C > 0$ such that the sets $\{U_a^C\}_{\# a = 2}$ defined by

$$U_a^C = \{x \| x| = 1, |x_i - x_j| > 2C \text{ for all } (ij) \not\subset a\}$$

covers the unit sphere, S. Since

$$\bigcup_{C>0} \bigcup_{\# a = 2} U_a^C = S ,$$

such a constant C exists by compactness. $\square$

One can use a little geometry in place of compactness and obtain an explicit value for C; see *Simon* [323].

The following proposition states properties of the Ruelle-Simon partition of unity that are crucial for the proof of the HVZ-theorem.

Proposition 3.6. (i) $(\nabla J_a)(H_0 + 1)^{-1}$ is compact.
(ii) $(J_a I_a)(H_0 + 1)^{-1}$ is compact.

Proof. (i) ∇J_a is continuous and homogeneous of degree -1 near infinity, so it tends to zero at infinity. Hence $\nabla J_a(H_0 + 1)^{-1}$ is compact (see e.g. *Reed* and *Simon* IV, XIII.4 [295]).

(ii) We prove (ii) for $f_{ij} \in C_0^\infty$; the general case follows by a straightforward approximation argument. For $f_{ij} \in C_0^\infty$, the function $J_a I_a$ has compact support (while I_a itself does not!). $\square$

HVZ-Theorem 3.7. For a cluster decomposition a, define $\Sigma(a) := \inf \sigma(H(a))$ and $\Sigma := \min_{\# a=2} \Sigma(a)$. Then

$$\sigma_{\text{ess}}(H) = [\Sigma, \infty) .$$

The HVZ-theorem was proven by *Zhislin* [382], *van Winter* [359] and *Hunziker* [171] (with increasing generality).

Proof. "Easy part": $[\Sigma, \infty) \subset \sigma_{\text{ess}}(H)$. This inclusion can be shown using the Weyl's criterion (see e.g. *Reed* and *Simon* I, VII.12 [292]) by construction of an appropriate sequence of trial functions; see *Reed* and *Simon* IV, XIII [295].

"Hard part": $\sigma_{\text{ess}}(H) \subset [\Sigma, \infty)$. By the IMS-localization formula

$$H = \sum J_a H(a) J_a + \sum I_a J_a^2 - \sum |\nabla J_a|^2$$

where $\{J_a\}_{\# a=2}$ is a Ruelle-Simon partition of unity. By Proposition 3.6, we know that both $I_a J_a$ and $|\nabla J_a|$ are relatively compact with respect to H_0. Therefore Weyl's theorem (see e.g. *Reed* and *Simon* IV, XIII.14 [295]) tells us that

$$\sigma_{\text{ess}}(H) = \sigma_{\text{ess}}(\sum J_a H(a) J_a) .$$

By definition of Σ, we have

$$H(a) \geq \Sigma(a) \geq \Sigma .$$

Hence,

$$\sum_{\# a=2} J_a H(a) J_a \geq \sum_{\# a=2} \Sigma J_a^2 = \Sigma .$$

Thus, $\sigma_{\text{ess}}(H) = \sigma_{\text{ess}}(\sum J_a H(a) J_a) \subset [\Sigma, \infty)$. □

We will present a second geometric proof of the HVZ-theorem. We need the following result which will be used again in the next chapter.

Proposition 3.8. Let $\{J_a\}$ denote a Ruelle-Simon partition of unity. For any $f \in C_\infty(\mathbb{R})$, the continuous functions vanishing at infinity, we have

(i) $[f(H(b)), J_a]$ is compact,
(ii) $[f(H(a)) - f(H)] J_a$ is compact.
(iii) If, furthermore, f has compact support, then both

$$H_0(f(H(b)) - f(H)) J_a \text{ and } H_0([f(H(b)), J_a]) \text{ are compact} .$$

Proof. We prove (i) and (ii) for the functions $f_z(x) = (x - z)^{-1}$ for $z \in \mathbb{C} \backslash \mathbb{R}$ and use the Stone-Weierstrass gavotte (see the Appendix to Chap. 3) to obtain the results for all of $C_\infty(\mathbb{R})$.

(i) We compute

$$
\begin{aligned}
[(H(b)-z)^{-1}, J_a] &= (H(b)-z)^{-1}[J_a, H_0](H(b)-z)^{-1} \\
&= (H(b)-z)^{-1}(\Delta J_a + 2\nabla J_a \nabla)(H(b)-z)^{-1} \\
&= \{(H(b)-z)^{-1}(\Delta J_a)\}(H(b)-z)^{-1} \\
&\quad + \{(H(b)-z)^{-1} 2\nabla J_a\} \nabla \cdot (H(b)-z)^{-1} \ . \qquad (3.16)
\end{aligned}
$$

Since ΔJ_a and ∇J_a are homogeneous of degree -2 and -1, respectively, near infinity, the terms in curly brackets are compact. Moreover, $\nabla \cdot (H(b)-z)^{-1}$ is bounded. Hence the whole expression is compact.

(ii) $\{(H(a)-z)^{-1} - (H-z)^{-1}\} J_a$

$$
\begin{aligned}
&= (H(a)-z)^{-1} I_a (H-z)^{-1} J_a \\
&= (H(a)-z)^{-1} I_a J_a (H-z)^{-1} + (H(a)-z)^{-1} I_a [(H-z)^{-1}, J_a] \ .
\end{aligned}
$$

The first term is compact by Proposition 3.6; the commutator in the second terms is compact by the argument in (i); thus, the whole expression is compact.
(iii) Set $g(x) := (x+i) f(x)$, then $g \in C_\infty$

$$
\begin{aligned}
H_0[f(H(b)), J_a] &= H_0(H(b)+i)^{-1}(H(b)+i)[f(H(b)), J_a] \\
&= H_0(H(b)+i)^{-1}\{[g(H(b)), J_a] - [H_0, J_a] f(H(b))\} \ . \qquad (3.17)
\end{aligned}
$$

By (i), $[g(H(b)), J_a]$ is compact. Furthermore

$$
\begin{aligned}
[H_0, J_a] f(H(b)) &= (-\Delta J_a - 2\nabla J_a \cdot \nabla)(H_0+i)^{-1}(H_0+i) \\
&\quad \times (H(b)+i)^{-1} g(H(b)) \ .
\end{aligned}
$$

Since J_a is homogeneous of degree zero, both $\Delta J_a (H_0+i)^{-1}$ and $\nabla J_a \cdot \nabla (H_0+i)^{-1}$ are compact. Hence the right-hand side of (3.17) is compact, which proves the first part of (iii). The second one can be proven in an analogous way. □

We now give a second proof of the HVZ-theorem: Let f be a continuous function on $\mathbb{R}$ with compact support below Σ. By Proposition 3.8, we know that

$$
C := \sum_{\# a = 2} [f(H) - f(H(a))] J_a^2
$$

is a compact operator. But

$$
f(H) = \sum [f(H) - f(H(a))] J_a^2 + \sum f(H(a)) J_a^2 = C \ ,
$$

since $\operatorname{supp} f \cap \sigma(H(a)) = \phi$. Thus, $f(H)$ is compact. By an operator inequality, it follows that the spectral projections $E(\lambda)$ of H are compact for $\lambda < \Sigma$, hence $\sigma_{\mathrm{ess}}(H) \subset [\Sigma, \infty)$. □

3.4 More on the Essential Spectrum

It is "general wisdom" that the essential spectrum of Schrödinger operators comes from "what happens very far away." The two theorems of this section make the above statement precise. They determine the essential spectrum, but not as explicitly as the HVZ-theorem does. On the other hand, they apply to potentials that do not decay at infinity even in the very weak sense of (3.4) and (3.5). Thus, for example, periodic, almost periodic and random potentials are included in those theorems while they are not in the HVZ-theorem.

The crucial property of Schrödinger operators that makes the above "general wisdom" true is local compactness, a concept particularly emphasized in the work of *Enss* (see e.g. [100]).

Definition 3.9. A Schrödinger operator $H = H_0 + V$ is said to have the *local compactness property* if $f(x)(H + i)^{-1}$ is compact for any bounded function f with compact support.

Virtually all Schrödinger operators of physical interest obey the local compactness property. For example, if V is operator bounded (or merely form bounded) with respect to H_0, then H has the local compactness property.

We will assume throughout this section that V is operator bounded. Notice, however, that we do not require any decay conditions at infinity. For those operators, we have the easy lemma:

Lemma 3.10. Suppose that V is operator bounded with respect to H_0. Let f be a bounded function with compact support. Then both $f(x)(H + i)^{-1}$ and $f(x)V(H + i)^{-1}$ are compact operators.

Proof.

$$f(x)V(H + i)^{-1} = (f(x)(H_0 + i)^{-1/2})[V(H_0 + i)^{-1/2}](H_0 + i)(H + i)^{-1} .$$

The first term in the above expression is compact, the others are bounded. The proof that $f(x)(H + i)^{-1}$ is compact is obvious from the above. □

We now state and prove the first of the announced theorems. We denote, by $B_n := \{x \mid |x| \leq n\}$, the ball around the origin, of radius n.

Theorem 3.11. Let V be operator bounded with respect to H_0. $H := H_0 + V$. Then $\lambda \in \sigma_{ess}(H)$ if and only if there exists a sequence of functions $\varphi_n \in C_0^\infty(\mathbb{R}^\nu \backslash B_n)$ with $\|\varphi_n\|_2 = 1$ such that

$$\|(H - \lambda)\varphi_n\| \to 0 . \tag{3.18}$$

Remark. (1) By the Weyl criterion, we know that $\lambda \in \sigma_{ess}(H)$ is equivalent to the existence of a sequence of trial functions $\{\varphi_n\}$ obeying $\|\varphi_n\|_2 = 1$ and $\varphi_n \xrightarrow{w} 0$ with

(3.18). Thus, Theorem 3.11 tells us that the weak convergence of the φ_n actually takes place in a particular way.

(2) The theorem can be proven under much weaker conditions on V. All we have to ensure is that the conclusions of Lemma 3.10 remain true and that C_0^∞ is an operator core.

Proof. By remark (1), the "$\Leftarrow$"-direction is trivial. Let us assume that $\lambda \in \sigma_{\text{ess}}(H)$. By the Weyl criterion there exists a sequence $\psi_n \in C_0^\infty(\mathbb{R}^\nu)$, $\|\psi_n\|_2 = 1$, $\psi_n \xrightarrow{w} 0$ such that

$$\|(H-\lambda)\psi_n\| \to 0 \ . \tag{3.19}$$

For any n choose a function $\chi_n \in C^\infty$, $0 \le \chi_n(x) \le 1$ such that $\chi_n(x) = 1$ for $|x| \ge n+1$, and $\chi_n(x) = 0$ for $|x| \le n$. We claim that, for any n, there exists an $i = i(n) > n$ such that

$$\|(1-\chi_n)\psi_{i(n)}\| < 1/n \ , \tag{3.20}$$

$$\|(\Delta\chi_n)\psi_{i(n)}\| < 1/n \ , \quad \text{and} \tag{3.21}$$

$$\|\nabla\chi_n\nabla\psi_{i(n)}\| < 1/n \ . \tag{3.22}$$

Assuming this for the moment, we set

$$\varphi_n := \frac{\chi_n\psi_{i(n)}}{\|\chi_n\psi_{i(n)}\|} \ .$$

We have

$$\|(H-\lambda)\varphi_n\| \le \frac{1}{\|\chi_n\psi_i\|}\{\|(H-\lambda)\psi_i\| + 2\|\nabla\chi_n\nabla\psi_i\| + \|(\Delta\chi_n)\psi_i\|\} \ ,$$

which goes to zero by (3.19–22). Thus, it remains to prove (3.20–22). For this, let χ be a C_0^∞-function, then:

$$\begin{aligned}\|\chi\psi_n\| &= \|\chi(H+\mathrm{i})^{-1}[(H-\lambda)+(\mathrm{i}+\lambda)]\psi_n\| \\ &\le \|\chi(H+\mathrm{i})^{-1}\|\,\|(H-\lambda)\psi_n\| \\ &\quad + |\lambda+\mathrm{i}|\,\|\chi(H+\mathrm{i})^{-1}\psi_n\| \ .\end{aligned}$$

The first term goes to zero because of (3.19); the second one goes to zero since $\psi_n \xrightarrow{w} 0$ and $\chi(H+\mathrm{i})^{-1}$ is compact by Lemma 3.10. This proves (3.20) and (3.21). A similar proof applies to (3.22). □

There are numerous related results, such as

$$\sigma_{\text{ess}}(H) = \bigcap_n \sigma(H + n\chi_{\{x \mid |x| \le n\}}) \ .$$

We now prove a result due to *Persson* [282] that gives the infimum of the essential spectrum in terms of a "min-max"-type expression:

Theorem 3.12 (Persson). Let V be operator bounded. Then

$$\inf \sigma_{\text{ess}}(H) = \sup_{\substack{K \subset \mathbb{R}^\nu \\ \text{compact}}} \inf_{\substack{\varphi \in C_0^\infty(\mathbb{R}^\nu \backslash K) \\ \|\varphi\|=1}} \langle \varphi, H\varphi \rangle \ .$$

Remarks. 1) Persson's theorem says that $\inf \sigma_{\text{ess}}$ is not effected by "what happens" in any compact set.

2) For any fixed K, the term

$$\inf_{\varphi \in C_0^\infty(\mathbb{R}^\nu \backslash K)} \langle \varphi, H\varphi \rangle$$

is just the ground-state energy for the Hamiltonian H on $L^2(\mathbb{R}^\nu \backslash K)$ with Dirichlet boundary conditions at ∂K. (See *Reed* and *Simon* IV [295]). Thus, $\inf \sigma_{\text{ess}}(H)$ is the sup over all these ground state energies.

3) Theorem 3.12 can be proven under weaker assumptions. See the book of *Agmon* [3] for another proof.

Proof. "$\geq$": Let λ_0 be the infimum of $\sigma_{\text{ess}}(H)$. Then $\lambda_0 \in \sigma_{\text{ess}}(H)$, and by Theorem 3.11, we can find a sequence $\varphi_n \in C_0^\infty(\mathbb{R}^\nu \backslash B_n)$ with $\|\varphi_n\| = 1$ and $(H - \lambda_0)\varphi_n \to 0$. Thus,

$$\sup_{\substack{K \\ \text{compact}}} \inf_{\substack{\varphi \in C_0^\infty(\mathbb{R}^\nu \backslash K) \\ \|\varphi\|=1}} \langle \varphi, H\varphi \rangle = \overline{\lim_n} \inf_{\substack{\varphi \in C_0^\infty(\mathbb{R}^\nu \backslash B_n) \\ \|\varphi\|=1}} \langle \varphi, H\varphi \rangle \leq \overline{\lim_n} \langle \varphi_n, H\varphi_n \rangle$$

$$\leq \overline{\lim} \langle \varphi_n, (H - \lambda_0)\varphi_n \rangle + \overline{\lim} \langle \varphi_n, \lambda_0 \varphi_n \rangle = \lambda_0$$

"$\leq$": Define

$$\mu_n := \sup_{\psi_1, \ldots, \psi_{n-1}} \inf_{\substack{\varphi \in C_0^\infty \\ \varphi \perp \psi_1, \ldots, \psi_{n-1};\, \|\varphi\|=1}} \langle \varphi, H\varphi \rangle \ .$$

By the min-max theorem (see *Reed* and *Simon* IV, XIII.1 [295]), μ_n is the nth eigenvalue of H from below counting multiplicity. If there are only n_0 eigenvalues below $\lambda_0 := \inf \sigma_{\text{ess}}(H)$, then $\mu_{n_0+1} = \mu_{n_0+2} = \cdots = \lambda_0$. Moreover, if there are infinitely many eigenvalues below $\sigma_{\text{ess}}(H)$, then $\mu_n \to \inf \sigma_{\text{ess}}(H)$. Thus, to show

$$\lambda_0 := \inf \sigma_{\text{ess}}(H) \leq \sup_K \inf_{\substack{\varphi \in C_0^\infty(\mathbb{R}^\nu \backslash K) \\ \|\varphi\|=1}} \langle \varphi, H\varphi \rangle =: \nu_0 \ ,$$

it suffices to show that $\nu_0 \geq \mu_n$ for all n. Since

$$\mu_1 = \inf_{\substack{\varphi \in C_0^\infty(\mathbb{R}^\nu) \\ \|\varphi\|=1}} \langle \varphi, H\varphi \rangle \ ,$$

obviously $\nu_0 \geq \mu_1$.

Suppose now $n > 1$ and $\nu_0 \geq \mu_{n-1}$. If $\mu_{n-1} = \lambda_0$, we are done. If $\mu_{n-1} < \lambda_0$, then

$$\mu_n = \inf_{\substack{\varphi \perp \rho_1, \ldots, \rho_{n-1} \\ \|\varphi\|=1}} \langle \varphi, H\varphi \rangle ,$$

where the ρ_i are normalized eigenfunctions corresponding to the eigenvalues μ_i, and moreover, $\rho_1, \ldots, \rho_{n-1}$ span the eigenspaces of the μ_i. Now choose $\varepsilon > 0$ and K_0 so large that

$$\int_{\mathbb{R}^\nu \setminus K_0} |\rho_i(x)|^2 \, dx < \varepsilon \quad \text{for } i = 1, \ldots, n-1 . \tag{3.23}$$

Define for any φ the function $\tilde{\varphi}(x) := \varphi(x) - \sum_{i=1}^{n} \langle \varphi, \rho_i \rangle \rho_i(x)$. Then $\tilde{\varphi} \perp \rho_i$. We have

$$\begin{aligned} \nu_0 &\geq \inf_{\substack{\varphi \in C_0^\infty(\mathbb{R}^\nu \setminus K_0) \\ \|\varphi\|=1}} \langle \varphi, H\varphi \rangle \\ &= \inf_{\substack{\varphi \in C_0^\infty(\mathbb{R}^\nu \setminus K_0) \\ \|\varphi\|=1}} \{ \langle \tilde{\varphi}, H\tilde{\varphi} \rangle + \langle \tilde{\varphi}, \Sigma \langle \varphi, \rho_i \rangle \mu_i \rho_i \rangle + \langle \Sigma \langle \varphi, \rho_i \rangle \mu_i \rho_i, \varphi \rangle \} . \end{aligned}$$

For any $\varphi \in C_0^\infty(\mathbb{R}^\nu \setminus K_0)$, $|\langle \varphi, \rho_i \rangle| < \varepsilon^{1/2} \|\varphi\|$ by (3.23), thus

$$\begin{aligned} \nu_0 &\geq \inf_{\substack{\varphi \in C_0^\infty(\mathbb{R}^\nu \setminus K_0) \\ \|\varphi\|=1}} \langle \tilde{\varphi}, H\tilde{\varphi} \rangle - C\varepsilon \\ &\geq (1 - C'\varepsilon) \inf_{\substack{\varphi \in C_0^\infty(\mathbb{R}^\nu \setminus K_0) \\ \|\varphi\|=1}} \frac{\langle \tilde{\varphi}, H\tilde{\varphi} \rangle}{\langle \tilde{\varphi}, \tilde{\varphi} \rangle} - C\varepsilon \\ &\geq (1 - C'\varepsilon) \inf_{\tilde{\varphi} \perp \rho_1, \ldots, \rho_{n-1}} \frac{\langle \tilde{\varphi}, H\tilde{\varphi} \rangle}{\langle \tilde{\varphi}, \tilde{\varphi} \rangle} - C\varepsilon \\ &\geq (1 - C'\varepsilon)\mu_n - C\varepsilon . \end{aligned}$$

Since ε was arbitrary, we have

$$\nu_0 \geq \mu_n \quad \text{and hence}$$

$$\nu_0 \geq \lambda_0 . \quad \square$$

3.5 A Theorem of Klaus: Widely Separated Bumps

Before we turn to applications of geometric ideas in atomic physics, we use geometric methods in a different context:

Theorem 3.13 (Klaus). Assume $V \in C_0^\infty(\mathbb{R})$, $V \leq 0$. Let $\{x_n\}_{n \in \mathbb{Z}}$ be a sequence of real numbers satisfying $x_n \leq x_{n+1}$ and $|x_n - x_{n+1}| \to \infty$ as $|n| \to \infty$.

Define $W := \sum_{n\in Z} V(x - x_n)$, $H := -d^2/dx^2 + W$ and $H' := -d^2/dx^2 + V$. Then

$$\sigma_{\text{ess}}(H) = \sigma(H') \ .$$

Remark. The above theorem is a special case of a theorem due to *Klaus* [214], who proved it by Birman-Schwinger techniques.

The negative eigenvalues of H' are isolated points of the essential spectrum of H, hence they cannot belong to the continuous spectrum of H. But they also cannot be eigenvalues of infinite multiplicity, because we have a one dimensional problem. Thus, they must be accumulation points of the discrete spectrum of H. Such a phenomenon is impossible if the potential decays at infinity (by the HVZ-theorem). In Chaps. 9 and 10, however, we will demonstrate other examples of "unexpected" spectral phenomena: singular continuous spectrum and dense point spectrum.

Proof. The direction "$\sigma_{\text{ess}}(H) \supset \sigma(H')$" can be proven by a standard application of Weyl's criterion. We only argue "$\sigma(H') \supset \sigma_{\text{ess}}(H)$". Let us define $V_n(x) = V(x - x_n)$ and $H_n = -d^2/dx^2 + V_n$. For notational convenience, we will assume that $\operatorname{supp} V_n \cap \operatorname{supp} V_m = \phi$ for $n \neq m$. Under this assumption, we may choose a partition of unity $\{j_n\}_{n\in\mathbb{Z}}$ with the following properties:

(i) $j_n W = j_n V_n$,
(ii) $j_n j_m = 0$ if $|n - m| > 2$,
(iii) $j_n \in C_0^\infty$ and $|\nabla j_n|_\infty \to 0$, $|\Delta j_n|_\infty \to 0$ as $|n| \to \infty$.

We define

$$A(z) := \sum j_n (H_n - z)^{-1} j_n \ . \tag{3.24}$$

It is not difficult to see that $A(z)$ is bounded and analytic as a function of z on $\Omega := \mathbb{C}\backslash\sigma(H')$. (The reader may adjust the proof of the lemma below.) We will show that

$$A(z) = (H - z)^{-1}[1 + B(z)] \tag{3.25}$$

for compact operators $B(z)$ analytic on Ω. Once we know (3.25), the analytic Fredholm theorem (see e.g. *Reed* and *Simon* I [292], Theorem VI.14) tells us that the inverse of $1 + B(z)$ exists on $\Omega\backslash D$ for a discrete set (in Ω), D. [From the definition of $B(z)$, we see that $\|B(z)\| \to 0$ as $z \to -\infty$, so $1 + B(z)$ is invertible for some z.] Moreover, the residues at the poles are finite rank operators. Thus, by (3.25), we can continue $(H - z)^{-1}$ to an analytic function on $\Omega\backslash D$, the residues of which are finite rank operators at the points of D. This implies that $\Omega \cap \sigma_{\text{ess}}(H) = \phi$, and we obtain the desired result.

To prove (3.25), we define

$$B_n(z) := [H_0, j_n](H_n - z)^{-1} j_n \ . \tag{3.26}$$

Here $B_n(z)$ is compact and analytic on Ω. We compute:

$$
\begin{aligned}
(H-z)^{-1}B_n(z) &= (H-z)^{-1}[(H-z),j_n](H_n-z)^{-1}j_n \\
&= (H-z)^{-1}[(H-z)j_n - j_n(H_n-z)](H_n-z)^{-1}j_n \\
&\quad \text{(we used: } j_nW = j_nV_n\text{)} \\
&= j_n(H_n-z)^{-1}j_n - (H-z)^{-1}j_n^2 \ .
\end{aligned}
$$

To prove that $B(z) = \sum B_n(z)$ is well defined and compact, we make use of the following lemma:

Lemma. Let C_n, $n \in \mathbb{Z}$ be bounded operators, and let f_n, g_n be bounded functions satisfying $\operatorname{supp} f_n \cap \operatorname{supp} f_m = \phi$, and $\operatorname{supp} g_n \cap \operatorname{supp} g_m = \phi$ for $n \neq m$. If $\| f_n C_n g_n \| \to 0$ as $|n| \to \infty$, then the series

$$\sum_{n=-\infty}^{+\infty} f_n C_n g_n$$

converges in norm.

Proof. Denote by χ_n and η_n the characteristic functions of $\operatorname{supp} f_n$ and $\operatorname{supp} g_n$, respectively. Then

$$
\begin{aligned}
&\left| \left\langle \psi, \sum_{|n|>M} f_n C_n g_n \varphi \right\rangle \right| = \left| \sum_{|n|>M} \langle \chi_n \psi, (f_n C_n g_n) \eta_n \varphi \rangle \right| \\
&\leq \sup_{|n|>M} \| f_n C_n g_n \| \sum_{|n|>M} \| \chi_n \psi \| \, \| \eta_n \varphi \| \\
&\leq \varepsilon \left(\sum_{|n|>M} \| \chi_n \psi \|^2 \right)^{1/2} \left(\sum_{|n|>M} \| \eta_n \varphi \|^2 \right)^{1/2} \\
&\leq \varepsilon \| \psi \| \, \| \varphi \| \quad \text{for } M \text{ large enough} \ . \quad \square
\end{aligned}
$$

Proof of the Theorem (*continued*). Now we write $B(z)$ as

$$
\begin{aligned}
B(z) &= \sum_{n=-\infty}^{+\infty} [H_0, j_n](H_n - z)^{-1} j_n \\
&= \sum_{n \text{ odd}} - j_n''(H_n - z)^{-1} j_n + \sum_{n \text{ odd}} - 2j_n'(\nabla (H_n - z)^{-1}) j_n \\
&\quad + \sum_{n \text{ even}} - j_n''(H_n - z)^{-1} j_n + \sum_{n \text{ even}} - 2j_n'(\nabla (H_n - z)^{-1}) j_n \ .
\end{aligned}
$$

We apply the lemma to any of the four terms separately. Since we have shown norm convergence, we conclude that $B(z)$ is compact. $\square$

3.6 Applications to Atomic Physics: A Warm-Up

For the rest of this chapter, we will be concerned with questions arising from atomic physics. We begin with a somewhat artificial example, which nevertheless will be illuminating for more realistic problems. Let us consider the Hamiltonian

$$H(A) = -\Delta_1 - \Delta_2 - \frac{1}{r_1} - \frac{1}{r_2} + \frac{A}{r_{12}} \tag{3.27}$$

acting on $L^2(\mathbb{R}^{2\cdot 3})$ where, as usual, $r_i = |x_i|$, $r_{12} = |x_1 - x_2|$. This operator describes two electrons moving under the influence of an infinitely heavy nucleus, with the repulsion strength between the electrons given by A.

By physical reasons, one expects that $H(A)$ has no bound states for very large repulsion between the electrons, i.e., for $A \gg 1$. We shall prove this here using the localization formula. By Lieb's method (see Sect. 3.8), one can prove there is no bound state once $A \geq 2$. Numerically (see *Reinhardt* [296]), the critical value seems to be about 1.03.

The HVZ-theorem tells us that $\sigma_{\text{ess}}(H(A)) = [-\frac{1}{4}, \infty)$, since $\inf\sigma(-\Delta_1 - 1/r_1) = -\frac{1}{4}$. Thus, the expected result is equivalent to

Proposition 3.14. For A sufficiently large, we have $H(A) \geq -\frac{1}{4}$.

Proof. We choose a partition of unity, j_0, j_1, j_2 with the following properties:

$$\operatorname{supp} j_0 \subset \{x \,|\, |x| < 1\} \ ,$$

$$\operatorname{supp} j_1 \subset \{x \,|\, |x_1| > \tfrac{1}{2}|x_2|; |x| > \tfrac{1}{2}\} \ ,$$

$$\operatorname{supp} j_2 \subset \{x \,|\, |x_2| > \tfrac{1}{2}|x_1|; |x| > \tfrac{1}{2}\} \ ,$$

and so that j_1 and j_2 are homogeneous of degree zero outside the unit sphere. To dominate the localization error $\sum |\nabla j_i|^2$, we may choose A_0 sufficiently large such that

$$\frac{A_0}{r_{12}} \geq \sum |\nabla j_i|^2 \ .$$

This choice of A_0 is possible because j_1 and j_2 are homogeneous of degree zero for r large, while j_0 has compact support. By the IMS-localization formula, we can write

$$\begin{aligned} H(A) &= \sum_{i=0}^{2} j_i H(A - A_0) j_i + \left(\frac{A_0}{r_{12}} - \sum_{i=0}^{2} |\nabla j_i|^2 \right) \\ &\geq \sum_{i=0}^{2} j_i H(A - A_0) j_i \ . \end{aligned}$$

Notice that it is the long-range nature of the Coulomb interaction that helps us to control the localization error.

Next we observe that, for any $\varepsilon > 0$ and sufficiently large A, we have

$$-\frac{1}{|x_1|} - \frac{1}{|x_2|} + \frac{A - A_0}{|x_1 - x_2|} > 0$$

for all x with $|x| < 1$ and $|x_1|, |x_2| > \varepsilon$. Thus, for large A

$$j_0 H(A - A_0) j_0 \geq -\Delta_1 - \Delta_2 - \frac{\chi_\varepsilon(x_1)}{|x_1|} - \frac{\chi_\varepsilon(x_2)}{|x_2|} , \tag{3.28}$$

where χ_ε denotes the characteristic function of the ball of radius ε. We know that the Hamiltonian on the right-hand side of (3.28) has no bound states if ε is small enough. This may be seen by (almost) any bound on the number of bound states (see e.g. [295], Theorem XIII.10). Hence,

$$j_0 H(A - A_0) j_0 \geq 0 ,$$

if A is large enough. Furthermore, on $\operatorname{supp} j_1$ we have

$$j_1\left(-\frac{1}{|x_1|} + \frac{3}{|x_1 - x_2|}\right) j_1 \geq 0 ,$$

since $|x_1 - x_2| \leq |x_1| + |x_2| \leq 3|x_1|$. Therefore, if $A \geq A_0 + 3$, we have

$$j_1 H(A - A_0) j_1 \geq j_1\left(-\Delta_1 - \Delta_2 - \frac{1}{|x_2|}\right) j_1 ,$$

and by symmetry

$$j_2 H(A - A_0) j_2 \geq j_2\left(-\Delta_1 - \Delta_2 - \frac{1}{|x_1|}\right) j_2 .$$

Since $-\Delta_i - \Delta_j - 1/|x_j| \geq -1/4$, we conclude

$$H(A) \geq j_1 H(A - A_0) j_1 + j_2 H(A - A_0) j_2 \geq -\tfrac{1}{4}$$

for sufficiently large A. □

In this argument, as well as in the next section, the separate region near zero is needed, because without it the localization error near zero is $O(r^{-2})$, which cannot be controlled by Coulomb potentials.

3.7 The Ruskai-Sigal Theorem

Now we come to an important application of geometric methods in atomic physics. We consider the Hamiltonian of an atom with nucleus charge Z, and N electrons:

$$H_N(Z) := \sum_{i=1}^{N}\left(-\Delta_i - \frac{Z}{|x_i|}\right) + \sum_{i<j} \frac{1}{|x_i - x_j|} . \tag{3.29}$$

By physical reasoning, one would expect that a nucleus of charge Z can bind only a limited number of electrons, because at some point the attraction of the nucleus should be dominated by the mutual repulsion of the electrons.

Let us give a more mathematical formulation of this expectation. Define

$$E(N,Z) := \inf \sigma(H_n(Z)) \; . \tag{3.30}$$

Then, by the HVZ-theorem

$$\sigma_{\text{ess}}(H_{N+1}(Z)) = [E(N,Z), \infty) \; .$$

Thus, our expectation can be formulated as

$$E(N+1,Z) = E(N,Z) \tag{3.31}$$

for large N. We emphasize that we are dealing with the discrete spectrum exclusively. Thus, we make no assertion on embedded eigenvalues.

Theorem 3.15 (Ruskai-Sigal Theorem). For any Z, there exists $N_{\max}(Z)$ such that

$$E(N+1,Z) = E(N,Z)$$

for all $N \geq N_{\max}(Z)$. Moreover, for fermionic particles, we have

$$\overline{\lim_{Z\to\infty}} \frac{N_{\max}(Z)}{Z} \leq 2 \; . \tag{3.32}$$

Remarks. (1) Theorem 3.15 was proven by *Ruskai* [302, 303] and by *Sigal* [310, 313].

(2) There exists an improved version of the Ruskai-Sigal theorem due to *Lieb* [231, 232] which gives $N_{\max}(Z) \leq 2Z$ for all integers Z. We present this theorem, as well as Lieb's elegant proof, in Sect. 3.8. Our proof below, however, follows *Sigal* [310]. Although Sigal's proof is much more lengthy than Lieb's, we present it here for two reasons. Firstly, we feel that it gives more physical insight into the phenomena, and secondly, there is another improvement of the Ruskai-Sigal theorem by *Lieb, Sigal, Simon* and *Thirring* [233] that states that the limit in (3.32) is actually 1. The proof of Lieb, Sigal, Simon and Thirring is a refinement of Sigal's proof which we give below.

(3) We will take into account the fermionic nature of our particles (for the second part of the proof) only by using the Pauli exclusion principle. A more careful investigation should check that one always discusses Hamiltonians restricted to antisymmetric states. For this, we refer to Appendix 4 in Sigal's paper.

Proof. Sketch of the Ideas. We divide the configuration space into $N+1$ pieces: $A_0, A_1, \ldots, A_N$. The first part, A_0, consists of the region where all the electrons are close to the nucleus, and A_i essentially consists of the region where the ith particle has larger distance to the nucleus than any other electron. We then construct a partition of unity, J_i, with $\operatorname{supp} J_i \subset A_i$ and with good control on the localization error $\sum_{i=0}^N |\nabla J_i(x)|^2$. On A_0, the strong repulsion between the electrons will dominate both the attraction by the nucleus and the localization error, provided N is sufficiently large. On A_i ($i \geq 1$), we split H_N into an $(N-1)$-body operator H_{N-1} corresponding to the electrons $1, 2, \ldots, i-1, i+1, \ldots, N$ and the additional terms due to the ith electron. Since that one is further from the nucleus than any other electron, the distance between the electrons i and j is at most twice the distance of the ith electron from the nucleus. Therefore, the repulsion between electron i and the other electrons dominates the attraction of the ith electron by the nucleus as well as the localization error if N is large enough.

Details of the Proof. Define

$$x_\infty(x) := \max_{i=1,\ldots,N} |x_i| .$$

$$A_0 := \{x \,|\, |x_j| < \rho \quad \text{for } j = 1, \ldots, N\}$$

$$A_i := \left\{x \,\Big|\, |x_i| > (1-\delta)x_\infty(x),\ x_\infty(x) > \frac{\rho}{2}\right\} ,$$

where ρ and $\delta < 1/2$ are positive numbers that will be fixed later on. We will eventually choose ρ in an N-dependent way.

We will construct a partition of unity, $\{J_i\}_{i=0}^N$, with $\operatorname{supp} J_i \subset A_i$. We single out some crucial estimates on the gradients of the J_i, and defer their proofs to the end of this section:

Lemma 3.16. There exists a partition of unity, $\{J_i\}_{i=0}^N$, with $\operatorname{supp} J_i \subset A_i$ such that the following estimates hold:

$$\sum_{i=0}^N |\nabla J_i(x)|^2 \leq \frac{AN^{1/2}}{\rho^2} \quad \text{on } A_0 \quad \text{and} \tag{3.33}$$

$$\sum_{i=0}^N |\nabla J_i(x)|^2 \leq \frac{AN^{1/2}}{x_\infty(x)\rho} \quad \text{on } A_j \quad j \geq 1 \tag{3.34}$$

for a suitable constant A.

Proof of Theorem 3.15 (continued). We set $L(x) = \sum_{i=0}^N |\nabla J_i(x)|^2$. By the IMS localization formula, we have

$$H_N = J_0(H_N - L(x))J_0 + \sum_{i=1}^N J_i(H_N - L(x))J_i . \tag{3.35}$$

Using (3.33), we estimate:

$$\begin{aligned}
J_0(H_N - L)J_0 &\geq J_0\left(\sum_{i=1}^{N}\left(-\Delta_i - \frac{Z}{|x_i|}\right) + \sum_{i<j}^{N}\frac{1}{|x_i - x_j|} - \frac{AN^{1/2}}{\rho^2}\right)J_0 \\
&\geq J_0\left(-\frac{1}{4}NZ^2 + \frac{N(N-1)}{4\rho} - \frac{AN^{1/2}}{\rho^2}\right)J_0 \\
&\geq 0 \quad \text{for large } N.
\end{aligned} \tag{3.36}$$

Observe that we used $-\Delta_i - Z/|x_i| \geq -\frac{1}{4}Z^2$, and $|x_i - x_j| \leq 2\rho$ on $\operatorname{supp} J_0$. For $i \neq 0$, we define

$$H_{N-1}^{(i)} := \sum_{\substack{j=1 \\ j\neq i}}^{N}\left(-\Delta_j - \frac{Z}{|x_j|}\right) + \sum_{\substack{k<j \\ k,j\neq i}}\frac{1}{|x_k - x_j|} \ . \tag{3.37}$$

We have, for $i \neq 0$, setting $E_{N-1} := E(N-1, Z)$:

$$\begin{aligned}
J_i(H_N - L)J_i &\geq J_i\left(H_N^{(i)} - \Delta_i - \frac{Z}{|x_i|} + \sum_{j\neq i}^{N}\frac{1}{|x_i - x_j|} - \frac{AN^{1/2}}{x_\infty(x)\rho}\right)J_i \\
&\geq J_i\left(E_{N-1} - \frac{Z}{|x_i|} + \frac{N-1}{2x_\infty(x)} - \frac{AN^{1/2}}{x_\infty(x)\rho}\right)J_i \\
&\geq J_i\left(E_{N-1} + \frac{1}{|x_i|}\left(\frac{N-1}{2}(1-\delta) - Z - \frac{AN^{1/2}}{\rho}\right)\right)J_i \\
&\geq J_i E_{N-1} J_i \quad \text{for large } N \ .
\end{aligned} \tag{3.38}$$

We used above that $x_i > (1-\delta)x_\infty(x)$ on A_i. Thus, we proved

$$H_N \geq \sum_{i=1}^{N} J_i E_{N-1} J_i \geq E_{N-1} \quad \text{if } N \text{ is sufficiently large} \ .$$

To obtain the additional result for fermions, we choose ρ N-dependent:

$$\rho := \eta N^{-1/3} \ .$$

Then the estimate of $J_i(H_N - L)J_i$, $i \neq 0$, reads

$$J_i(H_N - L)J_i \geq J_i\left(E_{N-1} + \frac{1}{|x_i|}\left(\frac{N-1}{2}(1-\delta) - Z - \frac{AN^{5/6}}{\eta}\right)\right)J_i \ .$$

The term

$$\left(\frac{N-1}{2}(1-\delta) - Z - \frac{AN^{5/6}}{\eta}\right)$$

will be eventually positive if $Z = \frac{1}{2}(1-2\delta)N$ and N is sufficiently large.

We have, however, to improve our estimate of $J_0(H_N - L)J_0$, because (3.36) is too rough for the asymptotics of $N_{\max}$. If we take into account the Pauli exclusion principle, we may estimate

$$\sum_{i=1}^{N}\left(-\Delta_i - \frac{Z}{|x_i|}\right) \geq -CN^{1/3}Z^2.$$

With this estimate, we obtain

$$J_0 H_N J_0 \geq J_0\left(-CN^{1/3}Z^2 + \frac{N^{4/3}(N-1)}{4\eta} - \frac{AN^{7/6}}{\eta^2}\right)J_0 \; .$$

Again, with the above choice of Z, the term in brackets is positive for appropriate η and sufficiently large N.

Thus, we have proved that

$$\overline{\lim_{Z\to\infty}} \frac{N_{\max}(Z)}{Z} < \frac{2}{1-2\delta}$$

which gives the desired result since $\delta > 0$ was arbitrary. □

Remark. Our boson proof yields $N_{\max}(Z) = O(Z^2)$. One can improve Lemma 3.16 to get $N_{\max}(Z) = O(Z^{1+\varepsilon})$ (see *Sigal* [313]), but that seems the best one can do with this method. Lieb's method (see Sect.3.8) shows that $N_{\max}(Z) < 2Z + 1$ for bosons.

Proof of Lemma 3.16. Let ψ be a C^∞-function on $\mathbb{R}$ satisfying $0 \leq \psi(t) \leq 1$ and $\psi(t) = 1$ for $t > 1 - \varepsilon$, $\psi(t) = 0$ for $t < 1 - \delta$; $0 < \varepsilon < \delta$. We define $\chi(t) := \psi(t)^2$, and set

$$F_0(x) = 1 - \chi\left(\frac{x_\infty(x)}{\rho}\right), \tag{3.39}$$

$$F_i(x) = \chi\left(\frac{x_\infty(x)}{\rho}\right)\chi\left(\frac{|x_i|}{x_\infty(x)}\right) \quad i = 1, \ldots, N \; . \tag{3.40}$$

We will show that

$$J_i(x) = \frac{F_i(x)}{\sum |F_j(x)|^2} \quad i = 0, \ldots, N$$

is a partition of unity with the required properties, except that it is not smooth. This is due to the fact that $x_\infty(x)$ is not differentiable at those points x where $x_i = x_j = x_\infty(x)$ for some $i \neq j$. However, ∇J_i in the distribution sense belongs locally to the domain of H_0, so that we could prove the localization formula for this more singular case without additional problems. An alternative way would be to smooth out $x_\infty(x)$ by convolution with a C^∞-function with small support around the origin, and to define F_i with this smoothed out version of x_∞. A third possibility (*Sigal* [310]) is to use $|x|_p := (\sum |x_i|^p)^{1/p}$ rather than x_∞. We leave these details to the reader, and argue the required estimates for the F_i defined in (3.39) and (3.40), eventually neglecting sets of measure zero where the gradients are not well defined.

First, it is easy to see that $\operatorname{supp} F_i \subset A_i$, $i = 0, \ldots, N$. For at least one i, we have $|x_i|/[x_\infty(x)] > 1 - \delta$, hence

$$\sum_{i=0}^{N} |F_i(x)|^2 = \left|1 - \chi\left(\frac{x_\infty(x)}{\rho}\right)\right|^2 + \left|\chi\left(\frac{x_\infty(x)}{\rho}\right)\right|^2 \sum_{i=1}^{N} \left|\chi\left(\frac{|x_i|}{x_\infty(x)}\right)\right|^2$$
$$\geq \left|1 - \chi\left(\frac{x_\infty(x)}{\rho}\right)\right|^2 + \left|\chi\left(\frac{x_\infty(x)}{\rho}\right)\right|^2 \geq \frac{1}{2}$$

(we used $|1 - x|^2 + |x|^2 \geq 1/2$). Therefore

$$J_i(x) := \frac{F_i(x)}{(\sum |F_j(x)|^2)^{1/2}}$$

is well defined. Moreover, $\{J_i\}$ is a partition of unity with $\operatorname{supp} J_i \subset A_i$ (but, as we emphasized above, J_i is not smooth!). Let us now prove the gradient estimates (3.33) and (3.34). By definition of χ, we have, for any $\gamma > 0$,

$$|\chi'(t)|^2 \leq 4|\psi'|_\infty^2 \psi(t)^2$$
$$\leq \gamma + \frac{4|\psi'|_\infty^4}{\gamma} \psi(t)^4$$
$$= \gamma + \frac{C}{\gamma} \chi(t)^2$$

[we used $2y \leq \gamma + (1/\gamma)y^2$]. Hence, for $i = 1, \ldots, N$

$$|\nabla F_i(x)|^2 = \sum_{j=1}^{N} \left\{\chi'\left(\frac{x_\infty(x)}{\rho}\right)\chi\left(\frac{|x_i|}{x_\infty(x)}\right)\frac{1}{\rho}\frac{\partial x_\infty}{\partial x_j}\right.$$
$$\left. + \chi\left(\frac{x_\infty(x)}{\rho}\right)\chi'\left(\frac{|x_i|}{x_\infty(x)}\right)\left[\frac{|x_i|}{x_\infty(x)^2}\frac{\partial x_\infty}{\partial x_j} + \delta_{ij}\frac{1}{x_\infty(x)}\right]\right\}^2$$

(the gradient $\partial x_\infty/\partial x_j$ is defined in L^2-sense, i.e. almost everywhere)

$$\leq \frac{\tilde{D}}{|x_\infty(x)|^2}\left\{\chi'\left(\frac{x_\infty(x)}{\rho}\right)^2 \chi\left(\frac{|x_i|}{x_\infty(x)}\right)^2\right.$$
$$\left. + \chi\left(\frac{x_\infty(x)}{\rho}\right)^2 \chi'\left(\frac{|x_i|}{x_\infty(x)}\right)^2\right\}\left(\sum_{j=1}^{N}\left(\frac{\partial x_\infty}{\partial x_j}(x)\right)^2 + 1\right),$$

since

$$\frac{1}{\rho} \leq \frac{1}{|x_\infty(x)|} \quad \text{if } \chi'\left(\frac{x_\infty(x)}{\rho}\right) \neq 0, \quad \text{and}$$

$$\frac{|x_i|}{x_\infty(x)} \leq 1 \quad \text{if } \chi'\left(\frac{|x_i|}{x_\infty(x)}\right) \neq 0 \ .$$

Furthermore, we read off from the definition of x_∞ that, for a given x, provided x_∞ is differentiable at x, there exists only one i with $\partial x_\infty(x)/\partial x_i \neq 0$, and for this i, we have $|\partial x_\infty(x)/\partial x_i| = 1$. Hence

$$\begin{aligned} |\nabla F_i(x)|^2 \leq \frac{2\tilde{D}}{|x_\infty(x)|^2} & \left\{\left[\gamma + \frac{C}{\gamma}\chi\left(\frac{x_\infty(x)}{\rho}\right)^2\right]\chi\left(\frac{|x_i|}{x_\infty(x)}\right)^2\right. \\ & \left. + \left[\gamma + \frac{C}{\gamma}\chi\left(\frac{|x_i|}{x_\infty(x)}\right)^2\right]\chi\left(\frac{x_\infty(x)}{\rho}\right)^2\right\} \\ \leq \frac{D}{|x_\infty(x)|^2} & \left(\gamma + \frac{C}{\gamma}F_i(x)^2\right) \ . \end{aligned}$$

It is easy to check that

$$|\nabla F_0(x)|^2 \leq \frac{D'}{\rho^2} \ .$$

Therefore

$$\sum |\nabla J_i(x)|^2 \leq \frac{\sum |\nabla F_j|^2}{\sum F_j^2} \leq \frac{\tilde{A}}{|x_\infty(x)|^2}\left(1 + \gamma N + \frac{C}{\gamma}\right) \ .$$

Inserting $\gamma = N^{-1/2}$, we obtain

$$\sum |\nabla J_i(x)|^2 \leq \frac{A}{|x_\infty(x)|^2} N^{1/2} \ .$$

Moreover, by enlarging A—if necessary—we have

$$\sum |\nabla J_i(x)|^2 \leq \frac{AN^{1/2}}{\rho^2} \quad \text{on supp } J_0 \quad \text{and}$$

$$\sum |\nabla J_i(x)|^2 \leq \frac{AN^{1/2}}{x_\infty(x)\rho} \quad \text{on supp } J_i \quad i \neq 0 \ . \quad \square$$

Remark. The reason why we get merely $\overline{\lim}\,[N_{\max}(Z)/Z] \leq 2$ is the estimate $|x_i - x_j| < 2x_\infty(x)$ in (3.38). Indeed, one might hope to improve the estimate (3.38) by a more clever choice of the J_i. This is actually what *Lieb, Sigal, Simon* and *Thirring* [233] do.

3.8 Lieb's Improvement of the Ruskai-Sigal Theorem

In this section, we present Lieb's simple proof of an improved version of Theorem 3.15. We use the notations of Sect. 3.7.

Theorem 3.17 [231, 232] For any Z: $N_{\max}(Z) < 2Z + 1$.

We single out

Corollary. If Z is an integer, then $N_{\max}(Z) \leq 2Z$.

In particular, the Corollary tells us that the ion H^{2-} has no bound states, i.e. it is unstable. To prove Theorem 3.17, we will use the following lemma:

Lemma 3.18. If $\varphi \in L^2(\mathbb{R}^3)$ and $\varphi \in D(-\Delta) \cap D(|x|)$, then $\operatorname{Re}\langle \varphi, |x|(-\Delta)\varphi\rangle \geq 0$.

Proof. If the function f is sufficiently regular, one has

$$\tfrac{1}{2}(fp^2 + p^2 f) = pfp + \tfrac{1}{2}[p,[p,f]] = pfp + \tfrac{1}{2}(-\Delta f) \ . \tag{3.41}$$

Choosing $f(x) = |x|^{-1}$ and multiplying (3.41) by $|x|$ from both sides, we obtain formally

$$\begin{aligned} \tfrac{1}{2}(p^2|x| + |x|p^2) &= |x|p|x|^{-1}p|x| - \tfrac{1}{2}|x|(\Delta|x|^{-1})|x| \\ &= |x|p|x|^{-1}p|x| + \tfrac{1}{2}|x|4\pi\delta_0|x| \\ &= |x|p|x|^{-1}p|x| \ , \end{aligned} \tag{3.42}$$

where δ_0 is the Dirac measure at the point $0 \in \mathbb{R}^3$. We used that $(4\pi|x|)^{-1}$ is a fundamental solution of $-\Delta$, i.e. $-\Delta(4\pi|x|)^{-1} = \delta_0$. From (3.42) we get

$$\operatorname{Re}\langle \varphi, |x|(-\Delta)\varphi\rangle = \int |x|^{-1}|\nabla(|x|\varphi)|^2\, dx \geq 0 \ . \tag{3.43}$$

However, we have to justify the above formal calculations. To do this, we approximate the function $|x|$ by $\rho_\varepsilon(x) := (|x|^2 + \varepsilon^2)^{1/2}$.

Doing the above calculations with ρ_ε instead of $|x|$, we arrive at

$$\begin{aligned} \operatorname{Re}\langle \varphi, \rho_\varepsilon(-\Delta)\varphi\rangle &= \int \rho_\varepsilon(x)^{-1}|\nabla(\rho_\varepsilon(x)\varphi(x))|^2\, dx \\ &\quad + \tfrac{3}{2}\int \rho_\varepsilon(x)^{-3}\varepsilon^2|\varphi(x)|^2\, dx \end{aligned} \tag{3.44}$$

[we used $\Delta(\rho_\varepsilon^{-1}) = -3\rho^{-5}\varepsilon^2$]. Since $\varepsilon\rho_\varepsilon^{-1}(x) \leq 1$ and

$$\int \left|\frac{1}{|x|}\varphi(x)\right|^2 dx \leq 4\int |\nabla\varphi(x)|^2\, dx$$

(see [293], X.2), the last term in (3.44) goes to zero as $\varepsilon \to 0$. Moreover,

$\langle\varphi, \rho_\varepsilon(-\Delta)\varphi\rangle \to \langle\varphi, |x|(-\Delta)\varphi\rangle$, and

$$\int \rho_\varepsilon^{-1}|\nabla(\rho_\varepsilon\varphi)|^2\,dx \to \int |x|^{-1}|\nabla(|x|\varphi)|^2\,dx \ .$$

(The reader may check that Lebesgue's theorem on dominated convergence applies, using $\nabla\varphi \in D(|x|^{1/2})$.) □

Proof of Theorem 3.18. Assume $E_N < E_{N-1}$. Thus, H_N has an isolated eigenvalue at the bottom of its spectrum. Therefore, the (normalized) ground state, ψ, of H_N decays fast as $|x| \to \infty$ (see [295], XIII.11), and consequently $\psi \in D(|x|)$. Moreover, we may assume that ψ is real. We have (x_i is the coordinate vector of the ith electron, and H_{N-1} the Hamiltonian of electrons $2, \ldots, N$):

$$\begin{aligned}
0 &= \langle |x_1|\psi, (H_N - E_N)\psi\rangle \\
&= \left\langle |x_1|\psi, \left(H_{N-1} - E_N - \Delta_1 - \frac{Z}{|x_1|} + \sum_{j=2}^{N} \frac{1}{|x_1 - x_j|}\right)\psi\right\rangle \\
&\geq \left\langle |x_1|\psi, \left(E_{N-1} - E_N - \Delta_1 - \frac{Z}{|x_1|} + \sum_{j=2}^{N} \frac{1}{|x_1 - x_j|}\right)\psi\right\rangle \\
&= \langle |x_1|\psi, (E_{N-1} - E_N)\psi\rangle + \langle |x_1|\psi, -\Delta_1\psi\rangle \\
&\quad - Z + \left\langle \psi, \sum_{j=2}^{N} \frac{|x_1|}{|x_1 - x_j|}\psi\right\rangle \\
&> -Z + \left\langle \psi, \sum_{j=2}^{N} \frac{|x_1|}{|x_1 - x_j|}\psi\right\rangle ,
\end{aligned}$$

where we used $\langle |x_1|\psi, (H_{N-1} - E_{N-1})\psi\rangle = \int |x_1| \langle \psi_{x_1}, (H_{N-1} - E_{N-1})\psi_{x_1}\rangle\, dx_1 \geq 0$ with $\psi_{x_1}(x_2, \ldots, x_N) = \psi(x_1, x_2, \ldots, x_N)$.

By symmetry of the above formulae, we obtain, replacing x_1 by x_i and summing over i,

$$\left\langle \psi, \sum_{\substack{i,j=1 \\ i \neq j}}^{N} \frac{|x_i| + |x_j|}{|x_i - x_j|}\psi\right\rangle < 2NZ \ .$$

Since $|x_i - x_j| \leq |x_i| + |x_j|$, we get

$$\left\langle \psi, \left(\sum_{\substack{i,j=1 \\ i \neq j}}^{N} 1\right)\psi\right\rangle = N(N-1) < 2NZ \ .$$

Thus $N < 2Z + 1$. □

Remarks. (1) One can show if $N \geq 2Z + 1$, then E_N is not an eigenvalue.

(2) *Lieb* [232] treats multi-center problems and various other refinements.

3.9 *N*-Body Systems with Finitely Many Bound States

The HVZ-theorem tells us that the infimum Σ of the essential spectrum of H is always defined by two cluster decompositions, i.e.

$$\Sigma = \Sigma_2 := \inf_{\#(a)=2} \sigma(H_a) .$$

In contrast to that, the question whether $\sigma_{\text{dis}}(H)$ is finite or infinite depends, in part, on

$$\Sigma_3 := \inf_{\#(a)=3} \sigma(H_a) .$$

We will show below that, in many cases, $\sigma_{\text{dis}}(H)$ is finite provided $\Sigma_3 > \Sigma$. On the other hand, if $\Sigma_3 = \Sigma$, the operator $H = -\Delta + \sum_{i<j} V_{ij}(x)$ may have infinitely many bound states even if the V_{ij} have compact support. This phenomenon is known as the *Efimov effect*, after its discoverer *Efimov* ([91], [92]). For rigorous treatments, see *Yafaev* [372], *Ovchinnikov* and *Sigal* [268] and references therein, and the discussion in *Reed* and *Simon* [295] after Theorem XIII.6.

In the following, we will show that $\Sigma_3 > \Sigma$ implies the finiteness of $\sigma_{\text{dis}}(H)$ for short-range potentials V_{ij}, as well as in the case of once negatively charged ions. These results will follow from an "abstract" theorem (Theorem 3.23) which we will prove first. The results of this chapter go back to *Zhislin* and his co-workers [16, 363, 382–384]. The form in which we state them, as well as the proofs we give are due to *Sigal* [310]. Additional references may also be found there.

We first introduce an appropriate partition of unity.

Definition 3.19. A partition of unity $\{j_a\}_a$ indexed by all cluster decompositions a of $\{1, 2, \ldots, N\}$ is called a *Deift-Agmon-Sigal partition* of unity if

(i) each j_a is homogeneous of degree zero outside the unit sphere,
(ii) $\{|x| > 1\} \cap \operatorname{supp} j_a \subset \{x = (x_1, \ldots, x_N) \,|\, |x_i - x_j| \geq C|x| \text{ whenever } (ij) \not\subset a\}$, with a suitable constant C,
(iii) for two distinct cluster decompositions a and a' with $\#a = \#a' = 2$, we have $\{|x| > 1\} \cap \operatorname{supp} j_a \cap \operatorname{supp} j_{a'} = \phi$.

Related partitions occurred first in *Deift* and *Simon* [78]; their importance in this context was noticed by *Sigal* [310]. An existence proof for a DAS partition of unity can be made by slightly changing the proof of Proposition 3.5 (existence of the Ruelle-Simon partition of unity). Like for the Ruelle-Simon partition of unity, we have that each $j_a I_a$ is relatively compact if the V_{ij} are (see Proposition 3.6). The following estimate for the localization error is crucial in our proof of the finiteness of σ_{dis}:

Proposition 3.20. For any $\varepsilon > 0$, there exists a C_ε such that outside the unit sphere

$$\sum |\nabla j_a(x)|^2 \le (1 + |x|^2)^{-1} \left(\varepsilon \sum_{\# a=2} j_a^2(x) + C_\varepsilon \sum_{\# a \ge 3} j_a^2(x) \right) . \tag{3.45}$$

Remark. It is well known that a two-body potential W, which decays at infinity like $a|x|^{-2}$, does not produce infinitely many bound states, provided the constant a is sufficiently small (*Reed* and *Simon* [295], XIII.3). The above proposition, therefore, ensures that the localization error will not produce an infinity of bound states for the two cluster Hamiltonians. Note that because of $\Sigma_3 > \Sigma$, the Hamiltonians with three or more clusters in any case have only finitely many bound states below $\Sigma := \inf \sigma_{\mathrm{ess}}(H)$ (see Lemma 3.22).

Proof. By (i), ∇j_a is homogeneous of degree minus one; therefore it suffices to show (3.45) on the unit sphere S. We consider the set

$$A = \left\{ x \in S \,\middle|\, \sum_{\# a=2} j_a^2(x) = 1 \right\} .$$

By Definition 3.19(iii), $x \in A$ implies $j_a(x) = 1$ for exactly one a with $\# a = 2$ and $j_{a'}(x) = 0$ for any other decomposition a'. Hence, $|\nabla j_b(x)| = 0$ for any cluster decomposition, b. Thus,

$$\sum |\nabla j_b(x)|^2 = 0 \quad \text{on } A . \tag{3.46}$$

Consider now $A^\delta = \{x \in S | \sum_{\# a=2} j_a^2(x) > 1 - \delta\}$; $0 < \delta < 1/2$. Taking $\delta = \delta(\varepsilon)$ small enough, we can assure, by (3.46), that

$$\sum |\nabla j_b(x)|^2 < \varepsilon/2 \quad \text{on } A^\delta , \quad \text{hence,}$$

$$\sum |\nabla j_b(x)|^2 < \varepsilon \sum_{\# a=2} j_a^2(x) \quad \text{on } A^\delta .$$

On the other hand

$$\sum_{\# a \ge 3} j_a^2(x) = 1 - \sum_{\# a=2} j_a^2(x) > \delta(\varepsilon) \quad \text{on } S \backslash A^\delta ,$$

so

$$\sum |\nabla j_b(x)|^2 < C_\varepsilon \sum_{\# a \ge 3} j_a^2(x) \quad \text{on } S \backslash A^\delta . \quad \square$$

For each cluster decomposition a let χ_a denote the characteristic function of $\operatorname{supp} j_a$. The IMS localization formula tells us that

$$\begin{aligned} H \ge & \sum_{\# a=2} j_a(H(a) + I_a \chi_a - \varepsilon(1 + |x|^2)^{-1}) j_a \\ & + \sum_{\# b \ge 3} j_b(H(b) + I_b \chi_b - C_\varepsilon(1 + |x|^2)^{-1}) j_b , \end{aligned} \tag{3.47}$$

where we used (3.45). To show that H has finitely many bound states, it suffices to prove that each of the terms on the right-hand side of (3.47) has finitely many bound states, because of the following lemma and the fact that the j's are a partition of unity.

Lemma 3.21. Let A, B be self-adjoint operators. (i) If $A \geq B$, then the number $N(A, \Sigma)$ of bound states of A below $\Sigma = \inf \sigma_{\text{ess}}(B)$ satisfies $N(A, \Sigma) \leq N(B, \Sigma)$.

(ii) If both A and B have a finite number of bound states below 0, $\inf \sigma_{\text{ess}}(A) \geq 0$, $\inf \sigma_{\text{ess}}(B) \geq 0$, and $A + B$ is essentially self-adjoint on $D(A) \cap D(B)$, then $A + B$ has finitely many bound states below 0.

Proof. (i) is easily proven using the min-max principle (see e.g. *Reed* and *Simon* [295], XIII.1).

(ii) Let P and Q denote the projections on the eigenspaces for eigenvalues below 0 of A and B, respectively. Then AP and BQ are finite rank operators. Moreover, $A + B \geq AP + BQ$, which is a finite rank operator, too. Applying (i) gives the desired result. □

Since we suppose $\Sigma_3 > \Sigma$, it is easy to see that the terms in (3.47) resulting from three and more clusters contribute only a finite number of bound states.

Lemma 3.22 Fix ε. If $\Sigma_3 > \Sigma$, the number of bound states of $H_b + I_b \chi_b - C_\varepsilon (1 + |x|^2)^{-1}$ below Σ is finite.

Proof. $I_b \chi_b - C_\varepsilon (1 + |x|^2)^{-1}$ is $H(b)$-compact, hence

$$\inf \sigma_{\text{ess}}(H(b) + I_b \chi_b - C_\varepsilon (1 + |x|^2)^{-1}) = \inf \sigma_{\text{ess}}(H(b)) \geq \Sigma_3 > \Sigma \ .$$

Hence the number of bound states below Σ is finite by the definition of σ_{dis}. □

Using the above considerations, H has finitely many bound states below Σ, if all the two cluster terms in (3.47) have. (Actually, only those with $\inf \sigma_{\text{ess}}(H(a)) = \Sigma$ have to be considered.) We therefore investigate now those terms more carefully.

Let a be a decomposition into two clusters. We saw already in Sect.3.2 that the Hilbert space $L^2(X)$ splits into

$$L^2(X) = L^2(X^a) \otimes L^2(X_a) \ ,$$

and moreover

$$H(a) = (h^a \otimes \mathbb{1}) \oplus (\mathbb{1} \otimes T_a) \ ,$$

see (3.13, 14). Since $\# a = 2$, we have $X_a \simeq \mathbb{R}^\mu$. Of course, h^a itself splits into two parts, corresponding to the two clusters in a.

Let us denote by ψ^a the normalized ground state of h^a. It is well known that the ground state is nondegenerate (see *Reed* and *Simon* [295], XIII.12). We denote, by P^a, the projection operator from $L^2(X^a)$ onto ψ^a. We set $P(a) := \mathbb{1}_{X_a} \otimes P^a$, where $\mathbb{1}_{X_a}$ denotes the projection on $L^2(X_a)$ onto the whole space, and we set $Q(a) := \mathbb{1}_X - P(a)$.

We will use below the brackets $\langle \cdot, \cdot \rangle$ to denote the scalar product in any of the spaces $L^2(X)$, $L^2(X_a)$, $L^2(X^a)$. It should be clear from the context which one is meant.

To state the "abstract" theorem, we introduce the potential W_a^δ on X_a:

$$W_a^\delta := \langle \psi^a, I_a \chi_a \psi^a \rangle + \delta^{-1}(\langle \psi^a (I_a \chi_a)^2 \psi^a \rangle - \langle \psi^a, I_a \chi_a \psi^a \rangle^2) \tag{3.48}$$

for any $\delta > 0$. The brackets $\langle \cdot, \cdot \rangle$ in (3.48) denote the scalar product in $L^2(X^a)$. The reader may notice that the operator $H(W_a^\delta) := -\Delta + W_a^\delta$ is a one-body operator acting on $L^2(X_a) \simeq L^2(\mathbb{R}^\mu)$. The following theorem reduces the question of finiteness of $\sigma_{\text{dis}}(H)$ to the investigation of $H(W_a^\delta)$.

Theorem 3.23. Suppose $\Sigma_3 > \Sigma$. If, for any two cluster decomposition a, the (one-body) operator $-(1-\eta)\Delta + W_a^\delta$ has finitely many bound states for all $\delta > 0$, and a suitable $\eta > 0$, then H has finitely many bound states.

Remarks. (1) It will become clear in the proof that the condition on $-(1-\eta)\Delta + W_a^\delta$ need only be required for those a with $\inf \sigma(H_a) = \Sigma$.

(2) For the treatment of Theorem 3.23 on the fermionic subspace, see *Sigal* [310].

(3) It is not easy to check the condition of the theorem for a given potential, which is the reason we called it "abstract". Later, we will present two important classes of examples.

Proof. Let j_a be a Deift-Agmon-Sigal partition of unity. By the IMS-localization formula and Proposition 3.20, we have

$$H \geq \sum_{\#a=2} j_a(H(a) + I_a\chi_a - \varepsilon(1+|x|^2)^{-1})j_a$$
$$+ \sum_{\#a\geq 3} j_a(H(a) + I_a\chi_a - C_\varepsilon(1+|x|^2)^{-1})j_a \ .$$

What remains to be proven is the finiteness of the discrete spectrum of each of the $H(a) + I_a\chi_a - \varepsilon(1+|x|^2)^{-1}$. We set

$$K(a) = H(a) + I_a\chi_a - \varepsilon(1+|x|^2)^{-1} \ .$$

We now write $K(a)$ as

$$K(a) = P(a)K(a)P(a) + P(a)K(a)Q(a)$$
$$+ Q(a)K(a)P(a) + Q(a)K(a)Q(a) \ . \tag{3.49}$$

It is clear that the term $Q(a)K(a)Q(a)$ contributes only finitely many bound states, since $\inf \sigma_{\text{ess}}(Q(a)K(a)Q(a)) > \Sigma$. To estimate the contribution of the mixed terms in (3.49), we use the following decoupling inequality:

Lemma 3.24 (Combes-Simon Decoupling Inequality [323]). Let A be a self-adjoint operator, let P be a projection, and set $Q = \mathbb{1} - P$. Then for any $\delta > 0$

$$A \geq PAP - \delta^{-1}PAQAP + Q(A - \delta)Q \; .$$

Before we prove Lemma 3.24, we continue the proof of the theorem. Applying the lemma to $K(a)$ and $P(a)$, we get

$$K(a) \geq P(a)K(a)P(a) - \delta^{-1}P(a)K(a)Q(a)K(a)P(a) + Q(a)[K(a) - \delta]Q(a) \; .$$

The last term still has the infimum of its essential spectrum above Σ provided we take δ small enough. Furthermore, take

$$\varphi = \varphi_1 \otimes \varphi_2; \quad \varphi_1 \in L^2(X^a), \quad \varphi_2 \in L^2(X_a) \; ,$$

then

$$\langle \varphi, P(a)K(a)P(a)\varphi \rangle = |\langle \varphi_1, \psi^a \rangle|^2 \langle \psi^a \otimes \varphi_2, K(a)(\psi^a \otimes \varphi_2) \rangle \; ,$$

and

$$\begin{aligned}\langle \psi^a \otimes \varphi_2, K(a)(\psi^a \otimes \varphi_2)\rangle &\geq \langle \varphi_2, -\Delta\varphi_2\rangle + \langle \varphi_2, \langle \psi^a, I_a\chi_a\psi^a\rangle\varphi_2\rangle \\ &\quad - \varepsilon\langle \varphi_2, \langle \psi^a, (1 + |x|^2)^{-1}\psi^a\rangle\varphi_2\rangle + \Sigma \\ &\geq \langle \varphi_2, [-\Delta + \langle \psi^a, I_a\chi_a\psi^a\rangle \\ &\quad - \varepsilon(1 + |x|_0^2)^{-1}]\varphi_2\rangle + \Sigma \; ,\end{aligned}$$

where $|x|^2$ is meant on $L^2(X)$, while $|x|_0^2$ is meant on $L^2(X_a)$. Therefore, we obtain

$$P(a)K(a)P(a) \geq -\Delta + \langle \psi^a, I_a\chi_a\psi^a\rangle - \varepsilon(1 + |x|_0^2)^{-1} + \Sigma P(a) \; .$$

In a similar way we get

$$P(a)K(a)Q(a)K(a)P(a) \geq \langle \psi^a, I_a^2\psi^a\rangle - \langle \psi^a, I_a\psi^a\rangle^2 - \varepsilon(1 + |x|_0^2)^{-1} \; .$$

In total, we have shown that

$$K(a) \geq (1 - \eta)\Delta + W_a^\delta - \eta\Delta - 2\varepsilon(1 + |x|_0^2)^{-1} + \Sigma \; .$$

By assumption, the Hamiltonian $-(1 - \eta)\Delta + W_a^\delta$ has finite discrete spectrum. Furthermore, $-\eta\Delta - 2\varepsilon(1 + |x|_0^2)^{-1}$ has finite discrete spectrum if ε is small enough (*Reed* and *Simon* [295], XIII.3). Therefore, using Lemma 3.24, we arrive at the conclusion of the theorem. □

We now prove the Combes-Simon decoupling inequality.

Proof (*of Lemma 3.24*).

$$A = PAP + PAQ + QAP + QAQ \; . \tag{3.50}$$

We estimate the mixed terms by the Schwarz inequality:

$$\begin{aligned}|\langle\varphi, QAP\varphi\rangle| &= |\langle\delta^{1/2}Q\varphi, \delta^{-1/2}QAP\varphi\rangle| \\ &\leq \langle\delta^{1/2}Q\varphi, \delta^{1/2}Q\varphi\rangle^{1/2}\langle\delta^{-1/2}QAP\varphi, \delta^{-1/2}QAP\varphi\rangle^{1/2} \\ &\leq \tfrac{1}{2}(\langle\delta^{1/2}Q\varphi, \delta^{1/2}Q\varphi\rangle + \langle\delta^{-1/2}QAP\varphi, \delta^{-1/2}QAP\varphi\rangle) \\ &= \tfrac{1}{2}(\delta\langle\varphi, Q\varphi\rangle + \delta^{-1}\langle\varphi, PAQAP\varphi\rangle) \ .\end{aligned}$$

Thus, we get

$$\mathrm{Re}(QAP) \geq -\tfrac{1}{2}\delta Q - \tfrac{1}{2}\delta^{-1}PAQAP \ .$$

Estimating $\mathrm{Re}(PAQ)$ in a similar way and inserting in (3.50), we obtain

$$A \geq PAP - \delta^{-1}PAQAP + Q(A-\delta)Q \ . \quad \square$$

We now present two applications of the above "abstract" theorem:

Theorem 3.25. Assume dimension $\mu \geq 3$. If the potentials V_{ij} belong to $L^{\mu/2}(\mathbb{R}^\mu)$ for $\mu > 3$, and to $L^2(\mathbb{R}^\mu)$ for $\mu = 3$, and if furthermore $\Sigma_3 > \Sigma$, then $H = H_0 + \sum V_{ij}$ has only finitely many bound states.

Proof. We show that the negative part of W_a^δ belongs to $L^{\mu/2}(\mathbb{R}^\mu)$. This implies that $-(1-\eta)\Delta + W_a^\delta$ has only finitely many bound states below zero by the Cwickel-Lieb-Rosenbljum bound (see e.g. *Reed* and *Simon* IV, [295], XIII.12). This implies the assertion by Theorem 3.23. Let us first consider

$$\langle\psi^a, I_a\chi_a\psi^a\rangle = \sum_{(ij)\not\subset a} \langle\psi^a, V_{ij}\chi_a\psi^a\rangle \ .$$

Let $a = \{A_1, A_2\}$. Define $M_k = \sum_{i\in A_k} m_i$. We can write any $x \in X$ as $(\tilde{x}_1 + y_1, \tilde{x}_2 + y_2, \ldots, \tilde{x}_n + y_n)$ with $(\tilde{x}_1, \ldots, \tilde{x}_n) \in X_a$ and

$$y_i = \frac{1}{M_k}\sum_{j\in A_k} m_j x_j \quad \text{for } i \in A_k \ .$$

Since $x \in X$, we have $M_1 y_1 = -M_2 y_2$ for $i \in A_1$, $j \in A_2$. Therefore $x_i - x_j = y_i - y_j + \tilde{x}_i - \tilde{x}_j = y + \tilde{x}_i - \tilde{x}_j$ with $y = y_i - y_j$. We see from this that

$$\langle\psi^a, V_{ij}\psi^a\rangle = \int_{X^a} |\psi^a(\tilde{x})|^2 V_{ij}(y + \tilde{x}_i - \tilde{x}_j)\, d\tilde{x}$$

is a convolution. It is well known that

$$\psi^a \in \bigcap_{1\leq p\leq\infty} L^p$$

(see *Reed* and *Simon* IV, [295], XIII.39). Thus, the Young inequality tells us that the convolution $\langle\psi^a, V_{ij}\psi^a\rangle \in L^{\mu/2}(\mathbb{R}^\mu)$. The term $\langle\psi^a, I_a^2\chi_a\psi^a\rangle$ can be handled by the estimate

$$\langle \psi^a, I_a^2 \chi_a \psi^a \rangle \le \left\langle \psi^a, \left(\sum_{(ij) \not\subset a} V_{ij} \right)^2 \psi^a \right\rangle \le C \sum_{(ij) \not\subset a} \langle \psi^a, V_{ij}^2 \psi^a \rangle .$$

As above, $\langle \psi^a, V_{ij}^2 \psi^a \rangle$ is a convolution. Assume first $\mu > 3$. Since $V_{ij} \in L^{\mu/2}$, we have $V_{ij}^2 \in L^{\mu/4}$. By the Young inequality this implies $\langle \psi^a, V_{ij}^2 \psi^a \rangle \in L^{\mu/2}$ (since $\psi^a \in L^q$ for any q). For $\mu = 3$, $V_{ij} \in L^2$, hence $V_{ij}^2 \in L^1$. Again we conclude that $\langle \psi^a, V_{ij}^2 \psi^a \rangle \in L^{\mu/2}$.

Since the third term, $\langle \psi^a, V_{ij} \psi^a \rangle^2$ is positive, we are done. □

Finally, we state without proof

Theorem 3.26. Once negatively charged bosonic ions have only a finite number of bound states.

Remarks. (1) The problem with fermions is that the corresponding ground state may be degenerate with a parity degeneracy producing a dipole term in the effective potential. If this ground state happens to be nondegenerate, Theorem 3.26 holds in this case, too.

(2) We can apply Theorem 3.23 only to negative ions of charge 1, since we do not know $\Sigma_3 > \Sigma$ for higher charges. For an ion of charge $-k$, $\Sigma_3 > \Sigma$ means $\Sigma = E(Z + k - 1, Z) \neq E(Z + k - 2, Z) = \Sigma_3$. So, what we cannot exclude is that an ion of charge $-k + 1$ has no bound states, while the corresponding ion of charge $-k$ has infinitely many bound states. The fact that we do not know how to exclude this physically absurd situation indicates how little we understand about atomic physics from a mathematical point of view.

Appendix: The Stone-Weierstrass Gavotte

It appears several times in this book that it is relatively easy to show an assertion for the resolvents $(H - z)^{-1}$ of an operator H, while a direct proof for $f(H)$ for an arbitrary function $f \in C_\infty(\mathbb{R})$ seems to be much harder.

However, it is in many cases easy to deduce this seemingly stronger assertion from the knowledge that it holds for resolvents, i.e. for the functions $f(x) = (x - z)^{-1}$.

One way to see this is the use of the abstract Stone-Weierstrass theorem, as follows: Suppose we know that $A := \{f \in C_\infty(\mathbb{R}) | f(H)$ has the desired property$\}$ obeys: (i) A is norm closed (ii) A is an involutative algebra, i.e. a vector space that contains, with f, g, also $\bar{f}$ and $f \cdot g$. Then the abstract Stone-Weierstrass theorem tells us that $A = C_\infty(\mathbb{R})$, i.e. the assertion holds for $f(H)$ with any $f \in C_\infty(\mathbb{R})$, provided we know it for $(H - z)^{-1}$ for a single $z \in \mathbb{C}$, $\operatorname{Re} z \neq 0$.

A second, more elementary way goes as follows: Suppose we know that a certain property holds for all $(H - z)^{-1}$ with z in an open subset, G, of $\mathbb{C}$. Suppose, furthermore, we know (i) above, i.e. if f_n has the property, $f_n \to f$ uniformly, then

f has the desired property, and (ii)$'$ A is a vector space. Then we conclude that $(H - z)^{-k}$, $k \in \mathbb{N}$, $z \in G$ belongs to A, since by Cauchy's integral formula

$$(H - z_0)^{-k} = \frac{1}{2\pi i} \int_\gamma \frac{(H - z)^{-1}}{(z_0 - z)^{-k}} dz \ , \tag{A.3.1}$$

where γ is a circle in G around z_0. The right-hand side of the above equation is a norm limit of linear combination of resolvents, and hence belongs to A.

"Mixed" polynomials of the type

$$(H - z_1)^{-1}(H - z_2)^{-1} \ldots (H - z_n)^{-1}$$

can be handled by the first resolvent inequality

$$(H - z_i)^{-1}(H - z_j)^{-1} = \frac{1}{z_i - z_j}[(H - z_i)^{-1} - (H - z_j)^{-1}]$$

for $z_i \neq z_j$, and by (A.3.1) for $z_i = z_j$. Thus, we know that all the polynomials in $(H - z)^{-1}$ with $z \in G$ belong to A. Since the polynomials in $(x - z)^{-1}$, $z \in G$ are dense in C_∞, we conclude by (i) that $A = C_\infty$.

4. Local Commutator Estimates

In this chapter, we will examine a number of theorems about operators H which follow from the Mourre estimate, an estimate which says that a commutator $[H, iA]$ is positive in some sense. The ideas in this chapter can be traced back to *Putnam* [289], *Kato* [191] and *Lavine* [225] for theorems on the absence of singular spectrum, and to *Weidmann* [367] and *Kalf* [189] for theorems on absence of positive eigenvalues.

All this earlier work applied to rather restricted classes of potentials. It was *Mourre* [256], in a brilliant paper, who realized that by only requiring localized estimates, one could deal with fairly general potentials. He developed an abstract theory which he was able to apply to 2- and 3-body Schrödinger operators. *Perry, Sigal* and *Simon* [281] showed that his ideas could handle N-body Schrödinger operators.

In Sect. 4.1, we prove Putnam's theorem on the absence of singular spectrum, and introduce the Mourre estimate. We then give some examples of Schrödinger operators for which a Mourre estimate holds, deferring the proof of the estimate for N-body Schrödinger operators until Sect. 4.5. In Sect. 4.2, we prove the virial theorem and show how this, together with a Mourre estimate, can give information about the accumulation of eigenvalues. In Sect. 4.3, we prove a variant of the theorem of *Mourre* [256] on absence of singular spectrum. In Sect. 4.4, we present theorems of *Froese* and *Herbst* [114], and *Froese, Herbst, Hoffmann-Ostenhof* and *Hoffmann-Ostenhof* [116] on L^2-exponential bounds for eigenfunctions of Schrödinger operators which imply that N-body Schrödinger operators have no positive eigenvalues.

4.1 Putnam's Theorem and the Mourre Estimate

Commutator methods appear in a simple form in Putnam's theorem, where positivity of a commutator is used to prove absolute continuity of spectrum. We first give a convenient criterion for the absolute continuity of spectrum.

Proposition 4.1. Suppose H is a self-adjoint operator, and $R(z) = (H - z)^{-1}$. Suppose for each φ in some dense set there exists a constant, $C(\varphi) < \infty$ such that

$$\varlimsup_{\varepsilon \downarrow 0} \sup_{\mu \in (a,b)} \langle \varphi, \operatorname{Im} R(\mu + i\varepsilon)\varphi \rangle \leq C(\varphi) .$$

Then H has purely absolutely continuous spectrum in (a, b).

Proof. By Stone's formula [292], if $E_\Delta = E_\Delta(H)$ denotes the spectral projection for H corresponding to Δ,

$$\frac{1}{2}\langle\varphi, (E_{[a',b']} + E_{(a',b')})\varphi\rangle = \lim_{\varepsilon\downarrow 0}\frac{1}{\pi}\int_{a'}^{b'}\langle\varphi, \operatorname{Im} R(\mu + \mathrm{i}\varepsilon)\varphi\rangle\, d\mu \ .$$

Since $E_{(a',b')} \le E_{[a',b']}$ this implies, if $(a',b') \subseteq (a,b)$

$$\begin{aligned}\langle\varphi, E_{(a',b')}\varphi\rangle &\le \frac{1}{\pi}\int_{a'}^{b'} C(\varphi)\, d\mu \\ &= \frac{1}{\pi} C(\varphi)|b' - a'|\end{aligned}$$

for a dense set of φ's. This implies

$$\langle\varphi, E_\Omega\varphi\rangle \le \pi^{-1}C(\varphi)|\Omega|$$

for every Borel set $\Omega \subseteq (a,b)$, which means that the spectral measures $d\mu_\varphi$ are absolutely continuous. Since the set of such φ's is assumed dense, the spectrum is purely absolutely continuous. □

Theorem 4.2 (Putnam's Theorem). Suppose H and A are bounded, self-adjoint operators. Assume

$$[H, \mathrm{i}A] = C^*C \ , \tag{4.1}$$

where $\operatorname{Ker}(C) = \{0\}$. Then H has purely absolutely continuous spectrum.

Proof. Set $R(z) := (H - z)^{-1}$. Then

$$\begin{aligned}\|CR(\mu \pm \mathrm{i}\varepsilon)\|^2 &= \|R(\mu \mp \mathrm{i}\varepsilon)C^*CR(\mu \pm \mathrm{i}\varepsilon)\| \\ &= \|R(\mu \mp \mathrm{i}\varepsilon)[H, \mathrm{i}A]R(\mu \pm \mathrm{i}\varepsilon)\| \\ &= \|R(\mu \mp \mathrm{i}\varepsilon)[H - \mu \mp \mathrm{i}\varepsilon, \mathrm{i}A]R(\mu \pm \mathrm{i}\varepsilon)\| \\ &\le \|AR(\mu \pm i\varepsilon)\| + \|R(\mu \mp i\varepsilon)A\| + 2\varepsilon\|R(\mu \mp \mathrm{i}\varepsilon)AR(\mu \pm \mathrm{i}\varepsilon)\| \\ &\le 4\varepsilon^{-1}\|A\| \ .\end{aligned}$$

Thus,

$$\begin{aligned}2\|C \operatorname{Im} R(\mu + \mathrm{i}\varepsilon)C^*\| &= \|CR(\mu + \mathrm{i}\varepsilon)(2\mathrm{i}\varepsilon)R(\mu - \mathrm{i}\varepsilon)C^*\| \\ &\le 8\|A\| \ .\end{aligned}$$

Since $\operatorname{ran}(C^*)$ is dense, the theorem now follows from Proposition 4.1. □

Remark. This proof shows that $[H, \mathrm{i}A] \ge \alpha I$ is impossible for bounded H and A, since this would imply that $R(z)$ is bounded for all z, i.e. that H has no spectrum.

The Mourre estimate can be thought of as a weak form of hypothesis (4.1). In the Mourre estimate, H and A can be unbounded, which is crucial for applications to Schrödinger operators. Moreover, the Mourre estimate is local in the spectrum of H. Thus, we will be able to prove absolute continuity of the spectrum of H away from eigenvalues without proving (as Putnam's theorem does) that eigenvalues do not exist.

Before describing the Mourre estimate, we need some definitions. We first define a scale of spaces associated with a self-adjoint operator H.

Definition 4.3. Given a self-adjoint operator H acting in a Hilbert space H, define $\mathrm{H}_{+2} := \mathrm{D}(H)$ with the graph norm

$$\|\psi\|_{+2} = \|(H + \mathrm{i})\psi\| \ .$$

Similarly, define $\mathrm{H}_{+1} := D(|H|^{1/2})$ with its graph norm. Define H_{-2} and H_{-1} to be the dual spaces of H_{+2} and H_{+1}, respectively, thought of as the closure of H in the norm $\|\varphi\|_{-j} = \|(|H| + 1)^{-j/2}\varphi\|$.

Thus, we have the inclusions

$$\mathrm{H}_{-2} \subset \mathrm{H}_{-1} \subset \mathrm{H} \subset \mathrm{H}_{+1} \subset \mathrm{H}_{+2} \ .$$

Remark. When $H = -\Delta$ or $-\Delta + V$ with V Δ-bounded with bound less than 1, these are just the usual Sobolev spaces.

We now give a list of hypotheses on a pair of self-adjoint operators H and A, to which we will refer later. In these hypotheses, $\{\mathrm{H}_k\}$ are the spaces associated with H.

Hypothesis 1. $\mathrm{D}(A) \cap \mathrm{H}_{+2}$ is dense in H_{+2}.

Hypothesis 2. The form $[H, \mathrm{i}A]$ defined on $\mathrm{D}(A) \cap \mathrm{H}_{+2}$ extends to a bounded operator from H_{+2} to H_{-1}.

Hypothesis 2'. The form $[H, \mathrm{i}A]$ defined on $\mathrm{D}(A) \cap \mathrm{H}_{+2}$ extends to a bounded operator from H_{+2} to H_{-2}.

Hypothesis 3. There is a self-adjoint operator H_0 with $\mathrm{D}(H_0) = \mathrm{D}(H)$ such that $[H_0, \mathrm{i}A]$ extends to a bounded map from H_{+2} to H and $\mathrm{D}(A) \cap \mathrm{D}(H_0 A)$ is a core for H_0.

Remark. In applications where $H = -\Delta + V$, H_0 will be $-\Delta$.

Hypothesis 4. The form $[[H, \mathrm{i}A], \mathrm{i}A]$ where $[H, \mathrm{i}A]$ is as in Hypothesis 2 extends from $\mathrm{H}_{+2} \cap \mathrm{D}(A)$ to a bounded map from H_{+2} to H_{-2}.

Definition 4.4 (The Mourre Estimate). We say that a self-adjoint operator H obeys a Mourre estimate on the interval Δ if there is a self-adjoint operator A, such that

(i) H and A satisfy hypotheses 1 and 2'

(ii) there exists a positive number α and a compact operator K such that

$$E_\Delta[H, \mathrm{i}A]E_\Delta \geq \alpha E_\Delta + K \ . \tag{4.2}$$

Here $E_\Delta = E_\Delta(H)$ is the spectral projection for H associated with the interval Δ. We say H satisfies a Mourre estimate at a point $\lambda \in \mathbb{R}$ if there exists an interval Δ containing λ such that H satisfies a Mourre estimate on Δ.

We close this section by giving four examples of Schrödinger operators which satisfy a Mourre estimate.

Example 1 (2-Body Potentials). The starting point for this example is the observation that if $H_0 = -\Delta$ acting in $L^2(\mathbb{R}^\nu)$, and A is the generator of dilations, i.e. $A = (x \cdot D + D \cdot x)/2\mathrm{i}$, where D is the gradient operator $Df = \nabla f$, then

$$[H_0, \mathrm{i}A] = 2H_0 \ .$$

Thus, it easily follows that H_0 obeys a Mourre estimate on any interval Δ not containing 0. We now show that the same is true for $H = H_0 + V$ if V satisfies

(i) $V(\Delta + 1)^{-1}$ is compact
(ii) $(-\Delta + 1)^{-1} x \cdot \nabla V(-\Delta + 1)^{-1}$ is compact

[see Remark 1 following Proposition 4.16 for the precise meaning of (ii)]. Since $C_0^\infty(\mathbb{R}^\nu) \subset \mathrm{D}(H) \cap \mathrm{D}(A)$, Hypothesis 1 is satisfied. Also

$$[H, \mathrm{i}A] = 2H_0 - x \cdot \nabla V \ , \tag{4.3}$$

so (ii) implies that Hypothesis 2′ holds. From (4.3) we see

$$E_\Delta[H, \mathrm{i}A]E_\Delta = 2E_\Delta H E_\Delta + E_\Delta W E_\Delta \ ,$$

where $W = 2V + x \cdot \nabla V$. By our assumptions, $E_\Delta W E_\Delta$ is compact for any finite interval Δ. If Δ lies below 0, then E_Δ and $E_\Delta H E_\Delta$ are also compact, since by (i) $\sigma_{\mathrm{ess}}(H) = \sigma_{\mathrm{ess}}(H_0) = [0, \infty)$, so the Mourre estimate is trivially satisfied. If $\Delta = (a, b)$ with $a > 0$, then $E_\Delta H E_\Delta \geq a E_\Delta$, so the Mourre estimate holds in this case also.

Example 2 (*Froese* and *Herbst* [114]). Consider $H = -d^2/dx^2 + V$ acting in $L^2(\mathbb{R})$, where

$$V(x) = \kappa_0 \frac{\sin(2x)}{x} + V_1(x) \ ,$$

with V_1 satisfying the conditions in Example 1. What we will show is that, with

$$A = \frac{1}{2\mathrm{i}}\left(\frac{d}{dx}x + x\frac{d}{dx}\right) \ ,$$

a Mourre estimate holds at all points except 0 and 1.

Since these Mourre estimates can be used to prove the non-existence of imbedded eigenvalues (see [114]), it is amusing to note that there exists a potential of this form, the Wigner-von Neumann potential [362, 295], which has an eigenvalue at 1. For notational simplicity, we set $\kappa_0 = 1$. The general case is proven in an identical way. It is easy to check that Hypothesis 1 and 2′ hold. Now let $f_\Delta \in C_0^\infty(\mathbb{R})$ be a smoothed out characteristic function of the interval Δ. Then, if $0 \notin \Delta$, it follows from Example 1 that, for $H_0 = -d^2/dx^2$

$$f_\Delta(H_0 + V_1)[H, \mathrm{i}A]f_\Delta(H_0 + V_1) \geq \alpha f_\Delta^2(H_0 + V_1) + K$$
$$+ f_\Delta(H_0 + V_1)\left[\frac{\sin(2x)}{x}, \mathrm{i}A\right]f_\Delta(H_0 + V_1) ,$$

where $\alpha > 0$ and K is compact. Now $(H_0 + 1)[f(H_0 + V_1) - f(H)]$ is compact by an argument similar to the one in Proposition 3.8. Also, $(H_0 + 1)^{-1}[H, \mathrm{i}A](H_0 + 1)^{-1}$ is bounded. Thus,

$$f_\Delta(H)[H, \mathrm{i}A]f_\Delta(H) \geq \alpha f_\Delta^2(H) + K'$$
$$+ f_\Delta(H_0 + V_1)\left[\frac{\sin(2x)}{x}, \mathrm{i}A\right]f_\Delta(H_0 + V_1) , \tag{4.4}$$

with K' compact. If we can show that the last term is compact for sufficiently small intervals about any point $\lambda \neq 1$, we will be done, since in that case, if $\lambda \notin \{0, 1\}$, we can choose f_Δ to be identically 1 in a neighborhood Δ' about λ. The Mourre estimate then follows upon multiplying (4.4) from both sides with $E_{\Delta'}$. Now $[x^{-1}\sin(2x), \mathrm{i}A] = 2\cos(2x) - x^{-1}\sin(2x)$ and $f_\Delta(H_0 + V_1)x^{-1}\sin(2x)f_\Delta(H_0 + V_1)$ is easily seen to be compact. Since $f(H_0 + V_1) - f(H_0)$ is compact, we need only show that $f_\Delta(H_0)\cos(2x)f_\Delta(H_0)$ is compact for any sufficiently small interval Δ about $\lambda \neq 1$. But this operator has an explicit integral kernel in momentum space:

$$[f_\Delta(H_0)\cos(2x)f_\Delta(H_0)]\ (p, p')$$
$$= \tfrac{1}{2}f_\Delta(p^2)[\delta(p - p' + 2) + \delta(p - p' - 2)]f_\Delta(p'^2) ,$$

which is identically zero for any small enough Δ interval about $\lambda \neq 1$, since for such Δ no $p, p' \in \operatorname{supp}(f_\Delta)$ obey $|p - p'| = 2$.

Example 3 (Electric fields; *Bentosela, Carmona, Duclos, Simon, Souillard* and *Weder* [45]). In the study of electric fields, the group of translations often plays a role analogous to the one played by dilations in the study of other Schrödinger operators (see Chap. 7). In this example, $A = \mathrm{i}d/dx$, the generator of translations and H is a one-dimensional Hamiltonian with an electric field

$$H = \frac{-d^2}{dx^2} + V(x) + Fx ,$$

where $F > 0$ is the field strength and V is assumed to be C^1 with bounded, uniformly continuous first derivative. Again, $C_0^\infty(\mathbb{R}) \subset \mathrm{D}(H) \cap \mathrm{D}(A)$ and $C_0^\infty(\mathbb{R})$ is dense in H_{+2}. Also $[H, \mathrm{i}A] = V' + F$ which is bounded. Thus, Hypotheses 1 and 2′ hold. Since

$$E_\Delta[H, \mathrm{i}A]E_\Delta = FE_\Delta + E_\Delta V' E_\Delta ,$$

we see that a Mourre estimate holds, provided the last term is compact. We will show in Sect. 7.2 that this is the case for any finite interval Δ.

The absolute continuity of the spectrum for operators of this form has been proven by other means (see e.g. *Titchmarsh* [357], *Naimark* [261], *Walter* [364]). The Mourre method actually proves at the same time that for suitable states φ, $|(\varphi, x(t)\varphi)|$ grows as t^2.

This example can be extended to $H = -\Delta + V + \boldsymbol{F}\cdot\boldsymbol{x}$ and $A = \mathrm{i}\boldsymbol{F}\cdot\boldsymbol{D}$ in $L^2(\mathbb{R}^\nu)$, provided $\nabla V \to 0$, at infinity.

Example 4 (N-Body Hamiltonians; *Perry, Sigal, Simon* [281]). Suppose H is an N-body Hamiltonian with center of mass removed, acting in $L^2(\mathbb{R}^{(N-1)\mu})$ as described in Sect. 3.2. Suppose the pair potentials V_{ij} each obey (i) and (ii) of Example 1 in their spaces $L^2(\mathbb{R}^\mu)$. Then, with $A = (x\cdot D + D\cdot x)/2\mathrm{i}$, H satisfies a Mourre estimate at every non-threshold point. The proof of this result is more involved than those in the previous examples. It is given in Sect. 4.5.

4.2 Control of Imbedded Eigenvalues

The first application of the Mourre estimate is a theorem of *Mourre* [256], which states that if H satisfies a Mourre estimate on an interval Δ, then the point spectrum of H in Δ is finite. The only tool we need to prove this result is the virial theorem, which says that if ψ is an eigenfunction of H, then $\langle\psi, [H, \mathrm{i}A]\psi\rangle = 0$. Formally, this is obvious (by expanding the commutator). However, when H and A are unbounded some care is required, since it might happen that $\psi \notin \mathrm{D}(A)$. The virial theorem has been proven by various authors [189, 367, 281]. The proof we give follows [281]. We will need the following lemma to regularize A.

Lemma 4.5. Assume that H and A satisfy Hypotheses 1 and 3, and let $\{\mathrm{H}_k\}$ be the spaces associated with H. For $\lambda \neq 0$, define $R_\lambda = \lambda(\mathrm{i}A + \lambda)^{-1}$. Then R_λ: $\mathrm{H}_k \to \mathrm{H}_k$ is uniformly bounded for large $|\lambda|$, and

$$s - \lim_{|\lambda|\to\infty} R_\lambda = I$$

in H_k for $k = -2, -1, 0, +1, +2$ (here $\mathrm{H}_0 = \mathrm{H}$).

Proof. We will prove this result for H_{+2}. By duality, we get that R_λ: $\mathrm{H}_{-2} \to \mathrm{H}_{-2}$ is also uniformly bounded for large $|\lambda|$. The uniform boundedness for the other

H_k's then follows by interpolation [293]. Since H_{+2} is dense in each H_j, uniform boundedness on each H_j and strong convergence in H_{+2} imply strong convergence in each H_j.

Let H_0 be as in Hypothesis 3. We will also regularize H_0. For $\varphi \in \mathrm{H}_{+2}$

$$(H_0 + \mathrm{i})(1 + \mathrm{i}\varepsilon H_0)^{-1} R_\lambda \varphi$$
$$= R_\lambda (H_0 + \mathrm{i})(1 + \mathrm{i}\varepsilon H_0)^{-1} \varphi + [(H_0 + \mathrm{i})(1 + \mathrm{i}\varepsilon H_0)^{-1}, R_\lambda]\varphi \; .$$

Since $\|R_\lambda\| = 1$,

$$\|R_\lambda (H_0 + \mathrm{i})(1 + \mathrm{i}\varepsilon H_0)^{-1} \varphi\| \leq \|(H_0 + \mathrm{i})\varphi\| \; ,$$

while

$$[(H_0 + \mathrm{i})(1 + \mathrm{i}\varepsilon H_0)^{-1}, R_\lambda] = (\mathrm{i}A + \lambda)^{-1}[(H_0 + \mathrm{i})(1 + \mathrm{i}\varepsilon H_0)^{-1}, \mathrm{i}A] R_\lambda$$
$$= (\mathrm{i}A + \lambda)^{-1}[-\mathrm{i}\varepsilon^{-1}(\mathrm{i}\varepsilon H_0 + 1 - \varepsilon - 1)(1 + \mathrm{i}\varepsilon H_0)^{-1}, \mathrm{i}A] R_\lambda$$
$$= \mathrm{i}\varepsilon^{-1}(1 + \varepsilon)(\mathrm{i}A + \lambda)^{-1}[(1 + \mathrm{i}\varepsilon H_0)^{-1}, \mathrm{i}A] R_\lambda$$
$$= \mathrm{i}\varepsilon^{-1}(1 + \varepsilon)(\mathrm{i}A + \lambda)^{-1}(-\mathrm{i}\varepsilon^{-1} + H_0)^{-1}[H_0, \mathrm{i}A](1 + i\varepsilon H_0)^{-1} R_\lambda \; .$$

Inserting a factor $(H_0 + \mathrm{i})^{-1}(H_0 + \mathrm{i})$ to the right of $[H_0, \mathrm{i}A]$, and using that $[H_0, A](H_0 + \mathrm{i})^{-1}$ is bounded (by Hypothesis 3), we find that for large $|\lambda|$

$$\|[(H_0 + \mathrm{i})(1 + \mathrm{i}\varepsilon H_0)^{-1}, R_\lambda]\varphi\| \leq C|\lambda|^{-1} \|(H_0 + \mathrm{i})(1 + \mathrm{i}\varepsilon H_0)^{-1} R_\lambda \varphi\| \; .$$

Thus,

$$(1 - C|\lambda|^{-1})\|(H_0 + \mathrm{i})(1 + \mathrm{i}\varepsilon H_0)^{-1} R_\lambda \varphi\| \leq C\|(H_0 + \mathrm{i})\varphi\| \; ,$$

so that for $|\lambda|$ large

$$\|(H_0 + \mathrm{i})(1 + \mathrm{i}\varepsilon H_0)^{-1} R_\lambda \varphi\| \leq C\|(H_0 + \mathrm{i})\varphi\| \; .$$

Taking $\varepsilon \downarrow 0$ and using that $\mathrm{D}(H_0) = \mathrm{D}(H) = \mathrm{H}_{+2}$, we see that $R_\lambda \colon \mathrm{H}_{+2} \to \mathrm{H}_{+2}$ is uniformly bounded for large $|\lambda|$.

Now $1 - R_\lambda = \lambda^{-1} R_\lambda \mathrm{i}A$. If $A\psi \in \mathrm{H}_{+2}$, this implies $\|(1 - R_\lambda)\psi\|_2 \to 0$ as $|\lambda| \to \infty$. Since $\mathrm{D}(A) \cap \mathrm{D}(H_0 A)$ is dense in H_{+2}, the uniform bound implies the strong convergence. □

Theorem 4.6 (The Virial Theorem). Assume Hypotheses 1, 2′ and 3 hold for H and A. If $E_{\{\mu\}}$ denotes the spectral projection for H corresponding to the point μ,

$$E_{\{\mu\}}[H, \mathrm{i}A]E_{\{\mu\}} = 0 \; . \tag{4.5}$$

In particular, $\langle \psi, [H, A]\psi \rangle = 0$ for any eigenfunction ψ of H.

Proof. Let $A_\lambda = AR_\lambda$ with R_λ as in Lemma 4.5. Then A_λ is bounded, and since $E_{\{\mu\}}H = \mu E_{\{\mu\}}$, we have

$$E_{\{\mu\}}[H, \mathrm{i}A_\lambda]E_{\{\mu\}} = \mu E_{\{\mu\}}\mathrm{i}A_\lambda E_{\{\mu\}} - \mu E_{\{\mu\}}\mathrm{i}A_\lambda E_{\{\mu\}}$$
$$= 0 \ .$$

By direct calculation, we find $[H, \mathrm{i}A_\lambda] = R_\lambda[H, \mathrm{i}A]R_\lambda$. Thus,

$$E_{\{\mu\}}R_\lambda[H, \mathrm{i}A]R_\lambda E_{\{\mu\}} = 0 \ .$$

Since $R_\lambda \to 1$ strongly in H_{+2} and H_{-2} as $|\lambda| \to \infty$, and $E_{\{\mu\}}$ maps H to H_{+2} and H_{-2} to H, this operator tends strongly to $E_{\{\mu\}}[H, A]E_{\{\mu\}}$ as an operator from H to H. This implies (4.5). □

We now can prove the theorem of *Mourre* [256] on finiteness of point spectrum.

Theorem 4.7. Assume Hypotheses 1, 2′ and 3 hold for H and A, and that H satisfies a Mourre estimate on the interval Δ. Then H has at most finitely many eigenvalues in Δ, and each eigenvalue has finite multiplicity.

Remark. This result shows that in the (open) set of points at which a Mourre estimate holds for H, eigenvalues cannot accumulate.

Proof. Suppose there are infinitely many eigenvalues of H in Δ, or that some eigenvalue has infinite multiplicity. Let $\{\psi_n\}_{n=1}$ be the corresponding orthonormal eigenfunctions. Then by virial theorem and the Mourre estimate

$$\begin{aligned} 0 &= \langle \psi_n, [H, \mathrm{i}A]\psi_n \rangle \\ &= \langle \psi_n, E_\Delta[H, \mathrm{i}A]E_\Delta\psi_n \rangle \\ &\geq \alpha\|\psi_n\|^2 + \langle \psi_n, K\psi_n \rangle \ . \end{aligned}$$

Now $\|\psi_n\| = 1$, and since $\psi_n \to 0$ weakly and K is compact, $\langle \psi_n, K\psi_n \rangle \to 0$ as $n \to \infty$. This is impossible, since $\alpha > 0$. □

Remark. For N-body Schrödinger operators, we will see that the Mourre estimate holds away from the set of thresholds, so that Theorem 4.7 says that eigenvalues can accumulate only at thresholds. *Perry* [280] has shown that, for N-body systems, eigenvalues can actually only accumulate at thresholds from below. There are examples of atomic Hamiltonians for which one knows (for reasons of symmetry) that there are infinitely many imbedded eigenvalues converging to a threshold. In Sect. 4.4, we will show that under suitable hypotheses, N-body systems cannot have positive eigenvalues.

4.3 Absence of Singular Continuous Spectrum

The purpose of this section is to prove that an operator H has no singular continuous spectrum in the set on which it obeys a Mourre estimate. Using this result, we can reduce the proof of the absence of singular continuous spectrum for a given operator H to the proof of the Mourre estimate for some choice of conjugate operator A. The strategy for proving this theorem is due to *Mourre* [256]. It was extended by *Perry, Sigal* and *Simon* in [281] to deal with more general operators.

Actually, what we will show is that H has a purely absolutely continuous spectrum on the set where H obeys the Mourre estimate (4.2) with $K = 0$. The following lemma with allow us to deduce the result on absence of singular continuous spectrum from this.

Lemma 4.8. Suppose H and A satisfy Hypotheses 1, 2′ and 3. If D is the (open) set of points at which H and A obey a Mourre estimate, then H and A obey a Mourre estimate with $K = 0$ at each point in $D \backslash \sigma_{pp}(H)$.

Proof. By the definition of D, there exists an interval Δ, about every point λ in $D \backslash \sigma_{pp}(H)$, such that a Mourre estimate (4.2) holds for some α and K. Multiplying this inequality from both sides with $E_{\Delta'}(H)$, where Δ' is an interval with $\lambda \in \Delta' \subseteq \Delta$, we obtain, for each such Δ',

$$E_{\Delta'}(H)[H, \mathrm{i}A]E_{\Delta'}(H) \geq \alpha E_{\Delta'}(H) + E_{\Delta'}(H)KE_{\Delta'}(H) \ . \tag{4.6}$$

Since $\lambda \notin \sigma_{pp}(H)$, $E_{\Delta'}(H)$ tends strongly to zero as Δ' shrinks about λ. Therefore, $E_{\Delta'}(H)KE_{\Delta'}(H)$ tends to zero in norm. If we choose Δ' such that $\|E_{\Delta'}(H)KE_{\Delta'}(H)\| < \alpha/2$, (4.6) implies

$$E_{\Delta'}(H)[H, A]E_{\Delta'}(H) \geq \alpha E_{\Delta'}(H) - \alpha/2$$

and the lemma follows upon multiplying this inequality from both sides with $E_{\Delta'}(H)$. □

We now come to the main theorem in this section.

Theorem 4.9. Suppose H and A satisfy Hypotheses 1, 2, 3 and 4. Then each point λ for which a Mourre estimate holds with $K = 0$ is contained in an open interval Δ, such that

$$\overline{\lim_{\delta \downarrow 0}} \sup_{\mu \in \Delta} \|(|A| + 1)^{-1}(H - \mu - \mathrm{i}\delta)^{-1}(|A| + 1)^{-1}\| \leq C \tag{4.7}$$

for some constant C.

Corollary 4.10. If H and A satisfy the hypotheses of Theorem 4.9, then H has a purely absolutely continuous spectrum in the (open) set where a Mourre estimate holds with $K = 0$.

Remark 1. Given the results of Sect. 4.5 (Example 4 above), Theorem 4.9 and Lemma 4.8 imply that N-body Schrödinger operators have no singular continuous spectrum.

Remark 2. For N-body Schrödinger operators, the conclusion (4.7) of Theorem 4.9 remains true when $(|A| + 1)^{-1}$ is replaced with $(|x| + 1)^{-1/2-\varepsilon}$, see [281]. More recently, *Jensen* and *Perry* [185] have improved this result, showing that $(H - \mu - i\delta)^{-1}$ remains bounded as a map between certain Besov spaces as $\delta \downarrow 0$.

Remark 3. The result of Perry, Sigal and Simon implies that $(1 + |x|)^{-1/2-\varepsilon}$ is a locally smooth perturbation of H (see *Reed* an *Simon* [295], XIII. 7 for the theory of smooth perturbations). This result immediately implies asymptotic completeness for two-body systems with potentials decaying like $|x|^{-1-2\varepsilon}$ and should be useful in studying N-body asymptotic completeness.

Remark 4. *Mourre* [257] has shown that $(1 + |A|)^{-1}$ can be replaced by spectral projections for A onto $\pm [0, \infty)$, yielding propagation estimates of use in scattering theory.

We will prove Theorem 4.9 in a sequence of lemmas. Let λ be a point where the Mourre estimate holds, with $K = 0$, i.e. for some interval Δ containing λ, and some $\alpha > 0$

$$E_\Delta(H)[H, iA]E_\Delta(H) \geq \alpha E_\Delta(H) \ .$$

Let $f \in C_0^\infty(\mathbb{R})$ be a smoothed characteristic function with support in Δ such that $f \equiv 1$ in some sub-interval Δ' containing λ. Then

$$f(H)[H, iA]f(H) \geq \alpha f^2(H) \ , \tag{4.8}$$

and we can define the nonnegative operator, $M^2 = f(H)[H, iA]f(H)$. The proof will center about the analysis of the operator

$$G_\varepsilon(z) = (H - i\varepsilon M^2 - z)^{-1}$$

which, as we show below, exists for $\varepsilon \geq 0$ and $\operatorname{Im} z > 0$. This operator is not as mysterious as it appears to be at first glance. If we ignore the $f(H)$ terms in M^2, G_ε is the resolvent of $H - \varepsilon[H, A]$, which is the first term in the formal power series expansion of the complex dilated Hamiltonian $\exp(\varepsilon A)H \exp(-\varepsilon A)$.

Remark 5. *Jensen, Mourre* and *Perry* in [184] have explored the idea of using more terms of this expansion. They establish a connection between the boundedness of higher-order terms and the smoothness of the resolvent in the limit $\delta \downarrow 0$.

Define the operators D and F_ε by

$$D = (|A| + 1)^{-1}$$

$$F_\varepsilon = F_\varepsilon(z) = DG_\varepsilon(z)D \ .$$

Then the strategy of the proof will be to show F_ε is C^1 in ε, and establish the following inequalities for small ε.

(a) $\|F_\varepsilon\| \leq C/\varepsilon$

(b) $\|dF_\varepsilon/d\varepsilon\| \leq C(\|F_\varepsilon\| + \varepsilon^{-1/2}\|F_\varepsilon\|^{1/2} + 1)$ with C independent of $\operatorname{Re} z = \mu$ for $\mu \in \Delta'$.

Proposition 4.11. The estimates (a) and (b) for small ε imply Theorem 4.9.

Proof. Inserting (a) into the right side of (b), we find that, for small ε,

$$\|dF_\varepsilon/d\varepsilon\| \leq C\varepsilon^{-1} ,$$

which implies

$$\|F_\varepsilon\| \leq C\log(\varepsilon) .$$

Using this new estimate in (b), we find

$$\|dF_\varepsilon/d\varepsilon\| \leq C\varepsilon^{-1/2}\log(\varepsilon)$$

near $\varepsilon = 0$, which shows that $\|F_\varepsilon\|$ stays bounded as $\varepsilon \downarrow 0$. □

In this proof and in what follows, C denotes a generic constant independent of $\mu = \operatorname{Re} z$ for $\mu \in \Delta'$, whose value might change from line to line. We prove next some technical lemmas which estimate quantities which will appear in the proof of (a) and (b). We remark that it is the need to control $[A, M^2]$ which forces us to assume Hypothesis 2 in place of 2′, and to assume Hypothesis 4.

Lemma 4.12. If $f \in C_0^\infty(\mathbb{R})$, then $[A, f(H)]$ is bounded from H_{-1} to H_{+1}.

Proof. To avoid domain difficulties, we regularize A. Let $R_\lambda = \lambda(\mathrm{i}A + \lambda)^{-1}$ as in Lemma 4.5. Then $A_\lambda := AR_\lambda$ is bounded, and

$$\begin{aligned} \mathrm{e}^{\mathrm{i}tH}A_\lambda - A_\lambda \mathrm{e}^{\mathrm{i}tH} &= (\mathrm{e}^{\mathrm{i}tH}A_\lambda \mathrm{e}^{-\mathrm{i}tH} - A_\lambda)\mathrm{e}^{\mathrm{i}tH} \\ &= \left(\int_0^t \mathrm{e}^{\mathrm{i}sH}[H, A_\lambda]\mathrm{e}^{-\mathrm{i}sH}\,ds\right)\mathrm{e}^{\mathrm{i}tH} . \end{aligned}$$

As in Lemma 4.6, $[H, A_\lambda] = R_\lambda[H, \mathrm{i}A]R_\lambda$ and R_λ is bounded uniformly in λ for large $|\lambda|$ from H_2 to H_2 and from H_{-1} to H_{-1}. On the other hand, by Hypothesis 2, $[H, A]$ is bounded from H_2 to H_{-1}. Thus,

$$\|[A_\lambda, \mathrm{e}^{\mathrm{i}tH}]\|_{2,-1} \leq Ct$$

with C independent of λ. Here $\|\cdot\|_{i,j}$ denotes the norm of maps from H_i to H_j. Now for $g \in C_0^\infty(\mathbb{R})$ we have

$$g(H) = (2\pi)^{-1/2}\int_{-\infty}^{\infty} \hat{g}(s)\mathrm{e}^{\mathrm{i}sH}\,ds ,$$

where $\hat{g}$ denotes the Fourier transform of g. Thus,

$$\|[A_\lambda, g(H)]\|_{2,-1} \leq C , \tag{4.9}$$

where C depends on g, but not on λ. Since

$$[A_\lambda, (H+\mathrm{i})^{-1}] = -(H+\mathrm{i})^{-1} R_\lambda [H, A] R_\lambda (H+\mathrm{i})^{-1} ,$$

we see that

$$\|[A_\lambda, (H+\mathrm{i})^{-1}]\|_{0,1} < C, \quad \|[A_\lambda, (H+\mathrm{i})^{-1}]\|_{-1,0} < C \tag{4.10}$$

for C independent of λ. Now for $f \in C_0^\infty(\mathbb{R})$, we write $f(H) = (H+\mathrm{i})^{-1} g(H)(H+\mathrm{i})^{-1}$ for $g \in C_0^\infty(\mathbb{R})$ and thereby obtain

$$\begin{aligned}[A_\lambda, f(H)] = [A_\lambda, (H+\mathrm{i})^{-1}] g(H)(H+\mathrm{i})^{-1} + (H+\mathrm{i})^{-1}[A_\lambda, g(H)](H+\mathrm{i})^{-1} \\ + (H+\mathrm{i})^{-1} g(H)[A_\lambda, (H+\mathrm{i})^{-1}] ,\end{aligned} \tag{4.11}$$

so, using (4.9) and (4.10), we find

$$\|[A_\lambda, f(H)]\|_{0,1} < C .$$

Using this estimate for $[A_\lambda, g(H)]$ in (4.11), we get

$$\|[A_\lambda, f(H)]\|_{-1,1} < C$$

with C independent of λ. Taking λ to ∞ completes the proof of the lemma. □

Lemma 4.13. $[A, M^2]$ is bounded from H to H.

Proof. We have

$$[A, M^2] = [A, f(H)]Bf(H) + f(H)[A, B]f(H) + f(H)B[A, f(H)] ,$$

where $B := [H, \mathrm{i}A]$, so this lemma follows from Lemma 4.12 and Hypothesis 4. □

Lemma 4.14. (a) For $\varepsilon \geq 0$ and $\operatorname{Im} z > 0$, $(H - \mathrm{i}\varepsilon M^2 - z)$ is invertible, and the inverse, G_ε, is C^1 in ε on $(0, \infty)$ and continuous on $[0, \infty)$.

(b) The following estimate holds for all ε with $0 < \varepsilon < \varepsilon_0$ for suitable ε_0, and for all z with $\operatorname{Re} z \in \Delta'$. (Recall that Δ' is an interval on which $f = 1$.)

$$\|f(H) G_\varepsilon(z)\varphi\| \leq C\varepsilon^{-1/2} |\langle \varphi, G_\varepsilon(z)\varphi \rangle|^{1/2} . \tag{4.12}$$

(c) For z and ε as in (a)

$$\|(1 - f(H)) G_\varepsilon(z)\| \leq C \tag{4.13}$$

$$\|G_\varepsilon(z)\| \leq C\varepsilon^{-1} . \tag{4.14}$$

(d) The estimates in (b) and (c) hold when the operator norm $\|\cdot\|$ on H is replaced with $\|\cdot\|_{0,2}$, the norm as operators from H to H_2.

(e) For ε and z as in (a)

$$\|G_\varepsilon(z)D\| \leq C(1 + \varepsilon^{-1/2}\|F_\varepsilon\|^{1/2}) .$$

Proof. (a) Write $z = \mu + \mathrm{i}\delta$. Then

$$\|(H - \mathrm{i}\varepsilon M^2 - z)\varphi\|^2 = \|(H - \mathrm{i}\varepsilon M^2 - \mu)\varphi\|^2 + \delta^2\|\varphi\|^2 + 2\delta\varepsilon\|M\varphi\|^2 .$$

Thus, for $\delta > \max(0, -2\varepsilon\|M\|)$ (ε may be negative), we conclude that $H - \mathrm{i}\varepsilon M^2 - z$ is invertible on its closed range. Since the adjoint operator $H + \mathrm{i}\varepsilon M^2 - z^*$ obeys a similar estimate, its null space is empty, which implies that the range of $H - \mathrm{i}\varepsilon M^2 - z$ is dense, and hence all of H. Since M^2 is bounded, $H - \mathrm{i}\varepsilon M^2 - z$, for fixed z, is an analytic family of type (A) in ε [196]. Thus, $G_\varepsilon(z)$, for fixed z, is analytic in a region surrounding $(-\delta/2\|M\|, \infty)$, which gives us the required smoothness and continuity. For future use, we note that, by differentiating $\psi = G_\varepsilon(z)(H - \mathrm{i}\varepsilon M^2 - z)\psi$ for $\psi \in \mathrm{D}(H) = \mathrm{H}_2$, and using the product rule, we find that

$$\frac{dG_\varepsilon}{d\varepsilon} = \mathrm{i}G_\varepsilon M^2 G_\varepsilon \tag{4.15}$$

(b) This is the only step where the Mourre estimate enters. By (4.8),

$$\begin{aligned}\|fG_\varepsilon\varphi\|^2 &= \langle\varphi, G_\varepsilon^* f^2 G_\varepsilon\varphi\rangle \\ &\leq (2\alpha\varepsilon)^{-1}\langle\varphi, G_\varepsilon^* 2\varepsilon M^2 G_\varepsilon\varphi\rangle \\ &\leq (2\alpha\varepsilon)^{-1}\langle\varphi, G_\varepsilon^*(2\varepsilon M^2 + 2\,\mathrm{Im}\,z)G_\varepsilon\varphi\rangle \\ &= (2\alpha\varepsilon)^{-1}\langle\varphi, \mathrm{i}(G_\varepsilon^* - G_\varepsilon)\varphi\rangle \\ &\leq (\alpha\varepsilon)^{-1}|\langle\varphi, G_\varepsilon\varphi\rangle| .\end{aligned}$$

(c) We can write

$$(1 - f)G_\varepsilon = (1 - f)G_0(1 + \mathrm{i}\varepsilon M^2 G_\varepsilon) ,$$

and for $\mathrm{Re}\, z \in \Delta'$, $(1 - f(H))G_0(z)$ is bounded. Thus,

$$\|(1 - f)G_\varepsilon\| \leq C(1 + \varepsilon\|G_\varepsilon\|) , \tag{4.16}$$

so that (4.13) follows from (4.14). To prove (4.14), we estimate

$$\begin{aligned}\|G_\varepsilon\| + 1 &\leq \|fG_\varepsilon\| + \|(1 - f)G_\varepsilon\| + 1 \\ &\leq C\varepsilon^{-1/2}\|G_\varepsilon\|^{1/2} + C_1(1 + \varepsilon\|G_\varepsilon\|) + 1 ,\end{aligned}$$

Here we used (4.12) and (4.16). Now if $C_1\varepsilon \leq \frac{1}{2}$ and $C_1 + \frac{1}{2} \leq C\varepsilon^{-1/2}$, we can continue estimating to conclude

$$\|G_\varepsilon\| + 1 \le C\varepsilon^{-1/2}(\|G_\varepsilon\|^{1/2} + 1) + \tfrac{1}{2}(\|G_\varepsilon\| + 1)$$
$$\le 2C\varepsilon^{-1/2}(\|G_\varepsilon\| + 1)^{1/2} + \tfrac{1}{2}(\|G_\varepsilon\| + 1) \ ,$$

which implies that

$$\|G_\varepsilon\| \le 16C^2\varepsilon^{-1} \ . \tag{4.17}$$

Thus, if $\varepsilon < \varepsilon_0 := \min\{(2C_1)^{-1}, C^2(C_1 + \frac{1}{2})^{-2}\}$, (4.14) holds.

(d) We remind the reader that for an operator S, $\|S\|_{0,2} = \|(H + \mathrm{i})S\|$. Since f has compact support in Δ, we have $(H + \mathrm{i})f(H) = (H + \mathrm{i})E_\Delta(H)f(H)$, so the required estimate follows easily for (4.12). Returning to the proof of (c), we note that in fact $(H + \mathrm{i})(1 - f(H))G_0(z)$ is bounded so that

$$\|(1 - f)G_\varepsilon\|_{0,2} \le C(1 + \varepsilon\|G_\varepsilon\|) \le C \ .$$

Here we used (4.14) with $\mathrm{H} \to \mathrm{H}$ norms. Since

$$\|fG_\varepsilon\|_{0,2} \le C\|(H + \mathrm{i})f(H)G_\varepsilon\| \le C\|G_\varepsilon\|$$

by the compact support of f we find, combining this estimate with the previous one, that (4.14) also holds for $\mathrm{H} \to \mathrm{H}_2$ norms.

(e) From (4.12), with $\varphi = D\psi$, it follows that

$$\|fG_\varepsilon D\psi\| \le C\varepsilon^{-1/2}|\langle\psi, F_\varepsilon\psi\rangle|^{1/2} \ , \quad \text{so that}$$
$$\|fG_\varepsilon D\| \le C\varepsilon^{-1/2}\|F_\varepsilon\| \ .$$

On the other hand,

$$\|(1 - f)G_\varepsilon D\| \le \|(1 - f)G_\varepsilon\|$$

is bounded by (4.13), so the result follows. □

The inequality (a) follows from (4.14), so the following lemma will complete the proof of Theorem 4.9.

Lemma 4.15. The differential inequality (b) holds.

Proof. From (4.15), we have

$$-\mathrm{i}dF_\varepsilon/d\varepsilon = DG_\varepsilon M^2 G_\varepsilon D = Q_1 + Q_2 + Q_3 \ , \quad \text{where}$$
$$Q_1 = -DG_\varepsilon(1 - f)[H, \mathrm{i}A](1 - f)G_\varepsilon D$$
$$Q_2 = -DG_\varepsilon(1 - f)[H, \mathrm{i}A]fG_\varepsilon D - DG_\varepsilon f[H, \mathrm{i}A](1 - f)G_\varepsilon D$$
$$Q_3 = DG_\varepsilon[H, \mathrm{i}A]G_\varepsilon D \ .$$

Now, from (4.13) with the $\|\cdot\|_{0,2}$ norm, $(1 - f)G_\varepsilon D$ is bounded from H to H_2, while $[H, \mathrm{i}A]$ is bounded from H_2 to H_{-2}. Thus,

$$\begin{aligned}
\|Q_1\| &\le C \quad \text{and} \\
\|Q_2\| &\le C\|(H+\mathrm{i})fG_\varepsilon D\| \\
&\le C\|G_\varepsilon D\| \\
&= C(1+\varepsilon^{-1/2}\|F_\varepsilon\|^{1/2})
\end{aligned}$$

by Lemma 4.14(e). We write

$$\begin{aligned}
Q_3 &= Q_4 + Q_5\ , \quad \text{where} \\
Q_4 &= DG_\varepsilon[H - \mathrm{i}\varepsilon M^2 - z, \mathrm{i}A]G_\varepsilon D \\
Q_5 &= \mathrm{i}\varepsilon DG_\varepsilon[M^2, \mathrm{i}A]G_\varepsilon D\ .
\end{aligned}$$

Expanding the commutator in Q_4, we find

$$\begin{aligned}
\|Q_4\| &\le 2\|DAG_\varepsilon D\| \\
&\le 2\|G_\varepsilon D\| \\
&\le C(1+\varepsilon^{-1/2}\|F_\varepsilon\|^{1/2})\ .
\end{aligned}$$

Here we used $\|DA\| \le 1$ and Lemma 4.14(e). Finally we estimate, using Lemmas 4.13 and 4.14(c)

$$\begin{aligned}
\|Q_5\| &\le \varepsilon\|G_\varepsilon D\|^2\|[M^2, \mathrm{i}A]\| \\
&\le C(\varepsilon^{1/2} + \|F_\varepsilon\|^{1/2})^2 \\
&\le C(1+\|F_\varepsilon\|)
\end{aligned}$$

for $\varepsilon < \varepsilon_0$. Combining the estimates for Q_1 through Q_5, we conclude that (b) holds. □

4.4 Exponential Bounds and Nonexistence of Positive Eigenvalues

In this section, we will describe the relationship between the decay rates of eigenfunctions of Schrödinger operators and the position of the eigenvalue relative to the points where the Mourre estimate fails to hold (by the results of the following section, these are the thresholds for N-body Hamiltonians). This will lead to a proof of the nonexistence of positive eigenvalues for N-body Schrödinger operators in two steps. First, we show that the eigenfunctions of N-body Hamiltonians corresponding to positive eigenvalues have to decay extremely rapidly. Then we show, in a sort of unique continuation theorem at infinity, that such rapid decay is impossible. The results of this section are less general than those of previous sections in that H is required to be of the form $-\Delta + V$, and A is always the dilation generator. This restriction arises because we use the special commutation properties of $-\Delta$, V and A.

The proof in this section follows *Froese* and *Herbst* [114] and *Froese, Herbst, Hoffmann-Ostenhof* and *Hoffmann-Ostenhof* [116]. That N-body Hamiltonians can have no positive eigenvalues was previously known for some special cases from the work of *Weidmann* [367], *Balslev* [35], and *Simon* [321]. For one-body systems, the absence of positive eigenvalues was known for quite general potentials from the work of *Kato* [190], *Agmon* [1] and *Simon* [318] (see also the recent book [90]). The exponential decay properties of eigenfunctions, which we use here as a tool, are interesting in their own right. We mention only the book of Agmon [3], which contains further references.

In this section, it will be convenient to use the antisymmetric dilation generator. Define

$$\tilde{A} = \tfrac{1}{2}(D\cdot x + x\cdot D) \ , \tag{4.18}$$

where D is the gradient operator, i.e. $Df := \nabla f$. Then $\tilde{A} = \mathrm{i}A$, where A is the dilation generator used above. We will also use the notation for $x \in \mathbb{R}^\nu$

$$\langle x\rangle = (1 + |x|^2)^{1/2} \ . \tag{4.19}$$

To give some idea of how commutators can give information about positive eigenvalues, we sketch a proof of *Weidmann*'s theorem [367], which applies in particular to potentials V which are homogeneous of degree -1, i.e. $V(\lambda x) = \lambda^{-1}V(x)$ for $\lambda > 0$ (e.g. atomic Hamiltonians). For these potentials, $[V, \tilde{A}] = -x\cdot \nabla V = V$. Thus, if ψ is a normalized eigenfunction of $H = -\Delta + V$ with eigenvalue E, we can apply the virial theorem to conclude

$$\begin{aligned} 0 &= \langle\psi, [H, \tilde{A}]\psi\rangle = \langle\psi, (-2\Delta + V)\psi\rangle \\ &= \langle\psi, H\psi\rangle + \langle\psi, -\Delta\psi\rangle \\ &\geq E \ . \end{aligned}$$

Here we used the commutation relation $[-\Delta, \tilde{A}] = -2\Delta$ and the positivity of $-\Delta$.

We will be dealing with exponential decay of eigenfunctions in the L^2 sense. A function ψ is said to satisfy an L^2 upper (lower) bound if $\exp(F)\psi$ is in (not in) L^2. Here F is a function which measures the decay rate. The next proposition lists some equations satisfied by $\psi_F := \exp(F)\psi$ when ψ is an eigenfunction of a Schrödinger operator, and $F(x)$ is an increasing function of $|x|$ alone. An important hypothesis in this proposition is that $\psi_F \in L^2$. This is how the L^2 decay properties of ψ will enter our proofs.

Proposition 4.16. Suppose $H = -\Delta + V$ in $L^2(\mathbb{R}^\nu)$, where V satisfies

(i) V is Δ-bounded with bound less than one,
(ii) $(-\Delta+1)^{-1}x\cdot\nabla V(-\Delta+1)^{-1}$ is bounded.
Suppose ψ is an eigenfunction of H, with eigenvalue E, i.e. $H\psi = E\psi$. Let F be a non-decreasing C^∞ function of $|x|$ alone, and assume that

$$|\nabla F| \leq C, (x\cdot\nabla)^2 g - x\cdot\nabla((\nabla F)^2) \leq C \ ,$$

where g is the nonnegative function defined by $\nabla F = xg$. Define $\psi_F := \exp(F)\psi$, and assume $\psi_F \in L^2$. Then

$$H(F)\psi_F = E\psi_F, \qquad \text{where } H(F) = H - (\nabla F)^2 + D\cdot\nabla F + \nabla F\cdot D \ , \tag{4.20}$$

$$\langle\psi_F, H\psi_F\rangle = \langle\psi_F, [(\nabla F)^2 + E]\psi_F\rangle \tag{4.21}$$

$$\begin{aligned}\langle\psi_F, [H, \tilde{A}]\psi_F\rangle &= -4\|g^{1/2}\tilde{A}\psi_F\|^2 \\ &\quad + \langle\psi_F, ((x\cdot\nabla)^2 g - (x\cdot\nabla)(\nabla F)^2)\psi_F\rangle \ .\end{aligned} \tag{4.22}$$

Here $\tilde{A}$ is given by (4.18) and D is the gradient operator.

Remark 1. Assumption (i) allows us to define H as a self-adjoint operator with domain $\mathrm{D}(\Delta)$. Assumption (ii) implies that Hypothesis 2′ holds for H and the generator of dilations. What (ii) really means is that form $Q(f_1, f_2)$ defined for f_1 and f_2 in the Schwartz space S by

$$Q(f_1, f_2) = \int V\left(\sum_i - \partial_i x_i[\overline{(-\Delta+1)^{-1}f_1}\cdot(-\Delta+1)^{-1}f_2]\right)d^\nu x$$

extends to the form of a bounded operator. Note the V need not have derivatives in the classical sense for this to hold. For example, if $(1+|x|)V$ is $-\Delta$ bounded, then Hypothesis (ii) will hold.

Remark 2. Although it might happen that $\psi_F \notin \mathrm{D}(\tilde{A})$, we will show that $\psi_F \in \mathrm{D}(g^{1/2}\tilde{A})$, so that (4.22) makes sense.

Remark 3. Formally, this proposition follows just from computing commutators.

Proof. Since ∇F is bounded, $H(-F)$ is a closed operator with domain $\mathrm{H}_2 = \mathrm{D}(\Delta)$, with C_0^∞ as a core, and with adjoint $H(F)$. For $\varphi \in C_0^\infty$, it follows from calculating commutators that $H(-F)\varphi = \exp(-F)H\cdot\exp(F)\varphi$. Thus, $\langle H(-F)\varphi, \psi_F\rangle = \langle\varphi, E\psi_F\rangle$ for $\varphi \in C_0^\infty$, which implies that $\psi_F \in \mathrm{D}(H(F))$, and that (4.20) holds. Equation (4.21) follows from (4.20), and the antisymmetry of $D\cdot\nabla F + \nabla F\cdot D$. Explicitly

$$\begin{aligned}E\|\psi_F\|^2 &= \langle\psi_F, H(F)\psi_F\rangle \\ &= \mathrm{Re}\langle\psi_F, H(F)\psi_F\rangle \\ &= \langle\psi_F, (H - (\nabla F)^2)\psi_F\rangle \ .\end{aligned}$$

To prove (4.22), we first verify the following identity for $\varphi \in C_0^\infty$ and $\xi := \exp(F)$

$$\langle \varphi, [\xi \tilde{A} \xi, -\Delta]\varphi\rangle = \langle \xi\varphi, [\tilde{A}, -\Delta]\xi\varphi\rangle - 4\|g^{1/2}\tilde{A}\xi\varphi\|^2 + \langle \xi\varphi, G\xi\varphi\rangle . \tag{4.23}$$

Here $G(x) = (x \cdot \nabla)^2 g - x \cdot \nabla((\nabla F)^2)$. Since $\varphi \in C_0^\infty$, this identity follows from the formal computation

$$\begin{aligned}
[\xi\tilde{A}\xi, -\Delta] &= \xi[\tilde{A}, -\Delta]\xi + \xi\tilde{A}[\xi, -\Delta] + [\xi, -\Delta]\tilde{A}\xi \\
&= \xi[\tilde{A}, -\Delta]\xi + \xi\tilde{A}(e^F gx \cdot D + D \cdot xge^F) \\
&\quad + (e^F gx \cdot D + D \cdot xge^F)\tilde{A}\xi \\
&= \xi[\tilde{A}, -\Delta]\xi + \xi(\tilde{A}(gx \cdot D + D \cdot xg) + (gx \cdot D + D \cdot xg)\tilde{A} \\
&\quad + [(\nabla F)^2, \tilde{A}])\xi \\
&= \xi[\tilde{A}, -\Delta]\xi + 4\xi\tilde{A}gA\xi + \xi((x \cdot \nabla)^2 g - x \cdot \nabla((\nabla F)^2))\xi .
\end{aligned}$$

Define the cut-off function $\chi_m(x) = \chi(x/m)$, where $\chi \in C_0^\infty$ and χ equals one in a neighborhood of the origin. Then it is not hard to see that (4.23) holds, with $\varphi = \chi_m\psi$. Adding $\langle \chi_m\psi, (x \cdot \nabla V)\chi_m\psi\rangle = \langle \chi_m\psi, [\tilde{A}, V]\chi_m\psi\rangle$ to each side, and introducing the constant E in the commutator on the left, we obtain

$$\begin{aligned}
\langle \chi_m\psi, [\xi\tilde{A}\xi, H - E]\chi_m\psi\rangle &= \langle \xi\chi_m\psi, [\tilde{A}, H]\xi\chi_m\psi\rangle \\
&\quad -4\|g^{1/2}\tilde{A}\xi\chi_m\psi\|^2 + \langle \xi\chi_m\psi, G\xi\chi_m\psi\rangle .
\end{aligned} \tag{4.24}$$

Using (4.20) and (4.21), it is possible to show that $\xi\chi_m\psi \to \xi\psi = \psi_F$ in H_{+2} as $m \to \infty$. Thus, the first and last terms on the right side of (4.24) converge. Here we use the boundedness of G. To handle the left side of (4.24), we write

$$\begin{aligned}
\langle \chi_m\psi, [\xi\tilde{A}\xi, (H - E)]\chi_m\psi &= -2\,\mathrm{Re}\langle \xi\tilde{A}\xi\chi_m\psi, (H - E)\chi_m\psi\rangle \\
&= -2\,\mathrm{Re}\langle \langle x\rangle^{-1}\tilde{A}\xi\chi_m\psi, \langle x\rangle\xi(H - E)\chi_m\psi\rangle .
\end{aligned} \tag{4.25}$$

[Recall that $\langle x\rangle = (1 + |x|^2)^{1/2}$]. Now

$$\langle x\rangle\xi(H - E)\chi_m\psi = -\langle x\rangle\xi(\Delta\chi_m - 2\nabla\chi_m \cdot D)\psi , \tag{4.26}$$

and $|\langle x\rangle\Delta\chi_m|$ and $|\langle x\rangle\nabla\chi_m|$ are bounded by a constant independent of m. Since $\xi\psi = \psi_F$ and $\xi\nabla\psi$ are both in L^2 (the latter follows from $\nabla\psi_F \in L^2$ and $\psi_F \in L^2$), the right side of (4.26) is bounded in absolute value by a fixed L^2 function. Moreover, it converges pointwise to zero. Thus, by Lebesgue's dominated convergence theorem

$$\|\langle x\rangle\xi(H - E)\chi_m\psi\| \to 0 \quad \text{as} \quad m \to \infty .$$

Since $\langle x\rangle^{-1}\tilde{A}$ is bounded from H_2 to H,

$$\|\langle x\rangle^{-1}\tilde{A}\xi\chi_m\psi\| \le C$$

with C independent of m. Thus, the left side of (4.25) converges to zero as $m \to \infty$.

Since all the other terms in (4.24) converge as $m \to \infty$, so must $\|g^{1/2}\tilde{A}\xi\chi_m\psi\|$. Thus, for $\varphi \in C_0^\infty$

$$\begin{aligned}|\langle\psi_F, \tilde{A}g^{1/2}\varphi\rangle| &= \lim_{m\to\infty} |\langle\xi\chi_m\psi, \tilde{A}g^{1/2}\varphi\rangle| \\ &\le \left(\lim_{m\to\infty} \|g^{1/2}\tilde{A}\xi\chi_m\psi\|\right)\|\varphi\| ,\end{aligned}$$

which shows that $\psi_F \in \mathrm{D}((-\tilde{A}g^{1/2})^*) = \mathrm{D}(g^{1/2}\tilde{A})$, and it follows easily that $\tilde{A}g^{1/2}\xi\chi_m\psi \to \tilde{A}g^{1/2}\psi_F$ as $m \to \infty$. Thus, all the terms in (4.24) converge to the corresponding ones in (4.22). $\square$

The next theorem relates the L^2 decay rate of an eigenfunction ψ with eigenvalue E to the set on which the Mourre estimate fails to hold. Consider $\psi_\alpha := \exp(\alpha\langle x\rangle)\psi$. When $\alpha = 0$, this function is in L^2 by hypothesis. If we increase α, it may happen that at some critical point, α_0, ψ_α leaves L^2. The next theorem says this can only happen if $\alpha_0^2 + E$ is a point where the Mourre estimate does not hold. This theorem does not rule out the possibility that ψ_α never leaves L^2. That such rapid decay cannot occur is proven in Theorem 4.18.

Theorem 4.17. Let $H = -\Delta + V$ in $L^2(\mathbb{R}^\nu)$, where V satisfies

(i) V is Δ-bounded with bound less than 1,
(ii) $(-\Delta + 1)^{-1}x\cdot\nabla V(-\Delta + 1)^{-1}$ is bounded.

Suppose $H\psi = E\psi$. Let $\mathrm{E}(H)$ be the complement of the set of points where a Mourre estimate (4.2) holds, with A the dilation generator. Define

$$\tau = \sup\{\alpha^2 + E\colon \alpha \ge 0, \exp(\alpha\langle x\rangle)\psi \in L^2(\mathbb{R}^\nu)\} .$$

Then $\tau \in \mathrm{E}(H) \cup \{+\infty\}$.

Proof. Suppose the theorem is false. Then $\tau = \alpha_0^2 + E \notin \mathrm{E}(H)$ for some $\alpha_0 < \infty$. If $\alpha_0 \ne 0$, choose α_1 and γ such that $\alpha_1 < \alpha_0 < \alpha_1 + \gamma$. If $\alpha_0 = 0$, let $\alpha_1 = 0$ and $\gamma > 0$. In both cases, $\exp(\alpha_1\langle x\rangle)\psi \in L^2$, while $\exp[(\alpha_1 + \gamma)\langle x\rangle] \notin L^2$. We will derive a contradiction for small γ. We assume that $0 \le \gamma \le 1$, so that all constants in the proof are independent of γ and α_1.

To begin, we define an interpolating function χ_s for $s \in \mathbb{R}$ by

$$\chi_s(t) = \int_0^t \langle sx\rangle^{-2}\,dx .$$

Then $\chi_s(t) \uparrow t$ as $s \downarrow 0$, and

$$\chi_s(t) \le c_s \quad \text{for } s > 0 , \tag{4.27}$$

$$\left|\left(\frac{d}{dt}\right)^n \chi_s(t)\right| \le ct^{-n+1} , \tag{4.28}$$

where the constant in (4.28) is independent of s. Define

$$F_s(x) = \alpha_1 \langle x \rangle + \gamma \chi_s(\langle x \rangle) \ .$$

By (4.27), $\exp(F_s)\psi \in L^2$ for all $s > 0$, but $\|\exp(F_s)\psi\| \to \infty$ as $s \downarrow 0$. Define

$$\Psi_s = \exp(F_s)\psi / \|\exp(F_s)\psi\| \ .$$

Then for any bounded set, B

$$\lim_{s\downarrow 0} \int_B |\Psi_s|^2 d^\nu x = 0 \ ,$$

$$\lim_{s\downarrow 0} \int_B |\nabla \Psi_s|^2 d^\nu x = 0 \ . \tag{4.29}$$

In particular, Ψ_s converges weakly to zero. In addition, we claim

$$\|\nabla \Psi_s\| < c \ , \tag{4.30}$$

$$\|(-\Delta + 1)\Psi_s\| < c \ . \tag{4.31}$$

To prove these inequalities, we need to use Proposition 4.16, and therefore must verify that F_s satisfies the hypotheses of that proposition. By direct calculation we find

$$\nabla F_s = (\alpha_1 + \gamma \chi_s'(\langle x \rangle))\langle x \rangle^{-1} x$$

$$g = (\alpha_1 + \gamma \chi_s'(\langle x \rangle))\langle x \rangle^{-1} \ , \quad \text{so that}$$

$$|x \cdot \nabla((\nabla F_s)^2)| \leq c\gamma(\alpha_1 + \gamma) + c(\alpha_1 + \gamma)^2 \langle x \rangle^{-2}$$

$$|(x \cdot \nabla)^2 g| \leq c(\alpha_1 + \gamma)\langle x \rangle^{-1} \ . \tag{4.32}$$

Thus, Proposition 4.16 holds. Using (4.21) of this proposition, together with the Δ-boundedness of V, we find

$$\|\nabla \Psi_s\|^2 \leq \langle \Psi_s, H\Psi_s \rangle + c\|\Psi_s\|^2$$

$$\leq c((\alpha_1 + \gamma)^2 + 1) \ .$$

This implies (4.30). Equation (4.31) now follows similarly from (4.30) and (4.20), together with the Δ-boundedness of V.

We wish to prove

$$\|g^{1/2} \tilde{A} \Psi_s\| \leq c \ . \tag{4.33}$$

To do this, we note that $\|g^{1/2}\tilde{A}\Psi_s\|$ is one of the terms in the equation obtained by dividing each term in (4.22) by $\|\exp(F_s)\psi\|^2$. Thus, it suffices to bound the remaining terms. By (4.31) and the boundedness of $(-\Delta + 1)^{-1}[H, \tilde{A}](-\Delta + 1)^{-1}$, which follows from (ii), we have

$$\langle \Psi_s, [H, \tilde{A}]\Psi_s \rangle \leq c \ .$$

The estimates (4.32) imply that

$$|\langle \Psi_s, ((x\cdot \nabla)^2 g - x\cdot \nabla((\nabla F_s)^2))\Psi_s\rangle| \le c\ .$$

Thus, (4.33) holds.

We now claim that

$$\lim_{s\downarrow 0} \|(H - E - (\nabla F_s)^2)\Psi_s\| = 0\ . \tag{4.34}$$

From (4.20), this is equivalent to

$$\lim_{s\downarrow 0} \|(D\cdot \nabla F_s + \nabla F_s\cdot D)\Psi_s\| = 0\ .$$

Now $D\cdot \nabla F_s + \nabla F_s\cdot D = 2g\tilde{A} + x\cdot \nabla g$. Let χ_N denote the characteristic function of $\{x\colon g(x) < N^{-1}\}$. Then

$$\begin{aligned}\overline{\lim_{s\downarrow 0}} \|g\tilde{A}\Psi_s\| &\le \overline{\lim_{s\downarrow 0}}\,(N^{-1/2}\|\chi_N g^{1/2}\tilde{A}\Psi_s\| + \|(1-\chi_N)g\tilde{A}\Psi_s\|)\\ &\le cN^{-1/2}\ .\end{aligned}$$

Here we used (4.29) and the fact that $1-\chi_N$ has support in a fixed, bounded set as $s\downarrow 0$. Since N is arbitrary, this shows that $\|g\tilde{A}\Psi_s\| \to 0$ as $s\downarrow 0$. Similarly, $\|x\cdot \nabla g \Psi_s\| \to 0$ as $s\downarrow 0$, and (4.34) is proven.

From the expression for ∇F_s, it is not hard to estimate

$$|(\nabla F_s)^2 - \alpha_1^2| \le c(\alpha_1\gamma + \gamma^2 + \alpha_1^2\langle x\rangle^{-2})$$

so that, from (4.34), it follows that

$$\overline{\lim_{s\downarrow 0}} \|(H - E - \alpha_1^2)\Psi_s\| \le c\gamma(\alpha_1 + \gamma)\ .$$

Now choose γ small enough so that the Mourre estimate holds in some interval Δ of width 2δ about $\alpha_1^2 + E$. This is possible since $\mathrm{E}(H)$ is closed and $\alpha_0^2 + E \notin \mathrm{E}(H)$. Then

$$\overline{\lim_{s\downarrow 0}} \|E(\mathbb{R}\backslash\Delta)\Psi_s\| \le \delta^{-1}\|E(\mathbb{R}\backslash\Delta)(H - E - \alpha_1^2)\Psi_s\| \le c\gamma \tag{4.35}$$

and

$$\begin{aligned}&\overline{\lim_{s\downarrow 0}} \|(H + \mathrm{i})E(\mathbb{R}\backslash\Delta)\Psi_s\|\\ &\quad\le \overline{\lim_{s\downarrow 0}} \|(E + \alpha_1^2 + \mathrm{i})E(\mathbb{R}\backslash\Delta)\Psi_s\| + \|(H - E - \alpha_1^2)E(\mathbb{R}\backslash\Delta)\Psi_s\|\\ &\quad\le c\gamma\ .\end{aligned} \tag{4.36}$$

Thus, we can insert spectral projections $E(\Delta)$ in the left side of the equation obtained by dividing (4.22) by $\|\exp(F_s)\psi\|^2$, and control the error terms to conclude that

$$\overline{\lim_{s\downarrow 0}}\,\langle \Psi_s, E(\Delta)[H, A]E(\Delta)\Psi_s\rangle \le c\gamma(\alpha_1 + \gamma)\ . \tag{4.37}$$

Here we use the negativity of the first term on the right of (4.22), the estimates for $|(x\cdot\nabla)^2 g - x\cdot\nabla(\nabla F)^2|$, the boundedness of $(H+\mathrm{i})^{-1}[H,\tilde{A}](H+\mathrm{i})^{-1}$ and the estimates (4.35) and (4.36). On the other hand, we know by the Mourre estimate

$$\langle \Psi_s, E(\Delta)[H,\tilde{A}]E(\Delta)\Psi_s\rangle \ge \alpha\|E(\Delta)\Psi_s\|^2 + \langle \Psi_s, K\Psi_s\rangle$$

for some $\alpha > 0$. Thus, since $\Psi_s \overset{W}{\to} 0$ and K is compact, we have, using (4.35)

$$\lim_{s\downarrow 0}\langle \Psi_s, E(\Delta)[H,\tilde{A}]E(\Delta)\Psi_s\rangle \ge a(1 - C\gamma^2)\ . \tag{4.38}$$

For small enough γ, (4.37) and (4.38) contradict each other, so the proof is complete. □

Remark. By making more careful estimates, one can prove this theorem without using weakly convergent sequences. This was done by *Perry* [280].

We now prove a theorem which eliminates the possibility $\tau = \infty$ in the result above. To prove this, we need to make an assumption on V that does not correspond to any of the hypotheses is Sect. 4.1. This assumption is not optimal; some alternative assumptions on V which imply the theorem are given in [114].

Theorem 4.18. Suppose $H = -\Delta + V$, where V satisfies (i) and (ii) of Proposition 4.1. Assume, in addition, that $x\cdot\nabla V$ is Δ-bounded with bound less than 2. Suppose $H\psi = E\psi$ and $\psi_\alpha := \exp(\alpha\langle x\rangle\psi) \in L^2$ for all α. Then $\psi = 0$.

Proof. The function $F = \alpha\langle x\rangle$ satisfies the hypotheses of Proposition 4.16. Thus, from (4.20) and the Δ-boundedness of V, we have

$$\begin{aligned}\langle\psi_\alpha, -\Delta\psi_\alpha\rangle &\ge \langle\psi_\alpha, H\psi_\alpha\rangle - C\|\psi_\alpha\|^2\\ &\ge \langle\psi_\alpha, (\nabla F)^2\psi_\alpha\rangle - C\|\psi_\alpha\|^2\\ &= \langle\psi_\alpha, \alpha^2 x^2\langle x\rangle^{-2}\psi_\alpha\rangle - C\|\psi_\alpha\|^2\ .\end{aligned}$$

On the other hand, we know that $x\cdot\nabla V \le a(-\Delta) + b$, with $a < 2$. Since $[H,\tilde{A}] = -2\Delta - x\cdot\nabla V$, this, together with (4.22), implies

$$\begin{aligned}\langle\psi_\alpha, -\Delta\psi_\alpha\rangle &\le C\langle\psi_\alpha, [H,\tilde{A}]\psi_\alpha\rangle + C\|\psi_\alpha\|^2\\ &\le C\langle\psi_\alpha, [(x\cdot\nabla)^2 g - x\cdot\nabla(\nabla F)^2 + 1]\psi_\alpha\rangle\\ &= C\langle\psi_\alpha, [\alpha(3x^2\langle x\rangle^{-5} - 2x^2\langle x\rangle^{-3}) - 2\alpha^2 x^2\langle x\rangle^{-4} + 1]\psi_\alpha\rangle\ .\end{aligned}$$

Combining these two inequalities, we have

$$\langle\psi_\alpha, (\alpha^2(x^2\langle x\rangle^{-2} + 2cx^2\langle x\rangle^{-4}) + C\alpha(2x^2\langle x\rangle^{-3} - 3x^2\langle x\rangle^{-5}) - 2C)\psi_\alpha\rangle \leq 0 \ .$$

But for large α, the expression in parentheses is increasing monotonically to ∞ at all points except 0 as $\alpha \to \infty$. This is impossible unless $\psi = 0$. □

We can now combine the theorems of this section with the results of the next section to prove that N-body Hamiltonians have no positive eigenvalues.

Theorem 4.19. Suppose H, acting in $L^2(\mathbb{R}^{(N-1)\mu})$, is an N-body Hamiltonian with center of mass removed (see Sect. 3.2), with pair potentials V_{ij} satisfying

(i) $V_{ij}(-\Delta + 1)^{-1}$ is compact in $L^2(\mathbb{R}^\mu)$,
(ii) $(-\Delta + 1)^{-1} x\cdot\nabla V_{ij}(-\Delta + 1)^{-1}$ is compact in $L^2(\mathbb{R}^\mu)$,
(iii) $x\cdot\nabla V_{ij}$ is Δ-bounded with bound zero in $L^2(\mathbb{R}^\mu)$.

Then H has no positive thresholds or eigenvalues.

Proof. The proof proceeds by induction. Suppose that for all $M < N$, M-body Hamiltonians have no positive eigenvalues. Then H has no positive thresholds, as thresholds are sums of eigenvalues of subsystem Hamiltonians. Now suppose $H\psi = E\psi$.

Since H is not of the form $-\Delta + V$ (unless all the masses are equal to $\frac{1}{2}$), Theorems 4.17 and 4.18 are not directly applicable to H and ψ. However, $H = H_0 + V$ where $H_0 = -D\cdot M^{-2}\cdot D$ for a symmetric positive definite matrix M, determined by the masses (see Chap. 3). Define the unitary operator U by $Uf(x) = \det(M)^{1/2} f(Mx)$. Then $\tilde{H} := U^*HU = -\Delta + V(M^{-1}x)$ does satisfy the hypotheses of these theorems. Moreover, $U^*\tilde{A}U = \tilde{A}$, which implies $\tilde{H}$ satisfies a Mourre estimate with $\tilde{A}$ if and only if H does. Thus, $E(\tilde{H}) = E(H)$ which, by Theorem 4.21, equals the set of thresholds of H. Applying Theorems 4.17 and 4.18 to $\tilde{H}$ and $U^*\psi = \det(M)^{-1/2}\psi(M^{-1}x)$, we find that

$$\tau := \sup\{\alpha^2 + E : \alpha \geq 0, \exp(\alpha\langle Mx\rangle)\psi \in L^2\}$$

is a threshold, and therefore nonpositive. But $E \leq \tau$. To start the induction effortlessly, we define an 0-body operator to be the zero operator on $\mathbb{C}$. □

4.5 The Mourre Estimate for *N*-Body Schrödinger Operators

The final topic in this chapter is a proof that N-body Schrödinger operators obey a Mourre estimate at all non-threshold points. The first proof of this result is due to *Perry, Sigal* and *Simon* [281]. It was previously proven for certain 3-body Hamiltonians by *Mourre* [256]. The proof given here follows *Froese* and *Herbst* [115]. Actually, in [115], this theorem is proven for a class of generalized N-body Hamiltonians whose geometric structure is explicit. To avoid introducing new notation, we will restrict ourselves to N-body operators.

The next lemma uses a Ruelle-Simon partition of unity to decompose $[H, \tilde{A}]$ for an N-body operator H into terms involving M-body operators with $M < N$. This is the key to an inductive proof of Theorem 4.21.

Lemma 4.20. Let H be an N-body Schrödinger operator (with center of mass removed) acting in $L^2(\mathbb{R}^{(N-1)\mu})$ as defined in Sect. 3.2. Suppose the pair potentials V_{ij} satisfy

(i) $V_{ij}(-\Delta + 1)^{-1}$ is compact on $L^2(\mathbb{R}^N)$,
(ii) $(-\Delta + 1)^{-1} x \cdot \nabla V_{ij}(-\Delta + 1)^{-1}$ is compact on $L^2(\mathbb{R}^N)$.
Let $\{J_a\}$ be a Ruelle-Simon partition of unity indexed by 2-cluster partitions a. Let $H(a)$ be the Hamiltonian corresponding to a (see Sect. 3.2), and $\tilde{A} = \frac{1}{2}(x \cdot D + D \cdot x)$, where x and D denote the variable and gradient operator in $L^2(\mathbb{R}^{(N-1)\mu})$, respectively. Then, if $f \in C_0^\infty(\mathbb{R})$, there exist compact operators K_1 and K_2 such that

$$f(H)^2 = \sum_a J_a f(H(a))^2 J_a + K_1 \ , \tag{4.39}$$

$$f(H)[H, \tilde{A}]f(H) = \sum_a J_a f(H(a))[H(a), \tilde{A}]f(H(a))J_a + K_2 \ . \tag{4.40}$$

Proof. Equation (4.39) follows immediately from Proposition 3.8. To prove (4.40), w note that $[H, \tilde{A}] = 2H_0 - W$, where H_0 is as in Sect. 3.2, and W has the form of an N-body potential with pair potentials $x \cdot \nabla V_{ij}$. Thus, we can apply the IMS localization formula to obtain

$$\begin{aligned} [H, \tilde{A}] &= \sum J_a [H, \tilde{A}] J_a - 2 \sum |\nabla J_a|^2 \\ &= \sum J_a [H(a), \tilde{A}] J_a + \sum J_a \tilde{I}_a J_a - 2 \sum |\nabla J_a|^2 \ , \end{aligned}$$

where $\tilde{I}_a$ is the interaction term

$$\sum_{(ij) \not\subset a} x \cdot \nabla V_{ij} \ .$$

The $|\cdot|$ in this equation is the norm associated with the mass weighted inner product on $\mathbb{R}^{(N-1)\mu}$ (see Chap. 3). Multiplying this equation from each side by $f(H)$, (4.40) now follows from Proposition 3.8 (which gives compactness of terms involving $[J_a, f(H(a))]$) and the fact that $f(H) J_a^2 \tilde{I}_a f(H)$ is compact. This compactness is easily seen to hold if the V_{ij} are in C_0^∞. Under condition (ii) on the V_{ij}, the compactness follows from an approximation argument (see [115]). □

Before turning to the statement and proof of the Mourre estimate, we will examine the structure of the intercluster Hamiltonians a bit more closely. Let $a = \{C_1, C_2\}$ be a 2-cluster partition. In Sect. 3.2, we saw that the space $X \subset \mathbb{R}^{N\mu}$ corresponding to H has the decomposition $X = X^a \oplus X_a$ [so that $L^2(X) = L^2(X^a) \otimes L^2(X_a)$], and $H(a)$ has the form $H(a) = h^a \otimes 1 \oplus 1 \otimes T_a$. In the case at hand where $a = \{C_1, C_2\}$, there is a further decomposition

$$X^a = X^{C_1} \oplus X^{C_2}, \quad L^2(X^a) = L^2(X^{C_1}) \otimes L^2(x^{C_2})$$
$$h^a = H(C_1) \otimes 1 + 1 \otimes H(C_2) \ ,$$

where $X^{C_k} := \{x \in X^a \colon x_i = 0 \text{ for } i \notin C_k\}$ (here x_i refers to the coordinates of x in the original space $\mathbb{R}^{N\mu}$, see Sect. 3.2). The operator $H(C_k)$ is the $|C_k|$-body Hamiltonian obtained by removing the center of mass from

$$-\Delta + \sum_{(i,j) \in C_k} V_{ij}(x_i - x_j) \ .$$

Thus,

$$x = X^{C_1} \oplus X^{C_2} \oplus X_a, \quad L^2(X) = L^2(X^{C_1}) \otimes L^2(X^{C_2}) \otimes L^2(X_a)$$

and

$$H(a) = H(C_1) \otimes 1 \otimes 1 + 1 \otimes H(C_2) \otimes 1 + 1 \otimes 1 \otimes T_a \ . \tag{4.41}$$

Whenever we decompose a space $Y = Y_1 \oplus Y_2, L^2(Y) = L^2(Y_1) \otimes L^2(Y_2)$, the generator of dilations $\tilde{A}$ in $L^2(Y)$ can be written

$$\tilde{A} = \tfrac{1}{2}(Y \cdot D + D \cdot Y) = \tfrac{1}{2}(Y_1 \cdot D_1 + D_1 \cdot Y_1) + \tfrac{1}{2}(Y_2 \cdot D_2 + D_2 \cdot Y_2) \ ,$$

where Y_k and D_k refer to the variables in Y_k, $k = 1, 2$. Thus, $\tilde{A} = \tilde{A}_1 \otimes I + I \otimes \tilde{A}_2$, where $\tilde{A}_k$ generates dilations in $L^2(Y_k)$. If we apply this to the decomposition $X = X^{C_1} \oplus X^{C_2} \oplus X_a$ above, it follows that

$$[H(a), \tilde{A}] = [H(C_1), \tilde{A}_1] \otimes I \otimes I + I \otimes [H(C_2), \tilde{A}_2] \otimes I + 2I \otimes I \otimes T_a \ . \tag{4.42}$$

Here we used the special commutation relation $[T_a, \tilde{A}_3] = 2T_a$.

Theorem 4.21. Suppose H is an N-body Schrödinger operator (with center of mass removed) acting in $X = L^2(\mathbb{R}^{(N-1)\mu})$ as described in Sect. 3.2. Suppose the pair potentials V_{ij} satisfy

(i) $V_{ij}(-\Delta + 1)^{-1}$ is compact on $L^2(\mathbb{R}^N)$,
(ii) $(-\Delta + 1)^{-1} x \cdot \nabla V_{ij}(-\Delta + 1)^{-1}$ is compact on $L^2(\mathbb{R}^\mu)$.

Suppose $\tilde{A}$ is the antisymmetric generator of dilations, i.e. $\tilde{A} = \frac{1}{2}(D \cdot x + x \cdot D)$, where x and D are the variable and gradient operator in $L^2(\mathbb{R}^{(N-1)\mu})$, respectively. Let $E_\Delta = E_\Delta(H)$ be the spectral projection for H corresponding to the interval Δ. Define

$$d(\lambda) := \text{distance}(\lambda, \{\text{thresholds of } H\} \cap (-\infty, \lambda]) \ .$$

Then

(a) For every $\varepsilon > 0$ and $\lambda \in \mathbb{R}$ there exists an open interval, Δ, containing λ, and a compact operator K such that

$$E_{\Delta}[H, \tilde{A}]E_{\Delta} \geq 2[d(\lambda) - \varepsilon]E_{\Delta} + K \ .$$

(b) The set τ of thresholds of H is closed and countable. Eigenvalues of H accumulate only at τ.

Proof. The proof is by induction on the number of particles. We have already proven the Mourre estimate for 2-body Hamiltonians in Example 1 of Sect. 4.1. This, and Theorem 4.7, imply (b) for these operators. In general, if we assume (b) for all subsystem Hamiltonians of an N-body operator H, then the first statement of (b) for H follows directly from the definition of thresholds. Once we have proven (a) for H, the second statement of (b) for H then follows from Theorem 4.7.

Thus, we assume that (a) and (b) hold for all M-body Hamiltonians satisfying our hypotheses with $M < N$, and will be done if we prove (a) for an N-body H.

The function d has the property that $d(\lambda) + \lambda' \geq d(\lambda + \lambda')$ for $\lambda' \geq 0$. The desire to have this inequality hold for λ' with $\lambda' \geq -\varepsilon$ motivates the following definition of d^{ε}:

$$d^{\varepsilon}(\lambda) = d(\lambda + \varepsilon) - \varepsilon \ .$$

We claim we can replace (a) in the theorem with

(a′) for every $\varepsilon > 0$ and $\lambda \in \mathbb{R}$, there exists an open interval Δ, containing λ, and a compact operator K, such that

$$E_{\Delta}[H, \tilde{A}]E_{\Delta} \geq 2(d^{\varepsilon}(\lambda) - \varepsilon)E_{\Delta} + K \ .$$

Certainly (a) implies (a′), since $d(\lambda) \geq d^{\varepsilon}(\lambda)$ for $\varepsilon > 0$. On the other hand, suppose (a′) holds. Then, given $\varepsilon > 0$ and $\lambda \in \mathrm{T}$, we can find Δ containing λ, such that

$$\begin{aligned} E_{\Delta}[H, \tilde{A}]E_{\Delta} &\geq 2(d(\lambda + \varepsilon/2) - \varepsilon/2 - \varepsilon/2)E_{\Delta} + K \\ &\geq -2\varepsilon E_{\Delta} + K \ . \end{aligned}$$

Since $d(\lambda) = 0$ for $\lambda \in \mathrm{T}$, this implies (a). Given $\lambda \notin \mathrm{T}$, there exists an interval about λ free of thresholds, since by the inductive hypothesis T is closed. Thus, for small enough ε, $d^{\varepsilon}(\lambda) = d(\lambda)$, and again (a′) $\Rightarrow$ (a).

The first step in the proof is to remove K in (a′) at the expense of including eigenvalues in the definition of d^{ε}. More precisely, let

$$\tilde{d}(\lambda) = \operatorname{distance}(\lambda, (\{\text{thresholds}\} \cup \{\text{eigenvalues}\}) \cap (-\infty, \lambda])$$

and $\tilde{d}^{\varepsilon}(\lambda) = \tilde{d}(\lambda + \varepsilon) - \varepsilon$. Then we claim (a′) implies

(c) for every $\lambda \in \mathbb{R}$ and $\varepsilon > 0$, there exists an open interval Δ about λ such that

$$E_{\Delta}[H, \tilde{A}]E_{\Delta} \geq 2(\tilde{d}^{\varepsilon}(\lambda) - \varepsilon)E_{\Delta} \ . \tag{4.43}$$

To prove (a′) implies (c), we assume (a′) and show that (c) holds. There are two cases. First, suppose λ is not an eigenvalue. From (a′) we know that

$$E_\Delta[H,\tilde{A}]E_\Delta \geq 2(d^{\varepsilon/2}(\lambda) - \varepsilon/2)E_\Delta + K$$

for some interval Δ. Now multiply this inequality from both sides by $E_{\Delta'}$ and let Δ' shrink about λ. Then, since λ is not an eigenvalue, $E_{\Delta'} \to 0$ strongly, and $\|E_{\Delta'}KE_{\Delta'}\| \to 0$ as Δ' shrinks. Since $d(\lambda + \varepsilon/2) \geq d(\lambda + \varepsilon) - \varepsilon/2 \geq \tilde{d}(\lambda + \varepsilon) - \varepsilon/2$, we have $d^{\varepsilon/2}(\lambda) \geq \tilde{d}^\varepsilon(\lambda)$. Thus, for small Δ', we have

$$E_{\Delta'}[H,\tilde{A}]E_{\Delta'} \geq 2(\tilde{d}^\varepsilon(\lambda) - \varepsilon/2)E_{\Delta'} - \varepsilon\ .$$

Multiplying again from both sides with $E_{\Delta'}$ we see that (4.43) holds.

Now suppose λ is an eigenvalues. Then $\tilde{d}^\varepsilon(\lambda) \leq 0$, so it suffices to show

$$E_\Delta[H,\tilde{A}]E_\Delta \geq -2\varepsilon E_\Delta\ . \tag{4.44}$$

Let $P = E_{\{\lambda\}}$. We will show that, for some compact operator K_1

$$E_\Delta[H,\tilde{A}]E_\Delta \geq -\varepsilon E_\Delta + (1-P)K_1(1-P)\ . \tag{4.45}$$

Since $E_{\Delta'}(1-P) \to 0$ strongly as Δ' shrinks about λ, an argument similar to the one above shows that (4.45) implies (4.44). By (a′)

$$E_\Delta[H,\tilde{A}]E_\Delta \geq -\varepsilon/2E_\Delta + K \tag{4.46}$$

for some interval Δ. Using compactness of K, pick a finite rank projection F, with Range $F \subseteq$ Range P, so that

$$\|(1-P)K(1-P) - (1-F)K(1-F)\| \leq \varepsilon/2\ . \tag{4.47}$$

Then, multiplying (4.46) from both sides with $(1-F)$ and using (4.47), we find

$$(E_\Delta - F)[H,\tilde{A}](E_\Delta - F) \geq -\varepsilon(E_\Delta - F) + (1-P)K(1-P)\ .$$

By the virial theorem, $P[H,\tilde{A}]P = 0$. Thus, we need only show that

$$\begin{aligned} R := {}& F[H,\tilde{A}]E_\Delta(1-P) + (1-P)E_\Delta[H,\tilde{A}]F \\ &\geq -\varepsilon F + (1-P)K_2(1-P) \end{aligned} \tag{4.48}$$

for some compact K_2. Let $C = F[H,\tilde{A}]E_\Delta(1-P)$. Then

$$(\varepsilon^{-1/2}C + \varepsilon^{1/2}F)^*(\varepsilon^{-1/2}C + \varepsilon^{1/2}F) \geq 0\ ,$$

from which it follows that

$$R = C^*F + F^*C \geq -\varepsilon F^*F + \varepsilon^{-1}C^*C\ ,$$

which implies (4.48). Thus, (c) is proven.

We now show that (c) implies the following uniform statement.

(c′) For every $\varepsilon > 0$ and compact interval M, there exist $\delta > 0$ such that, for $\mu \in M$ and $\Delta_\mu^\delta = (\mu - \delta, \mu + \delta)$,

$$E_{\Delta_\mu^\delta}[H, \tilde{A}]E_{\Delta_\mu^\delta} \geq 2[\tilde{d}^\varepsilon(\mu) - \varepsilon]E_{\Delta_\mu^\delta} . \tag{4.49}$$

We thus assume (c) holds, and prove (c′). Fix $\varepsilon > 0$ and M. From (c) we know that, for $\mu \in M$, (4.49) holds for some δ depending on μ. Let $\delta(\tilde{\varepsilon}, \mu) = \sup\{\delta\colon (4.49)$ holds with $\tilde{\varepsilon}$ in place of $\varepsilon\}$. Now, for $\delta < \delta(\mu, \varepsilon/2)$ and $|\mu'| < \varepsilon/2$

$$\begin{aligned} E_{\Delta_\mu^\delta}[H, \tilde{A}]E_{\Delta_\mu^\delta} &\geq 2[\tilde{d}^{\varepsilon/2}(\mu) - \varepsilon/2]E_{\Delta_\mu^\delta} \\ &\geq 2[\tilde{d}^\varepsilon(\mu + \mu') - \mu' - \varepsilon/2]E_{\Delta_\mu^\delta} \\ &\geq 2[\tilde{d}^\varepsilon(\mu + \mu') - \varepsilon]E_{\Delta_\mu^\delta} . \end{aligned}$$

Since $\Delta_{\mu+\mu'}^{\delta-|\mu'|} \subset \Delta_\mu^\delta$, we find that $\delta(\mu + \mu', \varepsilon) \geq \delta(\mu, \varepsilon/2) - |\mu'|$ for $|\mu'| < \varepsilon/2$. Thus, $\delta(\mu, \varepsilon)$ is locally bounded below by a continuous, positive function. Since M is compact, this implies $\delta(\mu, \varepsilon) \geq \delta(\varepsilon) > 0$ for all $\mu \in M$, which is precisely the uniformity needed in (c′).

We now come to the inductive step in the proof. We want to prove (a′) for an N-body H. Fix $\varepsilon > 0$ and $\lambda \in \mathbb{R}$. Then for $f \in C_0^\infty(\mathbb{R})$ we have, from Lemma 4.20,

$$f(H)[H, \tilde{A}]f(H) = \sum_a J_a f(H(a))[H(a), \tilde{A}]f(H(a))J_a + K_2 ,$$

where a runs over 2-cluster partitions and K_2 is compact. What we will show is that for each a, and for all f with small enough support about λ

$$f(H(a))[H(a), \tilde{A}]f(H(a)) \geq 2[d^\varepsilon(\lambda) - \varepsilon]f^2(H(a)) . \tag{4.50}$$

Then, since the a's run over a finite set, we can find one f that will work for all a's. Furthermore, we can choose f to be identically 1 in an interval Δ about λ. Summing (4.50) over a and using (4.39) and (4.40), we obtain

$$f(H)[H, \tilde{A}]f(H) \geq 2[d^\varepsilon(\lambda) - \varepsilon]f^2(H) + K ,$$

where K is compact and the theorem follows upon multiplying from both sides by $E_\Delta(H)$.

It thus remains to show (4.50). Suppose $a = \{C_1, C_2\}$. We have, from (4.41) and (4.42),

$$\begin{aligned} H(a) &= H(C_1) + H(C_2) + T_a \\ [H(a), \tilde{A}] &= [H(C_1), \tilde{A}] + [H(C_2), \tilde{A}] + 2T_a , \end{aligned}$$

where we have abused (and will continue to abuse) notation by writing $H(C_1)$ for $H(C_1) \otimes I \otimes I$, and $\tilde{A}$ for $\tilde{A}_1$, etc. We now can decompose our Hilbert space into a direct integral

$$L^2(\mathbb{R}^{(N-1)\mu}) = \int^{\oplus}_{\sigma(H(C_2))\times\sigma(T_a)} \mathrm{H}_{e_2,t}\, d\mu(e_2)\, dt$$

so that on each fibre $H(a)$ is represented by $H(C_1) + e_2 + t$. Now suppose $f \in C_0^\infty(\mathbb{R})$ has support in a small neighborhood of λ. Then $f(\cdot + e_2 + t)$ has support near $\lambda - e_2 - t$. Since $H(C_1)$, $H(C_2)$ and T_a are all bounded from below, $f(H(C_1) + e_2 + t)$ is only nonzero in a compact set. Thus, we can apply (c′), which holds for $H(C_1)$ by our inductive hypothesis, to conclude that for $\varepsilon_1 = \varepsilon/5$, if the support of f is small enough,

$$\begin{aligned} &f(H(C_1) + e_2 + t)[H(C_1), \tilde{A}]f(H(C_1) + e_2 + t) \\ &\quad \geq 2[\tilde{d}^{\varepsilon_1}_{C_1}(\lambda - e_2 - t) - \varepsilon_1]f(H(C_1) + e_2 + t)^2 \ , \end{aligned} \tag{4.51}$$

where $\tilde{d}^{\varepsilon_1}_{C_1}$ is the modified distance function for $H(C_1)$. Now assume, in addition, that $\mathrm{supp}(f) \subset (\lambda - \varepsilon_1, \lambda + \varepsilon_1)$. Then (4.51) implies

$$\begin{aligned} f(H(a))[H(C_1), \tilde{A}]f(H(a)) &\geq 2[\tilde{d}^{\varepsilon_1}_{C_1}(\lambda - H(C_2) - T_a) - \varepsilon_1]f(H(a))^2 \\ &= 2[\tilde{d}^{\varepsilon_1}_{C_1}(H(C_1) + (\lambda - H_a)) - \varepsilon_1]f(H(a))^2 \\ &\geq 2[\tilde{d}^{\varepsilon_1}_{C_1}(H(C_1)) - 2\varepsilon_1]f(H(a))^2 \ . \end{aligned}$$

Here we used that $|\lambda - \xi| < \varepsilon_1$ for $\xi \in \mathrm{supp}(f)$ and that $\tilde{d}^{\varepsilon_1}_{C_1}(a + b) \geq \tilde{d}^{\varepsilon_1}_{C_1}(a) - b$ for $|b| < \varepsilon_1$. Since there is a similar inequality for $H(C_2)$, we have, for $\mathrm{supp}(f)$ small enough about λ,

$$\begin{aligned} &f(H(a))[H(a), A]f(H(a)) \\ &\quad \geq 2[\tilde{d}^{\varepsilon_1}_{C_1}(H(C_1)) + \tilde{d}^{\varepsilon_1}_{C_2}(H(C_2)] + T_a - 4\varepsilon_1)f^2(H(a)) \ . \end{aligned}$$

We claim that, for $a \in \sigma(H(C_1))$, $b \in \sigma(H(C_2))$ and $c \in \sigma(T_a)$, (i.e. $c \geq 0$)

$$\tilde{d}^{\varepsilon_1}_{C_1}(a) + \tilde{d}^{\varepsilon_1}_{C_2}(b) + c \geq d^{2\varepsilon_1}(a + b + c) \ .$$

To see this, note that $\tilde{d}^{\varepsilon_1}_{C_1}(a) = a - \tau_1$, where τ_1 is a threshold or eigenvalue of $H(C_1)$ in $(-\infty, a + \varepsilon_1]$. Similarly, $\tilde{d}^{\varepsilon_1}_{C_2}(b) = b - \tau_2$. Thus, the left side is equal to $a + b + c - (\tau_1 + \tau_2)$. But by the definition of thresholds, $\tau_1 + \tau_2$ is a threshold of H, and since $c \geq 0$, $\tau + \tau_2 \in (-\infty, a + b + c + 2\varepsilon_1]$. Now the inequality follows from the definition of $d^{2\varepsilon_1}$. Combining the last two inequalities, we obtain

$$\begin{aligned} f(H(a))[H(a), A]f(H(a)) &\geq 2\{d^{2\varepsilon_1}[H(a)] - 4\varepsilon_1\}f^2(H(a)) \\ &= 2\{d^{2\varepsilon_1}(\lambda + [H(a) - \lambda]) - 4\varepsilon_1\}f^2(H(a)) \\ &\geq 2[d^{2\varepsilon_1}(\lambda) - 5\varepsilon_1]f^2(H(a)) \ . \end{aligned}$$

Since $5\varepsilon_1 = \varepsilon$ and d^ε is non-increasing in ε, this implies (4.50), and completes the proof. □

5. Phase Space Analysis of Scattering

In this chapter, we present an introduction into quantum mechanical scattering theory by geometric methods. Those methods were introduced for two-body scattering by *Enss* in his celebrated paper [95], and further developed by *Enss* [96–98, 100, 101], *Simon* [326], *Perry* [277, 279], *Davies* [75], *Mourre* [255], *Ginibre* [135], *Yafaev* [374, 375], *Muthuramalingam* [258, 259] *Isozoki* and *Kitada* [178, 180], and others.

The core of the Enss method is a careful comparison of the time evolution $\exp(-itH)$ of a given system with the "free" time evolution $\exp(-itH_0)$, thereby making rigorous the physicists' way of thinking about quantum mechanical scattering. One of the biggest advantages of the Enss method is its intimate connection to physical intuition, not only with regard to the general idea, but also even in single steps of the proof. Furthermore, there is a recent extension due to *Enss* [99, 102–105] of his method to three-body scattering, and there is hope that it may be possible to extend it to the N-body case.

We will give a complete proof of the Enss theory in the two-body case, with special emphasis on the new elements that Enss brought in from the three-body case. Moreover, we will discuss some of the features of the three-body case. We do not give a complete proof of the asymptotic completeness in the three-body case. Our intention is to present and discuss some of Enss' new ideas for three bodies, and to whet the reader's appetite for further reading.

Besides the research papers mentioned above, there are various expository works on the Enss theory, among them, Enss' lecture notes on the Erice Summer School [100], which is a self-contained introduction to the theory and which, at the same time, introduced the method of asymptotics of observables (which we will discuss in Sect. 5.5). There is the comprehensive monograph by *Perry* [279] on the Enss method, as well as chapters in the books by *Amrein* [13], *Berthier* [48] and *Reed* and *Simon* III [294] dedicated to that method. For other approaches to scattering theory (time-independent methods), see *Reed* and *Simon* III [294], where further references can also be found.

5.1 Some Notions of Scattering Theory

In typical (2-body) scattering experiments, we have a (test) particle and a scatterer (target) that are separated far away from each other at the beginning. As time evolves, the particle gets close to the scatterer and interacts with it. One expects, on the basis of physical experience, that after a sufficiently long time, the particle will again be far away from the scatterer. The particle should then move almost "freely", i.e. almost without influence by the scatterer.

In quantum mechanics, this expectation can be formulated in terms of the interacting time evolution $\exp(-itH)$ and the free time evolution $\exp(-itH_0)$. That a state φ looked in the remote past like a "free" state can be expressed by

$$e^{-itH}\varphi \approx e^{-itH_0}\varphi_- \quad \text{as } t \to -\infty$$

for some φ_-, while

$$e^{-itH}\varphi \approx e^{-itH_0}\varphi_+ \quad \text{as } t \to +\infty$$

expresses that φ looks, after a long time, asymptotically like a free state φ_+.

We therefore should have

$$\varphi = \lim_{t\to-\infty} e^{itH}e^{-itH_0}\varphi_- = \lim_{t\to\infty} e^{itH}e^{-itH_0}\varphi_+ \ .$$

This leads to the definition

$$\Omega^{\pm} := s\text{-}\lim_{t\to\mp\infty} e^{itH}e^{-itH_0} \ . \tag{5.1}$$

The operators $\Omega^{\pm}$ are called wave operators. Note the funny convention with respect to the signs $\pm$. This is due to the (historically earlier) definition of $\Omega^{\pm}$ in time-independent scattering theory.

We are interested in the correspondence, $\varphi \mapsto \varphi_+$, so one might think the limit one really wanted to consider is

$$\lim_{t\to\mp\infty} e^{itH_0}e^{-itH}\varphi \ , \tag{5.1'}$$

yet in (5.1), we have H and H_0 reversed. This is for several reasons: (a) the limit (5.1) tends to exist for all vectors, while (5.1′) will not exist for vectors φ which are eigenvectors of H; (b) It is much harder to control (5.1′) than (5.1); (c) Once one shows that (5.1) exists, it is not hard to show that the limit (5.1′) exists if and only if $\varphi = \Omega^{\pm}\varphi_{\mp}$, and the limit is then just $\varphi_{\mp}$.

Suppose we have proven existence of $\Omega^{\pm}$. Then $\varphi = \Omega^{+}\varphi_-$ is a state that developed backwards in time with the interacting dynamics looks asymptotically like the state φ_- developed with the free dynamics. A similar interpretation can be given to Ω^{-}.

It is therefore reasonable to call $\mathrm{H}_{\text{in}} = \operatorname{Ran}\Omega^{+}$ the incoming, and $\mathrm{H}_{\text{out}} = \operatorname{Ran}\Omega^{-}$ the outgoing states. If

$$\mathrm{H}_{\text{in}} = \operatorname{Ran}\Omega^{+} = \operatorname{Ran}\Omega^{-} = \mathrm{H}_{\text{out}} \tag{5.2}$$

any incoming state will be outgoing in the far future, and any outgoing state was incoming in the remote past. This is what we expect for scattering experiments. We will call a system obeying (5.2) *weakly asymptotic complete* (we usually adopt notations from *Reed* and *Simon* [294]; note that *Enss* [100] uses a slightly different terminology at this point). For weakly asymptotic complete systems, we will call $\mathrm{H}_{\mathrm{in}} = \mathrm{H}_{\mathrm{out}} = \mathrm{Ran}\, \Omega^{\pm}$ the scattering states. It is clear that bound states show a very different behavior than scattering states. Physically, one would expect that there are no other states than bound states and scattering states (and superpositions of them). Since the bound states correspond to H_{pp} (this can be justified by the RAGE theorem discussed in Sect. 5.4 below), the above physical expectation can be expressed by

$$\mathrm{Ran}\, \Omega^{\pm} = \mathrm{H}_{pp}^{\perp} = \mathrm{H}_{\mathrm{cont}} \ . \tag{5.3}$$

Property (5.3) is called *asymptotic completeness.* It is one of the main goals of scattering theory to prove asymptotic completeness for a wide class of interacting dynamics.

The above considerations, however, are only correct if no long range forces have to be considered. Roughly speaking, “short range” means decay of the potential at infinity, like $|x|^{-\alpha}$ for some $\alpha > 1$. Thus, the Coulomb potential is of long range nature. For long range potentials, the scattered particle will not move asymptotically freely. A correction to the free motion is needed to describe the asymptotic behavior of this motion. This is already true in classical mechanics (see *Reed* and *Simon* III, XI.9 [294]). This correction has to be considered also for the wave operators. Therefore, the definition (5.1) is not appropriate to long-range potentials, so “modified wave operators” are required. In the following, we will restrict ourselves to the case of short-range potentials, and refer, for the long-range case, to *Enss* [100, 105], *Isozaki* and *Kitada* [179] and *Perry* [278] and references given there. We will also restrict ourselves to the Enss time-dependent method; for the time-independent approach, see *Reed* and *Simon* III [294] and the works cited there.

We now state a few properties of wave operators before we turn to existence questions. Thus, let us assume that $\Omega^{\pm}$ exist. Since $\Omega^{\pm}$ are strong limits of unitary operators, they are isometries from L^2 to $\mathrm{Ran}\, \Omega^{\pm}$. Therefore the ranges $\mathrm{Ran}\, \Omega^{\pm}$ are closed subspaces of L^2. From the definition of $\Omega^{\pm}$, it is easy to see that $\exp(-\mathrm{i}Ht)\Omega^{\pm} = \Omega^{\pm} \exp(-\mathrm{i}H_0 t)$ and hence $H\Omega^{\pm} = \Omega^{\pm} H_0$. Therefore, $H \upharpoonright \mathrm{Ran}\, \Omega^{\pm}$ is unitarily equivalent to H_0. This implies

$$\mathrm{Ran}\, \Omega^{\pm} \subset \mathrm{H}_{\mathrm{ac}}(H) \ . \tag{5.4}$$

Because of (5.4), asymptotic completeness implies that the singular continuous spectrum is empty.

Now we turn to the question of existence of $\Omega^{\pm}$. We present a general strategy known as Cook's method, that will enable us to prove the existence of $\Omega^{\pm}$ for a wide class of short-range potentials.

Theorem 5.1 (Cook's Method). Let V be a Kato-bounded potential with relative bound $a < 1$. If there exists a set $D_0 \subset D(H_0)$ dense in L^2, such that, for all $\varphi \in D_0$

$$\int_T^\infty \| V e^{\mp itH_0} \varphi \| \, dt < \infty$$

for some T, then $\Omega^\pm$ exists.

Proof. We prove that, for $\varphi \in D_0$, $\eta(t) := \exp(\mathrm{i}Ht)\exp(-\mathrm{i}H_0 t)\varphi$ is a Cauchy sequence as $t \to -\infty$. By density of D_0 and $\|\exp(\mathrm{i}Ht)\exp(-\mathrm{i}H_0 t)\| \leq 1$, this suffices to prove the existence of Ω^-. We estimate

$$\|\eta(t) - \eta(s)\| \leq \int_s^t \|\eta'(u)\| \, du \leq \int_s^t \|e^{\mathrm{i}Hu}(H_0 - H)e^{-\mathrm{i}Hu}\varphi\| \, du$$

$$= \int_s^t \| V e^{-\mathrm{i}H_0 u} \varphi \| \, du \to 0 \text{ as } s, t \to \infty \text{ by hypothesis.} \quad \square$$

Corollary (Cook's Estimate). If $\Omega^\pm$ exists, then

$$\|(\Omega^\pm - 1)\varphi\| \leq \int_0^\infty \| V e^{\mp \mathrm{i}tH_0} \varphi \| \, dt \ . \tag{5.5}$$

5.2 Perry's Estimate

To apply Cook's method to some given class of potentials, we apparently need some control on the unitary group $\exp(-\mathrm{i}tH_0)$. In this section, we present a useful estimate due to *Perry* [277] that will enable us to apply Cook's method to a wide class of "short-range" potentials, and that is, furthermore, interesting by itself.

Perry's estimates were motivated by work of *Mourre* [255] on the Enss method. If we only wished to obtain existence of $\Omega^\pm$ using Cook's method, one could obtain an estimate on H_0 much more easily [e.g. take D_0 to be finite sums of Gaussians and compute $\exp(-\mathrm{i}tH_0\varphi)$ exactly]. The point of Perry's estimate is that it allows a uniformity in suitable φ that is critical to the Enss approach. Other methods of obtaining such uniform control are the original, direct phase space approach of *Enss* [95] (see also *Simon* [326]), the coherent vectors analysis of *Davies* [75] and *Ginibre* [135], and an approach due to *Yafaev* [374, 375] close in spirit to that of Mourre and Perry. We first introduce some technical tools concerning the *dilation generator* A (see also Chap. 4). We define

$$A := \tfrac{1}{2}(x \cdot p + p \cdot x)$$

(where $p = -\mathrm{i}\partial/\partial x$ in x-space representation). By $P_\pm$ we denote the spectral projection associated to A on the positive (resp. negative) half axis. Since

$$(Af)\hat{\ }(p) = -\frac{1}{2}\left(p\left(-\mathrm{i}\frac{\partial}{\partial p}\right) + \left(-\mathrm{i}\frac{\partial}{\partial p}\right)p\right)\hat{f}(p) = -A(\hat{f})(p)$$

($\hat{}$ denoting Fourier transform), we have

$$(P_\pm f)\hat{} = P_\mp \hat{f} \; . \tag{5.6}$$

We now introduce the *Mellin transform* that "diagonalizes" the operator A. Denote by $C_0^\infty(\mathbb{R}^n \backslash \{0\})$ the functions in C_0^∞ with support bounded away from 0. For $\varphi \in C_0^\infty(\mathbb{R}^n \backslash \{0\})$, we write $\varphi(x) = \varphi(|x|\omega)$ with $\omega \in S^{\nu-1}$ the sphere of radius one, and define, for $\lambda \in \mathbb{R}$ and $\omega \in S^{\nu-1}$

$$\varphi^{\#}(\lambda, \omega) := \frac{1}{(2\pi)^{1/2}} \int_0^\infty |x|^{\nu/2} |x|^{-i\lambda} \varphi(|x|\omega) \frac{d|x|}{|x|} \; . \tag{5.7}$$

The Mellin transform $^{\#}$ maps $C_0^\infty(\mathbb{R}^n \backslash \{0\})$ into $L^2(\mathbb{R} \times S^{\nu-1}, d\lambda \times d^{\nu-1}\omega)$. Moreover

Lemma 5.2. The Mellin transform preserves the L^2-norm, i.e.

$$\int |\varphi^{\#}(\lambda, \omega)|^2 \, d\lambda \, d^{\nu-1}\omega = \int |\varphi(x)|^2 \, d^\nu x \; . \tag{5.8}$$

Therefore, the Mellin transform extends to an isometric mapping, $^{\#}: L^2(\mathbb{R}^\nu) \to L^2(\mathbb{R} \times S^{\nu-1})$.

Remark. The Mellin transform can be viewed as the Fourier transform on the group $\mathbb{R}_+$ equipped with multiplication as group structure. R_+ is a locally compact, Abelian group with $d|x|/|x|$ as Haar measure. Moreover, the characters on $\mathbb{R}_+$ (dual group) are given by $|x|^{-i\lambda}$. Thus, the lemma above is nothing but the Plancherel theorem on $\mathbb{R}_+$.

Proof. We define, for $g \in C_0^\infty(\mathbb{R}^\nu \backslash \{0\})$

$$Ug(t, \omega) := \exp\left(\frac{\nu}{2} t\right) g(e^t \omega) \; .$$

It is easily verified that $\|Ug\|_{L^2(\mathbb{R} \times S^{\nu-1})} = \|g\|_{L^2(\mathbb{R}^\nu)}$. Now

$$\varphi^{\#}(\lambda, \omega) = \frac{1}{(2\pi)^{1/2}} \int_{\mathbb{R}} e^{-i\lambda t} Ug(t, \omega) \, dt \; .$$

Thus, $\varphi^{\#}$ is actually the Fourier transform in the t-variable of Ug. Since both U and the Fourier transform are unitary, the lemma holds. $\square$

Since $A(|x|^{\nu/2}|x|^{-i\lambda}) = \lambda |x|^{\nu/2}|x|^{-i\lambda}$, it is not difficult to show that the Mellin transform "diagonalizes" A, i.e. $(A\varphi)^{\#}(\lambda, \omega) = \lambda \varphi^{\#}(\lambda, \omega)$ for $\varphi \in D(A)$. It follows that

$$(P_\pm \varphi)^{\#}(\lambda, \omega) = \chi_{(0,\infty)}(\pm\lambda) \varphi^{\#}(\lambda, \omega) \; . \tag{5.9}$$

We adopt the following notation of Enss: Let S be a self-adjoint operator, M a subset of $\mathbb{R}$. Then $F(S \in M)$ denotes the spectral projection of S on the set M. In particular, $F(x \in M)$ is nothing but multiplication with the characteristic function of M in x-space. We are now prepared to prove Perry's estimate.

Theorem 5.3. Let g be a C^∞-function with support contained in $[a^2/2, b^2/2]$ for some $a, b > 0$. Furthermore, let $\delta < a$ be fixed. Then, for any N, there is a constant C_N, such that

$$\|F(|x| < \delta|t|)e^{-itH_0}g(H_0)P_\pm\| \leq C_N(1 + |t|)^{-N}$$

for $\pm t > 0$ (i.e. $t > 0$ for $P_+, t < 0$ for P_-).

Proof. Define $K_{x,t}(p) := \exp[i(p^2t/2 - px)]g(p^2/2)/(2\pi)^{\nu/2}$ and $\psi_t(x) = \exp(-itH_0)g(H_0)P_\pm\psi(x)$. Then $\psi_t(x) = \langle K_{x,t}, (P_\pm\psi)\hat{}\,\rangle = \langle K_{x,t}, P_\mp\hat{\psi}\rangle$ [by (5.6)], hence $|\psi_t(x)| \leq \|P_\mp K_{x,t}\|\,\|\psi\|$ (all norms are L^2-norm). Therefore, it suffices to prove

$$\|P_\mp K_{x,t}\| \leq \tilde{C}_N(1 + |t|)^{-N} \quad \text{for } |x| \leq \delta t \text{ and } \pm t > 0 \ .$$

We treat only the case of P_-, the P_+-case being similar. By Lemma 5.2 and (5.9), we have

$$\begin{aligned}\|P_- K_{x,t}\|^2 &= \|(P_- K_{x,t})^\#\|^2 \\ &= \int_{-\infty}^{0}\left(\int_{S^{\nu-1}} |K^\#_{x,t}(\lambda, \omega)|^2 d^{\nu-1}\omega\right)d\lambda \ .\end{aligned}$$

Thus, the theorem is proven if we show that

$$|K^\#_{x,t}(\lambda, \omega)| \leq C'_N(1 + |\lambda| + |t|)^{-N}$$

for $|x| < \delta t$ and $\lambda < 0$. Now

$$\begin{aligned}K^\#_{x,t}(\lambda, \omega) &= \frac{1}{(2\pi)^{1/2}}\int_0^\infty |p|^{\nu/2}|p|^{-i\lambda}K_{x,t}(|p|\,\omega)\frac{d|p|}{|p|} \\ &= \frac{1}{(2\pi)^{(\nu+1)/2}}\int_0^\infty |p|^{\nu/2}g(p^2/2)e^{i\alpha(p)}\frac{d|p|}{|p|}\end{aligned}$$

with $\alpha(p) = tp^2/2 - |p|\omega\cdot x - \lambda\log|p|$. Since $\lambda < 0$, $t > 0$ and $|x| < \delta t < at$,

$$\begin{aligned}\frac{\partial\alpha}{\partial|p|} &= t|p| - \omega\cdot x - \frac{\lambda}{|p|} \geq |p|t - \omega\cdot x \geq (|p| - \delta)t \\ &> (|p| - a)t \ .\end{aligned}$$

Thus, $\partial\alpha/\partial|p|$ is strictly positive on $K := \{p | g(p^2/2) \neq 0\}$, and we may estimate

$$\left(\frac{\partial\alpha}{\partial|p|}\right)^{-1} \leq C(1 + |t| + |\lambda|)^{-1} \quad \text{on } K \ .$$

Moreover, we see similarly that

$$\frac{\partial^k}{\partial|p|^k}\left(\frac{\partial\alpha}{\partial|p|}\right)^{-1} \leq C_k(1 + |t| + |\lambda|)^{-1} \quad \text{on } K \ . \tag{5.10}$$

Writing

$$e^{i\alpha(p)} = \left[-i\left(\frac{\partial\alpha}{\partial p}\right)^{-1}\frac{\partial}{\partial p}\right]^N e^{i\alpha(p)}$$

and integrating by parts N times, we get

$$\begin{aligned} |K^{\#}_{x,t}(\lambda, \omega)| &\leq \left|\int_0^\infty p^{\nu/2-1} g(p^2/2)\left[\left(\frac{\partial\alpha}{\partial p}\right)^{-1}\frac{\partial}{\partial p}\right]^N e^{i\alpha(p)}\, dp\right| \\ &= \left|\int_0^\infty \left[\frac{\partial}{\partial p}\left(\left(\frac{\partial\alpha}{\partial p}\right)^{-1}\frac{\partial}{\partial p}\right)^{N-1}\left\{\left(\frac{\partial\alpha}{\partial p}\right)^{-1} p^{\nu/2-1} g(p^2/2)\right\}\right] e^{i\alpha(p)}\, dp\right| \\ &\leq C'_N(1 + |t| + |\lambda|)^{-N} \end{aligned}$$

by (5.10). □

Remarks. (1) The last part of the above proof is a version of (the easy part of) the method of stationary phase (see e.g. *Reed* and *Simon* III [294]).

(2) We can replace $P_\pm$ by $P_{[-\alpha,\infty)}$ (resp. $P_{(-\infty,\alpha]}$ in the above argument, possibly increasing the constant C_N.

5.3 Enss' Version of Cook's Method

We remarked already that the wave operators $\Omega^\pm$ are appropriate only for "short-range" interactions, while for long-range potentials (e.g. the Coulomb potential), modified wave operators are required. Various definitions of "short range" are used in the literature (see e.g. *Reed* and *Simon* III [294], *Enss* [100], *Perry* [279]). Throughout the rest of this chapter, we will make the following assumption:

$$(1 + |x|)^{1+\varepsilon} V(H_0 + 1)^{-1} \quad \text{is compact for some } \varepsilon > 0 \ . \tag{5.11}$$

By the term "short-range potential" we will always mean a function V that satisfies (5.11).

In this section, our aim is to show the existence of the wave operators for short-range potentials using Cook's method.

Let us define $S(R) := \| V(H_0 + 1)^{-1} F(|x| > R)\|$. We first prove the following lemma:

Lemma 5.4. For short range potentials

$$\int_0^\infty S(R)\, dR < \infty \ . \tag{5.12}$$

Remark. Kato-bounded potentials (with relative bound $a < 1$) that satisfy (5.12) are called *Enss potentials* in *Reed* and *Simon* III [294]. To prove asymptotic completeness, it is enough to assume that V is an Enss potential.

Proof. Define $T(R) := \|F(|x| > R) V(H_0 + 1)^{-1}\|$. Then $\int T(R)\, dR < \infty$ since

$$\begin{aligned} T(R) &\le \|(1 + |x|)^{-1-\varepsilon} F(|x| > R)\| \, \|(1 + |x|)^{1+\varepsilon} V(H_0 + 1)^{-1}\| \\ &\le (1 + R)^{-1-\varepsilon} \|(1 + |x|)^{1+\varepsilon} V(H_0 + 1)^{-1}\| \ . \end{aligned}$$

Let j be a C^∞-function with $0 \le j(x) \le 1$ and $j(x) = 0$ for $|x| \le \frac{1}{2}$, $j(x) = 1$ for $|x| \ge 1$. Set $j_R(x) = j(x/R)$. Then

$$\begin{aligned} \|V(H_0 + 1)^{-1} F(|x| > R)\varphi\| &= \|V(H_0 + 1)^{-1} j_R F(|x| > R)\varphi\| \\ &\le \|V(H_0 + 1)^{-1} j_R\| \, \|\varphi\| \ , \end{aligned}$$

thus, $S(R) \le \|V(H_0 + 1)^{-1} j_R\| \le S(R/2)$, and similarly $T(R) \le \|j_R V(H_0 + 1)^{-1}\| \le T(R/2)$. Using the commutator $[(H_0 + 1)^{-1}, j_R] = (H_0 + 1)^{-1}(\Delta j_R + 2\nabla j_R \cdot \nabla)\cdot (H_0 + 1)^{-1}$ and, for $R > 1$, $|\nabla j_R|, |\Delta j_R| \le C/R\, j_{R/2}$, we have

$$\begin{aligned} S(R) &\le \|V(H_0 + 1)^{-1} j_R\| \\ &\le \|j_R V(H_0 + 1)^{-1}\| + \|V(H_0 + 1)^{-1}(\Delta j_R + 2\nabla j_R \cdot \nabla)(H_0 + 1)^{-1}\| \\ &\le T(R/2) + \frac{C'}{R} S(R/4) \ . \end{aligned}$$

Iterating and using the fact that $S(R)$ is bounded, we find that

$$S(R) \le T(R/2) + \frac{C'}{R} T(R/8) + \frac{(C')^2}{R^2} d \ .$$

Since T is integrable, it follows that S is integrable. □

Theorem 5.5 (Enss' Version of Cook's Method). For short-range potentials V

$$\int_0^\infty \|V e^{\mp itH_0} \varphi\|\, dt < \infty$$

for $\varphi \in D_0$, a dense set in $L^2(\mathbb{R}^\nu)$. Consequently, the wave operators $\Omega^\pm$ exist, and Cook's estimate (5.5) holds.

Proof. We take $D_0 = \{g(H_0)P_{(-\infty,a)}\psi | g \in C_0^\infty(\mathbb{R})$, $\operatorname{supp} g \subset [\alpha^2, \beta^2]$ for some α, $\beta > 0$, $a \in \mathbb{R}$, $\psi \in L^2(\mathbb{R}^\nu)\}$, $P_{(-\infty,a)}$ being the spectral projections corresponding to A. For $\varphi \in D_0$, i.e. $\varphi = g(H_0)P_{(-\infty,a)}\psi$,

$$\begin{aligned}\|V\mathrm{e}^{-itH_0}\varphi\| &\leq \|V(H_0+1)^{-1}F(|x| > \delta t)\| \, \|(H_0+1)\varphi\| \\ &\quad + \|V(H_0+1)^{-1}\| \, \|F(|x| < \delta t)\mathrm{e}^{-itH_0}(H_0+1)\varphi\| \ . \end{aligned} \tag{5.13}$$

The first term is integrable by Lemma 5.4. The second one can be estimated by

$$C\|F(|x| < \delta t)\mathrm{e}^{-itH_0}(H_0+1)g(H_0)P_{(-\infty,a)}\psi\| \leq C'(1+|t|)^{-2}$$

by Perry's estimate (Theorem 5.3); hence it is integrable. □

The following rather technical looking result will be a key to our proof of asymptotic completeness in Sect.5.6.

Proposition 5.6. Let φ_n be a sequence of vectors converging weakly to zero, with $\|\varphi_n\| = 1$. Then

$$\|(\Omega^- - 1)g(H_0)P_+\varphi_n\| \to 0 \ .$$

As usual, g denotes a C_0^∞-function with support on the (strictly) positive half-axis.

Proof. By Cook's estimate (5.5), we have

$$\|(\Omega^- - 1)g(H_0)P_+\varphi_n\| \leq \int_0^\infty \|V\mathrm{e}^{-itH_0}g(H_0)P_+\varphi_n\| \, dt$$

$$\|V\mathrm{e}^{-itH_0}g(H_0)P_+\varphi_n\| = \|V(H_0+1)^{-1}\mathrm{e}^{-itH_0}(H_0+1)g(H_0)P_+\varphi_n\|$$

goes to zero since $\varphi_n \overset{w}{\to} 0$, and by our short-range assumption, $V(H_0+1)^{-1}$ is compact. By (5.13), the integrand is bounded by an L^1-function. Therefore the assertion of the proposition follows from Lebesgue's theorem on dominated convergence. □

Proposition 5.6 says that $(\Omega^- - 1)g(H_0)P_+$ is compact. From this fact, one can prove asymptotic completeness fairly quickly (*Mourre* [255], *Perry* [277]). We will give a longer proof which is more intuitive, and which will serve as an introduction to the work of Enss on the three-body problem. We require two detours before returning to Proposition 5.6 in Sect. 5.6.

5.4 RAGE Theorems

In this section, we will prove three versions of the celebrated RAGE theorem. The theorem was originally proven by *Ruelle* [300], and extended by *Amrein* and *Georgescu* [14] and *Enss* [95] (hence the name "RAGE" theorem). The RAGE theorem states that the time mean of certain observables will tend to zero on the continuous subspace $\mathrm{H}_{\mathrm{cont}}$.

The theorems are based on the following result on time mean of Fourier transforms:

Theorem 5.7 (Wiener's Theorem). Let μ be a finite (signed) measure on $\mathbb{R}$, and let

$$F(t) = \int \mathrm{e}^{-\mathrm{i}xt}\, d\mu(x)$$

be its Fourier transform. Then

$$\lim_{T\to\infty} \frac{1}{T}\int_0^T |F(t)|^2\, dt = \sum_{x\in\mathbb{R}} |\mu(\{x\})|^2 \ .$$

We remark that the sum $\sum |\mu(\{x\})|^2$ is finite, since μ is finite. Since we will, in essence, give the proof of Wiener's theorem while proving Theorem 5.8 below, we do not give it now.

Theorem 5.8 (RAGE). Let A be a self-adjoint operator.

(1) If C is a compact operator and $\varphi \in \mathrm{H}_{\mathrm{cont}}$, then

$$\frac{1}{T}\int_0^T \|C\mathrm{e}^{-\mathrm{i}tA}\varphi\|^2\, dt \to 0 \quad \text{as } T \to \infty \ .$$

(2) If C is bounded and $C(A+\mathrm{i})^{-1}$ is compact, and $\varphi \in \mathrm{H}_{\mathrm{cont}}$, then still

$$\frac{1}{T}\int_0^T \|C\mathrm{e}^{-\mathrm{i}tA}\varphi\|^2\, dt \to 0 \ .$$

(3) If C is compact, then

$$\left\| \frac{1}{T}\int_0^T \mathrm{e}^{+\mathrm{i}tA} C P_{\mathrm{cont}}(A) \mathrm{e}^{-\mathrm{i}tA}\, dt \right\| \to 0 \quad \text{as } T \to \infty \ .$$

The integral in (3) is meant in the strong sense.

If we take $C = F(|x| \leq R)$ [in (2)], then the RAGE theorem tells us that any state in $\mathrm{H}_{\mathrm{cont}}$ will "infinitely often leave" the ball of radius R. This is indeed what we expect physically.

Proof. We first prove that (1) and (2) follow from (3). Let $\varphi \in \mathrm{H}_{\mathrm{cont}}$. Then

$$\frac{1}{T}\int_0^T \|C\mathrm{e}^{-\mathrm{i}tA}\varphi\|^2\, dt = \frac{1}{T}\int_0^T \langle \varphi, \mathrm{e}^{\mathrm{i}tA} C^* C \mathrm{e}^{-\mathrm{i}tA}\varphi\rangle\, dt$$

$$= \left\langle \varphi, \frac{1}{T}\int_0^T e^{-itA}C^*CP_{\text{cont}}(A)e^{-itA}\,dt\varphi \right\rangle$$

$$\leq \left\| \frac{1}{T}\int_0^T e^{-itA}C^*CP_{\text{cont}}(A)e^{-itA}\,dt \right\| \|\varphi\|^2 \to 0$$

by (3), since C^*C is compact. For $\varphi \in D(A) \cap \mathrm{H}_{\text{cont}}(A)$, we write $\varphi = (A + i)^{-1}\psi$ $[\psi \in \mathrm{H}_{\text{cont}}(A)]$. Therefore,

$$\frac{1}{T}\int_0^T \|Ce^{-itA}\varphi\|^2\,dt = \frac{1}{T}\int_0^T \|C(A+i)^{-1}e^{-itA}\psi\|^2\,dt$$

converges to zero, given (1). This implies (2), since C is bounded and $D(A) \cap \mathrm{H}_{\text{cont}}(A)$ is dense in $\mathrm{H}_{\text{cont}}(A)$.

We now come to the proof of (3). Since the compact operator C can be approximated in norm by finite rank operators, it suffices to prove (3) for those operators. Since any operator of finite rank is a (finite) sum of rank 1 operators, we may restrict ourselves to rank 1 operators. Thus, let $C\varphi = \langle \rho, \varphi \rangle \psi$ (the most general operator of rank 1). Then $C^*\varphi = \langle \psi, \varphi \rangle \rho$. Define

$$Q(T) := \frac{1}{T}\int_0^T e^{itA}CP_{\text{cont}}(A)e^{-itA}\,dt$$

$$= \frac{1}{T}\int_0^T \langle e^{itA}P_{\text{cont}}(A)\rho, \cdot \rangle e^{itA}\psi\,dt \;,$$

we have

$$Q(T)^* = \frac{1}{T}\int_0^T \langle e^{itH}\psi, \cdot \rangle e^{itA}P_{\text{cont}}(A)\rho\,dt \;,$$

and therefore

$$Q(T)Q(T)^*\varphi = \frac{1}{T}\int_0^T \langle e^{itA}P_{\text{cont}}(A)\rho, Q(T)^*\varphi \rangle e^{itA}\psi\,dt$$

$$= \frac{1}{T^2}\int_0^T\int_0^T \langle e^{itA}P_{\text{cont}}\rho, e^{isA}P_{\text{cont}}\rho \rangle \langle e^{isA}\psi, \varphi \rangle e^{itA}\psi\,ds\,dt \;.$$

Therefore,

$$\left\| \frac{1}{T}\int_0^T e^{itA}Ce^{-itA}P_{\text{cont}}(A)\,dt \right\|^2$$

$$= \|Q(T)\|^2 = \|Q(T)Q(T)^*\|$$

$$\leq \frac{1}{T^2}\int_0^T\int_0^T |\langle \mathrm{e}^{\mathrm{i}tA}\rho, \mathrm{e}^{\mathrm{i}sA}P_{\mathrm{cont}}(A)\rho\rangle|\, ds\, dt\, \|\psi\|^2$$

$$\leq \|\psi\|^2\left(\frac{1}{T^2}\int_0^T\int_0^T |\langle P_{\mathrm{cont}}\rho, \mathrm{e}^{-\mathrm{i}(t-s)A}P_{\mathrm{cont}}\rho\rangle|^2\, ds\, dt\right)^{1/2} .$$

Let μ denote the spectral measure for $P_{\mathrm{cont}}\rho$. Then

$$\frac{1}{T^2}\int_0^T\int_0^T |\langle P_{\mathrm{cont}}\rho, \exp[-\mathrm{i}(t-s)A]P_{\mathrm{cont}}\rho\rangle|^2\, ds\, dt$$

$$\leq \frac{1}{T^2}\int_0^T\int_0^T \left|\int \exp[-\mathrm{i}(t-s)\lambda]\, d\mu(\lambda)\right|^2 ds\, dt$$

$$= \frac{1}{T^2}\int_0^T\int_0^T \iint \exp[-\mathrm{i}(t-s)(\lambda-\kappa)]\, d\mu(\lambda)\, d\mu(\kappa)\, ds\, dt$$

$$= \iint\left[\frac{1}{T}\int_0^T \exp[-\mathrm{i}t(\lambda-\kappa)]\, dt\right.$$

$$\left.\times\frac{1}{T}\int_0^T \exp[+\mathrm{i}s(\lambda-\kappa)]\, ds\right] d\mu(\lambda)\, d\mu(\kappa) \quad \text{(by Fubini)} . \qquad (5.14)$$

Computing

$$\frac{1}{T}\int_0^T \exp[\mathrm{i}(\lambda-\kappa)s]\, ds \frac{1}{T}\int_0^T \exp[-\mathrm{i}(\lambda-\kappa)t]\, dt$$

$$= \frac{1}{T^2(\lambda-\kappa)^2}\{\exp[\mathrm{i}(\lambda-\kappa)T]-1\}\{\exp[-\mathrm{i}(\lambda-\kappa)T]-1\}$$

$$= \frac{1}{T^2(\lambda-\kappa)^2}\{\exp[\mathrm{i}(\lambda-\kappa)T/2]-\exp[-\mathrm{i}(\lambda-\kappa)T/2]\}$$

$$\cdot\{\exp[-\mathrm{i}(\lambda-\kappa)T/2]-\exp[\mathrm{i}(\lambda-\kappa)T/2]\}$$

$$= \frac{4\sin^2\{(\lambda-\kappa)T/2\}}{T^2(\lambda-\kappa)^2}$$

with the convention that $\sin 0/0 = 1$. Since

$$\frac{4\sin^2\{(\lambda-\kappa)T/2\}}{T^2(\lambda-\kappa)^2} \leq 1$$

[which is in $L^2(d\mu)$], and since furthermore

$$\frac{4\sin^2(\lambda-\kappa)T/2}{T^2(\lambda-\kappa)^2}$$

tends to zero for $\lambda \neq \kappa$, and to one for $\lambda = \kappa$ as $T \to \infty$, we have that (5.14) tends to

$$\int \mu(\{\kappa\})\, d\mu(\kappa) = \sum_{\kappa \in \mathbb{R}} \mu(\{\kappa\})^2$$

by Lebesgue's theorem on dominated convergence. Since the measure μ (the spectral measure for $P_{\text{cont}}\varphi$) is continuous, i.e. does not have atoms, we know that $\sum_{\kappa \in \mathbb{R}} \mu(\{\kappa\})^2 = 0$. □

We will make use of the RAGE theorem in Sect.5.5, as well as in the chapter on random Jacobi matrices.

5.5 Asymptotics of Observables

In this section, we are concerned with recent developments of time-dependent scattering theory due to *Enss* [98, 100]. These new ideas present, in the two-body case, more physical insight and simplify the proof of asymptotic completeness for long-range forces. Furthermore, they are an essential ingredient for Enss' three-body proof.

The main result of this section states that some observables, $B(t) = \exp(\mathrm{i}Ht)B\exp(-\mathrm{i}Ht)$, behave on H_{cont} asymptotically in time in a similar way as they would under the free time evolution, more precisely: $(x(t)/t)^2 \sim 2H$, $A(t)/t \sim 2H$, $H_0(t) \sim H$.

Theorem 5.9. For $f \in C_\infty(\mathbb{R})$ and any $\varphi \in \mathrm{H}_{\text{cont}}(H)$:

(i) $f\left(\left(\frac{x(t)}{t}\right)^2\right)\varphi \to f(2H)\varphi$

(ii) $f\left(\frac{A(t)}{t}\right)\varphi \to f(2H)\varphi$

(iii) $f(H_0(t))\varphi \to f(H)\varphi \quad$ as $t \to \pm\infty$.

Remark. The only assumptions on V we need for the proof below are $D(H) = D(H_0)$, and $[A, V]$ is a compact operator from H_{+2} to H_{-2}. For a proof under very weak assumptions allowing long-range forces, see [98]. Before proving the theorem, we first state and prove two of its consequences.

Corollary 1. For $\varphi \in \mathrm{H}_{\text{cont}}$

$$\|P_- \mathrm{e}^{-\mathrm{i}tH}\varphi\| \to 0 \quad \text{as } t \to \infty \quad \text{and}$$

$$\|P_+ \mathrm{e}^{-\mathrm{i}tH}\varphi\| \to 0 \quad \text{as } t \to -\infty\ .$$

Proof. By the usual density argument, it is enough to prove the Corollary for vectors $\varphi = g(H)\varphi$ where $g \in C^\infty$, $0 \le g \le 1$, $g(x) = 1$ for $x > 2\delta$, $g(x) = 0$ for $x \le \delta$ for some $\delta > 0$. Furthermore, let $f \in C^\infty$ satisfy $f(x) = 1$, $x < \delta/2$, $f(x) = 0$, $x \ge \delta$, $0 \le f \le 1$. For such a vector φ we have (for $t \ge 0$):

$$\|P_- \mathrm{e}^{-\mathrm{i}tH}\varphi\| = \|\mathrm{e}^{\mathrm{i}tH}P_- \mathrm{e}^{-\mathrm{i}tH}\varphi\| = \left\| \chi_{(-\infty,0)}\left(\frac{A(t)}{t}\right)\varphi \right\|$$

$$\le \left\| f\left(\frac{A(t)}{t}\right)\varphi \right\| \to \|f(2H)\varphi\|$$

by Theorem 5.9. Actually, Theorem 5.9 as stated is not applicable since $f \notin C_\infty(R)$, but an elementary argument ([294, p. 286]) allows one to extend the result to all bounded continuous f. Since $\varphi = g(H)\varphi$ and $fg = 0$, we know

$$\|P_- \mathrm{e}^{-\mathrm{i}tH}\varphi\| \to 0 \quad \text{as } t \to \infty \ .$$

$$\|P_+ \mathrm{e}^{-\mathrm{i}tH}\varphi\| \to 0 \quad \text{as } t \to -\infty$$

is proven in the same way, observing that $P_+ = \chi_{(-\infty,0)}(A/t)$ for $t < 0$. □

Remark. The corollary states that a $\varphi \in \mathrm{H}_{\mathrm{cont}}$ cannot have an incoming part in the far future, or an outgoing part in the remote past.

Corollary 2. For $\varphi \in \mathrm{H}_{\mathrm{cont}}(H)$: $\exp(-\mathrm{i}tH)\varphi \to 0$ weakly.

Proof. Let f, g be the functions defined in the proof of Corollary 1. Let $\varphi = g(H)\varphi$.

$$\|F(|x| \le a)\mathrm{e}^{-\mathrm{i}tH}\varphi\| = \left\| F\left(\left|\frac{x(t)}{t}\right|^2 \le \frac{a^2}{t^2}\right)\varphi \right\|$$

$$\le \left\| f\left(\left|\frac{x(t)}{t}\right|^2\right)\varphi \right\| \to 0 \quad \text{by Theorem 5.9} \ . \quad \square$$

Proof of Theorem 5.9. We first do some formal calculations explaining why the theorem is true.

$$\dot{A}(t) = [H, A](t) = 2H + W(t) \ ,$$

where $W := -2V - x\nabla V$ and $W(t) = \exp(\mathrm{i}Ht)W\exp(-\mathrm{i}Ht)$. Thus, neglecting any domain questions,

$$\frac{A(t)}{t} = \frac{A}{t} + \frac{1}{t}\int_0^t \dot{A}(t)\,dt = \frac{A}{t} + 2H + \frac{1}{t}\int_0^t W(t)\,dt \ .$$

Since we can reasonably expect that $1/t\int_0^t W(t)\,dt$ will go to zero when applied to (nice) $\varphi \in \mathrm{H}_{\mathrm{cont}}$ by a RAGE-type theorem, and also $(A/t)\varphi$ will tend to zero, we formally obtain the desired result.

Let us now make the above calculations rigorous. Define $N = p^2 + x^2$. Here N is a self-adjoint operator on $D(N) = D(H_0) \cap D(x^2)$. $D(N)$ has the advantage that all of the relevant operators are defined on this set. Since $\exp(-itH)$ leaves $D(N)$ invariant (cf. *Fröhlich* [117], *Kato* [192] and *Radin* and *Simon* [290]), $D(N)$ is also in the domain of $A(t)$, $x(t)$, etc. However, we do not know whether $D(N)$ is a core for HP_{cont}. This will cause some complications and requires an additional approximation argument.

By the Stone-Weierstrass gavotte (see the Appendix to Chap. 3), it suffices to prove the theorem for $f(x) = (x - z)^{-1}$; $z \notin \mathbb{R}$. For bounded operators M, N we have

$$\|(M - N)\varphi\|^2 = -\langle N\varphi, (M - N)\varphi\rangle - \langle (M - N)\varphi, N\varphi\rangle + \langle \varphi, (M^*M - N^*N)\varphi\rangle \ .$$

Inserting, for M, N, the resolvents $(A(t)/t - z)^{-1}$ and $(2H - z)^{-1}$, and using

$$(A - z)^{-1*}(A - z)^{-1} = (-2\,\mathrm{Im}\, z)^{-1}[(A - z)^{-1*} - (A - z)^{-1}] \ ,$$

we see that weak convergence of the resolvents implies strong convergence. Thus, it suffices to prove

$$\left\langle \eta, \left[\left(\frac{A(t)}{t} - z\right)^{-1} - (2H - z)^{-1}\right]\varphi\right\rangle \to 0 \tag{5.15}$$

for $\varphi \in \mathrm{H}_{\text{cont}}$ and $\eta \in \mathrm{H}$. By a density argument, it is enough to prove (5.15) for η, φ in suitable dense sets. We estimate, for $\eta \in D(H)$,

$$\begin{aligned}
&\left|\left\langle \eta, \left[\left(\frac{A(t)}{t} - z\right)^{-1} - (2H - z)^{-1}\right]\varphi\right\rangle\right| \\
&= \left|\left\langle \eta, \left(\frac{A(t)}{t} - z\right)^{-1}\left(\frac{A(t)}{t} - 2H\right)(2H - z)^{-1}\varphi\right\rangle\right| \\
&= \left|\left\langle (H - \bar{z})\eta, (H - z)^{-1} t(A(t) - tz)^{-1}(H - z)(H - z)^{-1}\left(\frac{A(t)}{t}\right.\right.\right. \\
&\qquad \left.\left.\left. - 2H\right)(2H - z)^{-1}\varphi\right\rangle\right| \\
&\le \|(H - \bar{z})\eta\| \, \|(H - z)^{-1} t(A(t) - tz)^{-1}(H - z)\| \\
&\qquad \left\|(H - z)^{-1}\left(\frac{A(t)}{t} - 2H\right)(2H - z)^{-1}\varphi\right\| \ .
\end{aligned} \tag{5.16}$$

The operator $(H - z)^{-1} t(A(t) - tz)^{-1}(H - z)$ is bounded uniformly in (large) t. This can be proven by exploiting the commutator relation $i[H_0, A] = 2H_0$ (see Chap. 4). For $\psi \in D(N) \cap \mathrm{H}_{\text{cont}}$, we will prove below that

$$\left\| (H - z)^{-1} \left(\frac{A(t)}{t} - 2H \right) \psi \right\|$$

tends to zero. (The proof is a rigorous version of the above formal calculations.)

From this, we can conclude that (5.16) converges to zero for those $\varphi \in \mathrm{H}_{\mathrm{cont}}$ for which $(2H - z)^{-1} \varphi \in D(N)$. This set, however, is only dense in $\mathrm{H}_{\mathrm{cont}}$ if $D(N)$ is a core for HP_{cont}. Since we do not see how to prove this, we use a "regularization" for $\psi := (2H - z)^{-1} \varphi$.

Define $\psi^{(t)}(x) := (1 + x^2/t)^{-1} \psi(x)$. Then $\psi^{(t)} \in D(N)$ for $\psi \in D(H)$, and furthermore

(a) $\psi^{(t)} \to \psi$ in L^2
(b) $H\psi^{(t)} \to H\psi$ in L^2
(c) $\|A\psi^{(t)}\| \leq Ct^{1/2}$.

The proofs of (a)–(c) are straightforward calculations. We now estimate (remember $\psi = (2H - z)^{-1} \varphi$):

$$\begin{aligned}
&\left| \left\langle \eta, \left[\left(\frac{A(t)}{t} - z \right)^{-1} - (2H - z)^{-1} \right] \varphi \right\rangle \right| \\
&\leq \left| \left\langle \eta, \left(\frac{A(t)}{t} - z \right)^{-1} \left(\frac{A(t)}{t} - 2H \right) (\psi^{(t)} - \psi) \right\rangle \right| \\
&\quad + \left| \left\langle \eta, \left(\frac{A(t)}{t} - z \right)^{-1} \left(\frac{A(t)}{t} - 2H \right) \psi^{(t)} \right\rangle \right| . \qquad (5.17)
\end{aligned}$$

The first term of the right-hand side of (5.17) tends to zero because of (a), (b) above. As in (5.16), the second term can be estimated by

$$C \left\| (H - z)^{-1} \left(\frac{A(t)}{t} - - 2H \right) \psi^{(t)} \right\| .$$

Since $\psi^{(t)} \in D(N)$, we have

$$\frac{1}{t} A(t)\psi^{(t)} = \frac{A\psi^{(t)}}{t} + \frac{1}{t} \int_0^t \mathrm{e}^{\mathrm{i}sH} [H, A] \mathrm{e}^{-\mathrm{i}sH} \psi^{(t)} \, dt .$$

Thus,

$$\begin{aligned}
&\left\| (H - z)^{-1} \left(\frac{A(t)}{t} - 2H \right) \psi^{(t)} \right\| \\
&\leq \frac{1}{t} \|A\psi^{(t)}\|
\end{aligned}$$

$$+\left\|\frac{1}{t}\int_0^t e^{isH}(H-z)^{-1}W(H-z)^{-1}e^{-isH}(H-z)\psi\,ds\right\|$$

$$+\left\|\frac{1}{t}\int_0^t e^{isH}(H-z)^{-1}W(H-z)^{-1}e^{-isH}(H-z)(\psi^{(t)}-\psi)\,ds\right\| .$$

The first term goes to zero because of (c). The second one tends to zero by the RAGE theorem [Theorem 5.8(3)], and the third one does because of (b) and the uniform boundedness of the operator

$$\frac{1}{t}\int_0^t e^{isH}(H-z)^{-1}W(H-z)^{-1}e^{-isH}\,ds .$$

This finishes the proof of part (ii) of the theorem. The proof of part (i) is similar, but uses a double integral and $d/dt\,[x(t)^2] = A(t)$. By the Stone-Weierstrass gavotte, it suffices to prove part (iii) for resolvents.

$$\|[(H-z)^{-1} - (H_0(t)-z)^{-1}]\varphi\| = \|[(H-z)^{-1}V(H-z)^{-1}]e^{itH}\varphi\| .$$

By Corollary 2 above, $\exp(itH)\varphi$ converges weakly to zero. Since the operator $(H_0-z)^{-1}V(H-z)^{-1}$ is compact, this implies that the norm goes to zero, thus giving (iii). The application of Corollary 2 is correct, since its proof makes no use of part (iii) of the theorem. □

5.6 Asymptotic Completeness

We now come to the proof of asymptotic completeness.

Theorem 5.10. $\operatorname{Ran}\Omega^- = \operatorname{Ran}\Omega^+ = \mathrm{H}_{\mathrm{cont}}$

Remark. Since $\operatorname{Ran}\Omega^\pm \subset \mathrm{H}_{\mathrm{ac}}$, it follows $\mathrm{H}_{\mathrm{sc}} = \{0\}$.

Proof. As usual, take $\varphi \in \mathrm{H}_{\mathrm{cont}}$ with $g(H)\varphi = \varphi$ ($0 \le g \le 1$, $g = 1$ on $[a^2, b^2]$, $\operatorname{supp} g \subset [a^2/2, 2b^2]$). Set $\varphi_s := \exp(-isH)\varphi$.

$$\begin{aligned}\|(\Omega^- - 1)\varphi_s\| &\le \|(\Omega^- - 1)g(H_0)\varphi_s\| + \|(\Omega^- - 1)[g(H_0) - g(H)]\varphi_s\| \\ &\le \|(\Omega^- - 1)g(H_0)P_+\varphi_s\| + 2\|g(H_0)P_-\varphi_s\| \\ &\quad + 2\|[g(H_0(s)) - g(H)]\varphi\| .\end{aligned}$$

The first term goes to zero as $s \to \infty$ by Proposition 5.6, since $\varphi_s \xrightarrow{w} 0$ by Corollary 2 of Theorem 5.9. The second term converges to zero by Corollary 1 of Theorem 5.9, and the third one goes to zero by Theorem 5.9. Thus,

$$\varphi = \lim_{s\to\infty} e^{isH}\Omega^- e^{-isH}\varphi = \lim_{s\to\infty} \Omega^- e^{isH_0}e^{-isH}\varphi$$

$$\in \overline{\operatorname{Ran}\Omega^-} = \operatorname{Ran}\Omega^- .$$

The assertion on Ω^+ is proven in the same way. □

Theorem 5.11. Positive eigenvalues are isolated and of finite multiplicity.

Proof. Suppose the assertion of the theorem is wrong. Then we find a C_0^∞-function g, such that $g(H)\varphi_n = \varphi_n$ for a sequence $\varphi_n \xrightarrow{w} 0$ with $\|\varphi_n\| = 1$ and $\varphi_n \in \mathrm{H}_{\mathrm{ac}}^\perp$.
Since $g(H_0) - g(H)$ is compact, $\varphi_n \xrightarrow{w} 0$ implies $\|[g(H_0) - g(H)]\varphi_n\| \to 0$. But

$$\begin{aligned}\varphi_n &= g(H_0)\varphi_n + [\varphi_n - g(H_0)\varphi_n] \\ &= g(H_0)\varphi_n + [g(H) - g(H_0)]\varphi_n \\ &= g(H_0)P_-\varphi_n + g(H_0)P_+\varphi_n + [g(H) - g(H_0)]\varphi_n \\ &= \Omega^- g(H_0)P_-\varphi_n + \Omega^+ g(H_0)P_+\varphi_n + (1 - \Omega^-)g(H_0)P_-\varphi_n \\ &\quad + (1 - \Omega^+)g(H_0)P_+\varphi_n + [g(H) - g(H_0)]\varphi_n \to \Omega^- g(H_0)P_-\varphi_n \\ &\quad + \Omega^+ g(H_0)P_+\varphi_n\end{aligned}$$

because of Proposition 5.6. Hence, the projection of φ_n to $\mathrm{H}_{\mathrm{ac}}^\perp$ tends to zero, which is impossible since $\varphi_n \perp \mathrm{H}_{\mathrm{ac}}$ and $\|\varphi_n\| = 1$. □

The above proof also excludes singular continuous spectrum (and did not use asymptotic completeness). Thus RAGE is not needed to prove $\sigma_{\mathrm{sc}} = \phi$. It was remarked by *Davies* [75] that, even for a geometric proof of asymptotic completeness, this celebrated theorem is not required.

For other methods to exclude positive eigenvalues, see *Eastham* and *Kalf* [90] (especially Theorem 4.19) and *Reed* and *Simon* IV, XIII.13 [295].

5.7 Asymptotic Completeness in the Three-Body Case

This paragraph concerns Enss' recent geometric proof of asymptotic completeness of three-body Hamiltonians. Here we make no attempt to give complete proofs, but rather discuss some of the main ideas in Enss' three-body proof, referring the reader to the papers [99, 102–105] for details. We follow, more or less, Enss' first three-body paper [99]. However, the articles [102, 105] are more self-contained, and we recommend them for further reading.

Asymptotic completeness for two-body systems in increasing generality was obtained by *Povzner* [287], *Ikebe* [175], *Agmon* [2] and *Kuroda* [223] (see *Reed* and *Simon* III [294] for further historical references), and then in *Enss*' famous paper [95]. In a celebrated monograph, *Faddeev* [106] proved asymptotic completeness for a class of three-body systems. There were extensions (and corrections of one gap) of Faddeev's work by *Sigal* [309], *Yafaev* [373], *Thomas* [354], *Ginibre* and *Moulin* [136] and *Kato* [194]. Faddeev's work had two severe limitations: (1) It required $r^{-2-\varepsilon}$ decay on potentials, and, in particular, systems with an infinity of channels were not allowed. (2) Two-body subsystems were not allowed to have quasibound states at threshold (this assumption was later removed under stronger decay hypotheses on V_{ij} by *Loss* and *Sigal* [236]). Recently, extending their ideas, which is the basis of this and the last chapter, *Enss* [99] and *Mourre* [257] treated three-body systems without any assumptions on bound states, and with sufficiently slow decay allowed to have an infinity of channels.

As the reader will expect, the three-body case—compared with the two-body case—shows new, physically interesting phenomena, as well as a variety of mathematical difficulties. First of all, there are more possibilities of "asymptotic configurations": All particles may move essentially free, or two of them are bound together and the third one is free, or all three particles are in a bound state. These three possibilities are the physically expected ones, and asymptotic completeness says that they are the only ones that can occur. However, a priori configurations such as one particle bouncing back and forth between the two others might occur.

Moreover, the asymptotic configurations in the far future and in the remote past may look "rather different," e.g. a configuration with two particles bounded together (and the third one free) in the remote past may have all three particles moving freely in the far future, etc. However, one certainly expects, on physical grounds (and we will indicate a proof in the sequel), that a state with one particle asymptotically free in the remote past has necessarily (at least) one particle free in the far future, and vice versa.

To formalize the above discussion, let us introduce some notation. As usual, we work in the center of mass frame. By α, we denote a two-cluster decomposition, i.e. $\alpha = \{(i,j), k\}$ with $\{i,j,k\} = \{1,2,3\}$. We use Jacobi coordinates to describe the positions of the particles. Let m_i, $i = 1, 2, 3$ be the masses of the particles. For $\alpha = \{(i,j), k\}$ we set

$$x^\alpha := x_i - x_j \quad \text{and} \quad y_\alpha := x_k - \frac{m_i x_i + m_j x_j}{m_i + m_j}, \tag{5.18}$$

where x^α describes the relative position in the pair (i,j), while y_α is the position of the third particle relative to the center of mass of the pair. We write

$$p^\alpha := -\mathrm{i}\frac{\partial}{\partial x^\alpha} \quad \text{and} \quad q_\alpha := -\mathrm{i}\frac{\partial}{\partial y_\alpha} \tag{5.19}$$

for the corresponding momentum operators. The reduced mass of the pair and the third particle are given by

$$\mu^\alpha := \frac{m_i m_j}{m_i + m_j} \quad \text{and} \quad \nu_\alpha := \frac{m_k(m_i + m_j)}{m_i + m_j + m_k}, \tag{5.20}$$

respectively. Defining

$$h_0^\alpha := \frac{1}{2\mu^\alpha}(p^\alpha)^2 \quad \text{and} \quad k_{0\alpha} := \frac{1}{2v_\alpha}(q_\alpha)^2 \ , \tag{5.21}$$

we have

$$H_0 = h_0^\alpha + k_{0\alpha} \ , \tag{5.22}$$

where we used the symbols p_α, h_0^α, etc. in (5.21) as operators on $L^2(\mathbb{R}^{2\nu}, d^\nu x^\alpha d^\nu y_\alpha)$ with the obvious understanding. As usual, we denote the internal potential by V^α (this is the interaction between the particles in the pair). We set

$$h^\alpha := h_0^\alpha + V^\alpha \quad \text{and} \tag{5.23}$$

$$H(\alpha) := H_0 + V^\alpha \ . \tag{5.24}$$

It seems reasonable that the dynamics of a state which is asymptotically free is properly described by the free time evolution $\exp(-iH_0 t)$, for large (resp. small negative) t. However, if one particle moves freely while the other two are bound together, $\exp(-iH_0 t)$ is obviously not the right description of the asymptotic behavior of the corresponding state. Rather, $\exp[-iH(\alpha)t]$ should play the role of $\exp(-iH_0 t)$. Note that $\exp[-iH(\alpha)t]$ actually gives a simple comparison dynamics for configurations for which the pair in a is in a bound state. Indeed, if E_0 is the bound state energy and P_0 the projector on the corresponding eigenspace for the pair, then

$$e^{-iH(\alpha)t}P_0 = e^{-iE_0 t}e^{-ik_{0\alpha}t}P_0 \ .$$

This leads to the definition of a set of wave operators, each of which is expected to give the correct asymptotic behavior of states in a suitable subspace

$$\Omega_\pm^0 = s\text{-}\lim_{t\to\mp\infty} e^{iHt}e^{-iH_0 t} \ , \tag{5.25}$$

the range of which consists of configurations of three asymptotically free particles, and

$$\Omega_\pm^\alpha = \lim_{t\to\mp\infty} e^{itH}e^{-itH(\alpha)}P_p(h^\alpha) \tag{5.26}$$

[$P_p(h^\alpha)$ denotes the projection onto the point subspace $\mathrm{H}_{pp}(h^\alpha)$ of h^α]. The range of those operators are given by states, with the particles in the pair (of α) asymptotically in a bound state and the third particle moving asymptotically free. We used $P_p(h^\alpha)$ in (5.26) to single out only states for which the pair actually is in a bound state asymptotically. The dynamics $\exp[-itH(\alpha)]$ does, of course, describe also configurations with three asymptotically free particles correctly. It is, however, only the range of $P_p(h^\alpha)$ where this dynamics is particularly simple.

We say that *asymptotic completeness* holds, if

$$\begin{aligned} \mathrm{H}_{\mathrm{cont}}(H) &= \operatorname{Ran}\Omega_-^0 \oplus \left(\bigoplus_\alpha \operatorname{Ran}\Omega_-^\alpha\right) \\ &= \operatorname{Ran}\Omega_+^0 \oplus \left(\bigoplus_\alpha \operatorname{Ran}\Omega_+^\alpha\right). \end{aligned} \tag{5.27}$$

In other words, any state is a superposition of the following types of states: (1) bound states; (2) a state with asymptotically two particles in a bound state and the other one moving freely; (3) a state with three asymptotically free particles.

Using the asymptotic completeness of two-body systems, it is easy to see that asymptotic completeness of the three-body system is equivalent to:

For any $\psi \in \mathrm{H}_{\mathrm{cont}}$ and $\varepsilon > 0$, there is a τ large enough such that

$$e^{-iH\tau}\psi = \psi_0(\tau) + \sum_a \psi_\alpha(\tau)$$

with the following, uniformly in $t \geq 0$:

$$\|(e^{-iHt} - e^{-iH_0t})\psi_0(\tau)\| < \varepsilon \tag{5.28}$$

$$\|(e^{-iHt} - e^{-iH(\alpha)t})\psi_\alpha(\tau)\| < \varepsilon \tag{5.29}$$

and a similar decomposition for the past. Knowing this one splits $\psi_\alpha(\tau)$ into $P_{\mathrm{p}}(h^\alpha)\psi_\alpha(\tau)$ and $P_{\mathrm{c}}(h^\alpha)\psi_\alpha(\tau)$. Then (5.29) says that $P_{\mathrm{p}}(h^\alpha)\psi_\alpha(\tau) \in \operatorname{Ran}\Omega_-^\alpha$, while one uses the asymptotic completeness of the two-body system to show that $P_{\mathrm{c}}(h^\alpha)\psi_\alpha(\tau) \in \operatorname{Ran}\Omega_-^0$.

In the following, we will assume, for simplicity, that the V_{ij} are continuously differentiable functions satisfying

(1) $(1 + |x|)^{1+\varepsilon} V_{ij}(x)$ is bounded, and
(2) $x \cdot \nabla V_{ij}(x) \to 0$ as $|x| \to \infty$.

However, Enss' proof works for a larger class of potentials.

In the two-body case, we restricted our considerations to those states with energy E between $0 < a \leq E \leq b < \infty$, eventually sending a to zero and b to infinity. The upper cut-off was for mathematical convenience, while the lower cut-off had the physical reason that particles with energy near to zero may travel extremely slowly, thus making problems for many estimates.

In the three-body case, we are faced with a new set of "trouble-makers", namely states with energy around thresholds. Recall that thresholds are eigenvalues of subsystems (Sect. 3.2). It is known (see Theorem 4.19 or *Reed* and *Simon* IV, XIII.58 [295]) that the operators h^α have no positive eigenvalues. Denote by e_i^α the nonzero eigenvalues of h^α in increasing order. The e_i^α can accumulate at most at zero. The set of thresholds T is given by $\mathrm{T} = \{e_i^\alpha\} \cup \{0\}$.

It might happen that the pair in α is in a bound state with energy e_i^α, or has energy almost e_i^α, while the third particle is traveling with very small velocity with respect to the pair. Those states obviously cause the same—if not a more difficult—problem as the states with small energy in the two-body case.

For $a, b > 0$, let us define

$$\mathrm{A} := \mathrm{A}(a,b) := \{E \in \mathbb{R} | d(E, \mathrm{T}) > a, E < b\} ,$$

where $d(x, A)$ denotes the distance of the point x from the set A.

It is clear that

$$\bigcup_{a,b>0} \mathrm{A}(a,b) \text{ is dense in } \mathbb{R} ,$$

and thus the set of states ψ, with $\psi = F(H \in \mathrm{A}(a,b))\psi$, is dense in L^2 if a, b run through $\mathbb{R}^+$. States obeying $\psi = F(H \in \mathrm{A}(a,b))P_c(H)\psi$ have the following property: If two particles are in a bound state, then the third particle has kinetic energy at least a with respect to the pair, provided it is far enough separated.

The strategy of the proof will be the following: We define $\mathrm{H}_s := \mathrm{H}_c(H) \cap (\bigoplus_\alpha \mathrm{Ran}\, \Omega_-^\alpha)^\perp$. We then show that any $\psi \in \mathrm{H}_s$ such that $\psi = F(H \in \mathrm{A}(a,b))\psi$, $a, b > 0$ arbitrary belongs to $\mathrm{Ran}\, \Omega_-^0$.

One of the crucial observations of Enss' three-body paper is the following remarkable proposition which we will not prove here:

Proposition 5.12. Suppose $\psi \in \mathrm{H}_s$ and $\psi = F(H \in \mathrm{A}(a,b))\psi$. Then for any α

$$\lim_{T\to\infty} \frac{1}{T} \int_0^T \|F(|x^\alpha| < \rho) e^{-iHt}\psi\| = 0 .$$

This proposition tells us that for the states under consideration, any two of the particles have to separate from each other in the time mean. This proposition has much in common with the RAGE theorems we discussed in Sect. 5.4. Note, however, that the assertion of the proposition is definitely not true on the whole of H_c! Proposition 5.12 indeed expresses our physical intuition. If, for a state ψ, it is not true that asymptotically two particles are in a bound state and the third one moves freely ($\psi \perp \Omega_-^\alpha$), and if, furthermore, one particle has strictly positive kinetic energy with respect to the pair of the two others, then all the particles will be separated from each other in the time mean.

It is now easy to obtain an analog of Theorem 5.9 (i) and (ii) for three particles. Let us define

$$X^2 = \mu^\alpha |x^\alpha|^2 + \nu_\alpha |y_\alpha|^2 \quad \text{and} \tag{5.30}$$

$$A = \tfrac{1}{2}\{p^\alpha \cdot x^\alpha + x^\alpha \cdot p^\alpha + q_\alpha \cdot y_\alpha + y_\alpha \cdot q_\alpha\} . \tag{5.31}$$

Proposition 5.13. For $f \in C_\infty(\mathbb{R})$ and $\varphi \in \mathrm{H}_s$

(i) $f\left(\left(\frac{X(t)}{t}\right)^2\right)\varphi \to f(2H)\varphi$

(ii) $f\left(\frac{A(t)}{t}\right)\varphi \to f(2H)\varphi$.

Proof. [Sketch of (ii)]. As in the two-body case (Theorem 5.9), we have (omitting domain questions)

$$\frac{A(t)}{t} = \frac{A}{t} + 2H + \sum_{\alpha} \frac{1}{t} \int_0^t W^\alpha(s)\, ds \tag{5.32}$$

with $W^\alpha = -2V^\alpha - x\nabla V^\alpha$.

However, this time we cannot use the RAGE theorem to prove that the third term in the right-hand side of (5.32) goes to zero, because W^α has no chance to be compact in the three-body case, even if sandwiched between resolvents.

However, by the decay assumptions on V^α, we have

$$\|W^\alpha F(|x^\alpha| < \rho)\| \to 0 \quad \text{as } \rho \to \infty$$

while we can control

$$\frac{1}{t}\int_0^t \mathrm{e}^{\mathrm{i}sH} W^\alpha F(|x^\alpha| > \rho)\mathrm{e}^{-\mathrm{i}sH}\, ds$$

by Proposition 5.12. □

An analog of Theorem 5.9 (iii) holds in a weak form:

Proposition 5.14. For $\varphi \in \mathrm{H}_s \cap D(H_0)$ with $\varphi = F(H \in \mathrm{A}(a, b))\varphi$, there exists a sequence $\tau_n \to \infty$ such that

$$H_0(\tau_n)\varphi \to H\varphi \ .$$

Proof (Sketch). By the assumptions on the potentials V^α we know that, for any sequence $\rho_n \to \infty$,

$$\left\|(H_0 - H)\prod_\alpha F(|x^\alpha| > \rho_n)\right\| \to 0 \ .$$

Proposition 5.12 enables us to choose a sequence $\tau_n \to \infty$, such that

$$\sum_\alpha \|F(|x^\alpha| < \rho_n)\mathrm{e}^{-\mathrm{i}H\tau_n}\varphi\| \to 0 \ .$$

Thus,

$$\|e^{i\tau_n H} H_0 e^{-i\tau_n H}\varphi - H\varphi\| \le \left\| (H_0 - H)\prod_\alpha F(|x^\alpha| > \rho_n) \right\| \|\varphi\| + \sum_\alpha \|V_\alpha\| \, \|F(|x^\alpha| < \rho_n) e^{-iH\tau_n}\varphi\|$$

converges to zero. □

Theorem 5.9 is a very powerful tool for the investigation of two-body systems. As they stand, the three-body analogs (Propositions 5.13 and 5.14) give much less information, since they do not say much about x^α, y_α. However, they enable us to prove the following proposition which turns out to be a cornerstone of the asymptotic completeness result:

Proposition 5.15. For $\psi \in \mathrm{H}_s$ with $\psi = F(H \in \mathrm{A}(a,b))\psi$, there exists a sequence $\tau_n \to \infty$, such that

(i) For any $f \in C_0^\infty(\mathbb{R}^\nu)$:

$$\left\| \left\{ f\left(\mu^\alpha \frac{x^\alpha}{\tau_n}\right) - f(p^\alpha) \right\} e^{-iH\tau_n}\psi \right\| \to 0 \quad \text{and}$$

$$\left\| \left\{ f\left(\nu_\alpha \frac{y_\alpha}{\tau_n}\right) - f(q_\alpha) \right\} e^{-iH\tau_n}\psi \right\| \to 0 \, .$$

(ii) For any $g \in C_0^\infty(\mathbb{R})$

$$\left\| \left[g\left(\frac{\mu^\alpha}{2}\left(\frac{x^\alpha}{\tau_n}\right)^2\right) - g(h^\alpha) \right] e^{-iH\tau_n}\psi \right\| \to 0 \, .$$

Proof (Sketch). We only indicate the main idea of the proof of (i).

Neglecting domain questions, we formally have

$$\frac{1}{2\mu^\alpha}\left[\frac{\mu^\alpha x^\alpha(\tau_n)}{\tau_n} - p^\alpha(\tau_n)\right]^2 + \frac{1}{2\nu_\alpha}\left[\frac{\nu_\alpha y_\alpha(\tau_n)}{\tau_n} - q_\alpha(\tau_n)\right]^2$$

$$= H_0(\tau_n) + \frac{X^2(\tau_n)}{\tau_n^2} - \frac{A(\tau_n)}{\tau_n}$$

$$\text{“}\to\text{”}\ H_0(\tau_n) + H - 2H \ \text{“}\to\text{”}\ 0 \, .$$

Therefore, both

$$\frac{\mu^\alpha x^\alpha(\tau_n)}{\tau_n} - p^\alpha(\tau_n), \quad \text{and} \quad \frac{\nu_\alpha y_\alpha(\tau_n)}{\tau_n} - q_\alpha(\tau_n)$$

go to zero.

It is not difficult to turn the above formal calculations into a rigorous proof, using resolvents and the Stone-Weierstrass gavotte (see Appendix to Chap. 3). □

We now state the asymptotic completeness result of Enss, and say a few words about the proof.

Theorem 5.16. For three-body Hamiltonians, asymptotic completeness holds.

Remark. Although we require the conditions (i) and (ii) above for the potentials, Enss' proof works under weaker assumptions.

Enss distinguishes the case where the energy of the pair, (h^α), is negative or small positive, and the case where the energy of the pair is positive and bounded away from zero.

For the first case, he proves that the (full) time evolution is well approximated by $\exp[-itH(\alpha)]$, thus showing (5.29) for those states.

For the second case, Enss proves that the time evolution is approximately given by $\exp(-itH_0)$, showing (5.28) for those states. The proof makes use of the asymptotic of observables as well as of Proposition 5.12.

During the preparation of this manuscript, *I.M. Sigal* and *A. Soffer* [314] have announced a proof of N-body asymptotic completeness under fairly general conditions.

6. Magnetic Fields

In this chapter, we discuss only a few aspects of Schrödinger operators with magnetic fields. We refer the reader to the review of *Hunziker* [172] for a more extensive survey and a more complete list of references.

The dimension (of the configuration space) here will always be $\nu = 2$ or $\nu = 3$. We will actually discuss two types of operators, i.e., the usual magnetic Hamiltonian (see Sect. 1.3), $H(\boldsymbol{a}, V) := (-\mathrm{i}\nabla + \boldsymbol{a})^2 + V$, which is defined on $\mathrm{H} := L^2(\mathbb{R}^\nu)$ and the Pauli Hamiltonian, which describes particles with spin (we only consider the spin-$\frac{1}{2}$ case [182, p. 249])

$$\tilde{H}(\boldsymbol{a}, V) := H(\boldsymbol{a}, V)\mathbb{1} + \boldsymbol{\sigma}\cdot\boldsymbol{B} \tag{6.1}$$

defined on $\mathrm{H} := L^2(\mathbb{R}^\nu) \otimes \mathbb{C}^2$, where $\mathbb{1}$ is the unit 2×2-matrix and $\boldsymbol{\sigma}$ is the matrix-vector $\boldsymbol{\sigma} := (\sigma_1, \sigma_2, \sigma_3)$, with

$$\sigma_1 := \begin{pmatrix} 0 & 1 \\ 1 & 0 \end{pmatrix}, \quad \sigma_2 := \begin{pmatrix} 0 & -\mathrm{i} \\ \mathrm{i} & 0 \end{pmatrix}, \quad \sigma_3 := \begin{pmatrix} 1 & 0 \\ 0 & -1 \end{pmatrix},$$

and $\boldsymbol{B}$ is the magnetic field associated with the vector potential $\boldsymbol{a}$, i.e. $\nabla \times \boldsymbol{a} = \boldsymbol{B}$.

Note that we always require assumptions on V and $\boldsymbol{a}$ which are stronger than the ones necessary for $H(\boldsymbol{a}, V)$ and $\tilde{H}(\boldsymbol{a}, V)$ to be essentially self-adjoint (on a suitable subspace; see Chap. 1). In the whole chapter, we disregard domain questions; the closure of the operators on C_0^∞ is always intended.

We will first discuss gauge invariance very briefly for smooth $\boldsymbol{a}$. Then we prove a result which says that if $\boldsymbol{B} \to 0$ at ∞ and V is short range in some sense, then both operators H and $\tilde{H}$ have essential spectrum $= [0, \infty)$. This is not a completely trivial result, because if $\boldsymbol{B} \to 0$ at ∞ only as $r^{-\alpha}[\alpha \in (0, 1)]$, then the corresponding $\boldsymbol{a}$ must go to ∞. But this does not change the essential spectrum.

We will use this fact to construct a rather striking example of a Hamiltonian with dense pure point spectrum, a type of spectrum which is known for random potential Hamiltonians (but there seems to be no connection).

In the third section, we give a set-up for supersymmetry (in a very restricted sense) and give some examples. Then we present a result of Aharonov and Casher which gives the number of zero energy eigenstates. This implies an index theorem, which we might understand as a physical example of the Atiyah-Singer index theorem on an unbounded space [57]. Section 6.5 contains a theorem of Iwatsuka yielding certain two-dimensional $H(\boldsymbol{a})$ with purely absolutely continuous spectrum. In the last section, we give an introduction to other phenomena of Schrödinger operators with magnetic fields.

Before we start with Sect. 6.1, we introduce a notation which allows us to write $\tilde{H}$ in a different, more convenient way. First, we look at the case of $v = 3$ dimensions. Consider the vector operator

$$\boldsymbol{O} := (O_1, O_2, O_3), \quad \text{where } O_j \text{ are operators in } L^2(\mathbb{R}^3) \quad (j \in \{1,2,3\}) .$$

Then we denote

$$\not{O} := \boldsymbol{O} \cdot \boldsymbol{\sigma} = \sum_{j=1}^{3} O_j \sigma_j .$$

Now we have

$$\sigma_k \sigma_l = \delta_{kl} \mathbb{1} + \mathrm{i} \sum_{m=1}^{3} \varepsilon_{klm} \sigma_m , \tag{6.2}$$

where ε_{klm} is the usual sign of permutations of (k, l, m), i.e., 0 if any two are equal, 1 for even, and -1 for odd permutations of $(1, 2, 3)$. Then using (6.2), an easy calculation yields

$$\not{O}^2 = \sum_{l=1}^{3} O_l^2 \mathbb{1} + \mathrm{i} \sum_{\substack{m=1 \\ (\varepsilon_{klm}=1)}}^{3} \sigma_m [O_k, O_l] .$$

If we use the relations

$$[(p_k - a_k), (p_l - a_l)] = -\mathrm{i} \left(\frac{\partial}{\partial x_k} a_l - \frac{\partial}{\partial x_l} a_k \right) ,$$

(note $p_k := -\mathrm{i}\partial/\partial x_k$) we get, in particular

$$(\not{p} - \not{a})^2 = (\boldsymbol{p} - \boldsymbol{a})^2 \mathbb{1} + \boldsymbol{\sigma} \cdot \boldsymbol{B} , \tag{6.3}$$

which gives

$$\tilde{H}(\boldsymbol{a}, V) = (\not{p} - \not{a})^2 + V \mathbb{1} \tag{6.4}$$

(we will drop the $\mathbb{1}$ in the following). The case of $v = 2$ dimensions can always be understood as a special case of the 3-dimensional one, in the sense that $\boldsymbol{B}$ has a constant direction, and one looks only at the motion projected on the orthogonal plane. Therefore, all the above relations are formally still true for the two-component vector operators. There are some simplifications, however, i.e.,

$$\nabla \times \boldsymbol{a} = \frac{\partial}{\partial x_1} a_2 - \frac{\partial}{\partial x_2} a_1 = B$$

is now a scalar field (since it has only one component) and (6.1) and (6.3) reduce to

$$\tilde{H}(\boldsymbol{a}, V) := H(\boldsymbol{a}, V)\mathbb{1} + \sigma_3 B \quad \text{and} \tag{6.1'}$$

$$(\not{p} - \not{a})^2 = (\boldsymbol{p} - \boldsymbol{a})^2\mathbb{1} + \sigma_3 B \ , \tag{6.3'}$$

respectively. A one-component B as in (6.3′) is a real simplification, since then

$$(\boldsymbol{p} - \boldsymbol{a})^2 + \sigma \cdot B + V$$

is just a direct sum of the pair of operators

$$(\boldsymbol{p} - \boldsymbol{a})^2 \pm B + V \ .$$

6.1 Gauge Invariance and the Essential Spectrum

If $\boldsymbol{a}$ is smooth, then gauge invariance is quite simple (there are subtleties, however, if $\boldsymbol{a}$ is nonsmooth), namely if we have two different vector potentials $\boldsymbol{a}_1$ and $\boldsymbol{a}_2$, where both of them are smooth and have the same curl, i.e., if $\boldsymbol{a}_1, \boldsymbol{a}_2 \in (C^\infty(\mathbb{R}^\nu))^\nu$ and

$$\nabla \times \boldsymbol{a}_1 = \nabla \times \boldsymbol{a}_2 = \boldsymbol{B} \quad \text{then}$$

$$\nabla \times (\boldsymbol{a}_1 - \boldsymbol{a}_2) = 0 \ ,$$

which means that there is a (gauge–) function, $\lambda \in C^\infty(\mathbb{R}^\nu)$, such that

$$\boldsymbol{a}_1 - \boldsymbol{a}_2 = \nabla\lambda \ .$$

This gives the gauge invariance of the "magnetic momentum"

$$\mathrm{e}^{\mathrm{i}\lambda}(-\mathrm{i}\nabla - \boldsymbol{a}_1)\mathrm{e}^{-\mathrm{i}\lambda} = (-\mathrm{i}\nabla - \boldsymbol{a}_2) \ ,$$

which implies

$$\mathrm{e}^{\mathrm{i}\lambda} H(\boldsymbol{a}_1, V)\mathrm{e}^{-\mathrm{i}\lambda} = H(\boldsymbol{a}_2, V) \ .$$

This expresses the important physical fact that "physics" depends only on B. So, if one has different vector potentials with the same B, then the operators are unitarily equivalent under a multiplication operator. Thus, not only spectral properties are the same, but also various other properties which are described by functions of the x-variable.

In the following, we will see that, in particular, the essential spectrum of the Hamiltonian is much more stable than general spectral properties. That is, we will show that the essential spectrum is always the positive real axis if $\boldsymbol{B}$ decays at ∞, provided V (depending on x) is not too weird. The first rigorous proof of this result is due to Miller (see *Miller* and *Simon* [244]). There are improvements for nonsmooth magnetic potentials due to *Leinfelder* [228].

The crucial idea of the proof is that one "adapts" the operator to a chosen

Weyl sequence by choosing a suitable gauge which transforms to a vector potential which is suitably small in the region where the Weyl vector lives. This is possible because $\boldsymbol{B} \to 0$ at ∞.

Theorem 6.1. Let $\nu = 2, 3$. Suppose $\boldsymbol{a} \in (C^\infty(\mathbb{R}^\nu))^\nu$ and $|\boldsymbol{B}(\boldsymbol{x})| \to 0$ as $|\boldsymbol{x}| \to \infty$, and suppose that V is H_0-compact. Then $\sigma_{\text{ess}}(H(\boldsymbol{a}, V)) = \sigma(\tilde{H}(\boldsymbol{a}, V)) = [0, \infty)$.

Proof:

Step 1. We reduce the problem to showing that $\sigma_{\text{ess}}(H(a, 0)) = [0, \infty)$. In the spinless case, we know by the diamagnetic inequality (1.8)

$$|\mathrm{e}^{-tH(\boldsymbol{a},0)}\varphi| \le \mathrm{e}^{-tH_0}|\varphi|, \quad t \in \mathbb{R}, \varphi \in \mathrm{H}$$

that

$$|V(H(\boldsymbol{a},0) + 1)^{-1}\varphi| \le |V|(H_0 + 1)^{-1}|\varphi| \tag{6.5}$$

(see [22, p. 851]). Thus, V is also $H(\boldsymbol{a}, 0)$-compact (see [285, 88]). Therefore, $H(\boldsymbol{a}, V)$ and $H(\boldsymbol{a}, 0)$ have the same essential spectrum [295, p. 113]. In the "spin"-case, $\tilde{H}(\boldsymbol{a}, 0)$, one uses the fact that $|\boldsymbol{\sigma} \cdot \boldsymbol{B}| \to 0$ as $|\boldsymbol{x}| \to \infty$. So $\boldsymbol{\sigma} \cdot \boldsymbol{B}$ is also an $H(\boldsymbol{a}, 0)$-compact perturbation.

Step 2. Since $H(\boldsymbol{a}, 0) \ge 0$, we know that $\sigma(H(\boldsymbol{a}, 0)) \subseteq [0, \infty)$.

Step 3. Weyl's theorem in a slightly strengthened version: Suppose A is self-adjoint and $A \ge 0$. If there is an orthonormal sequence, $\{\psi_n\}_{n \in \mathbb{N}} \subseteq \mathrm{H}$ such that $\psi_n \to 0$, $(n \to \infty)$ weakly and $\|(A + 1)^{-1}(A - z)\psi_n\| \to 0$, $(n \to \infty)$, then $z \in \sigma_{\text{ess}}(A)$. This can be seen by the spectral mapping theorem, i.e., $0 \in \sigma_{\text{ess}}((A + 1)^{-1}(A - z))$ if and only if $z \in \sigma_{\text{ess}}(A)$, and by applying the usual Weyl criterion to the operator $(A + 1)^{-1}(A - z)$ ([292, p. 237]).

Step 4. We show that there is a sequence $\{\boldsymbol{x}_n\}_{n \in \mathbb{N}} \subseteq \mathbb{R}^\nu$ such that $|\boldsymbol{x}_n| \to \infty$, $(n \to \infty)$ and a sequence of vector potentials, $\{\boldsymbol{a}_n\}_{n \in \mathbb{N}}$ with

$$\nabla \times \boldsymbol{a}_n = \boldsymbol{B} \quad (n \in \mathbb{N}) \quad \text{such that}$$

$$\sup_{|\boldsymbol{x} - \boldsymbol{x}_n| \le n} |\boldsymbol{a}_n(\boldsymbol{x})| \le cn^{-1} \quad (n \in \mathbb{N}) \tag{6.6}$$

for a suitable $c > 0$.

Proof. (Step 4). Let $n \in \mathbb{N}$, choose $\boldsymbol{x}_n \in \mathbb{R}^\nu$ such that

$$\sup_{|\boldsymbol{x} - \boldsymbol{x}_n| \le n} |\boldsymbol{B}(\boldsymbol{x})| \le cn^{-2},$$

for a suitable c which is possible due to the decay of $\boldsymbol{B}$. If $\nu = 2$, choose

$$\boldsymbol{a}_n(\boldsymbol{x}) := \left(0, \int_{x_n}^{x} B(t, y)\, dt\right),$$

where $\boldsymbol{x} = (x, y)$ and $\boldsymbol{x}_n = (x_n, y_n)$, and if $v = 3$, choose $\boldsymbol{a}_n(\boldsymbol{x}) = (t_n, s_n, 0)$ where

$$t_n(\boldsymbol{x}) := -\int_{y_n}^{y} B_3(x, t, 0)\, dt + \int_{z_n}^{z} B_2(x, y, s)\, ds$$

$$s_n(\boldsymbol{x}) := -\int_{z_n}^{z} B_1(x, y, t)\, dt \ ,$$

where $\boldsymbol{x} = (x, y, z)$ and $\boldsymbol{x}_n = (x_n, y_n, z_n)$. Now using $\nabla \cdot \boldsymbol{B} = 0$, one checks easily that $\nabla \times \boldsymbol{a}_n = \boldsymbol{B}$. Now (6.6) follows just by the estimate we assumed for $\boldsymbol{B}$ in the ball $\{|\boldsymbol{x} - \boldsymbol{x}_n| \leq n\}$.

Step 5. Given $\boldsymbol{k} \in \mathbb{R}^v$, we construct a sequence $\{\psi_n\}_{n \in \mathbb{N}} \subseteq \mathrm{H}$, for which Step 3 applies, i.e.,

$$\|(H(\boldsymbol{a}) + 1)^{-1}(H(\boldsymbol{a}) - \boldsymbol{k}^2)\psi_n\| \to 0, \quad (n \to \infty) \ . \tag{6.7}$$

First, note that since for any $n \in \mathbb{N}$, $\nabla \times (\boldsymbol{a} - \boldsymbol{a}_n) = 0$, there exists a gauge function $\lambda_n \in C^1(\mathbb{R}^v)$ such that

$$H(\boldsymbol{a}, 0) = e^{i\lambda_n} H(\boldsymbol{a}_n, 0) e^{-i\lambda_n} \ .$$

Now select a subsequence from $\{\boldsymbol{x}_n\}$ in Step 4 (also denoted by $\{\boldsymbol{x}_n\}$) such that $|\boldsymbol{x}_n - \boldsymbol{x}_{n-1}| > 2n$. Choose a $g \in C_0^\infty(\mathbb{R}^v)$ such that

$$g(\boldsymbol{x}) = \begin{cases} 1 & \text{if } |\boldsymbol{x}| < \frac{1}{2} \\ 0 & \text{if } |\boldsymbol{x}| > 1 \end{cases} \qquad \boldsymbol{x} \in \mathbb{R}^v$$

with $\|g\|_\infty = 1$, denote

$$g_n(\boldsymbol{x}) := \alpha_n g\left(\frac{1}{n}(\boldsymbol{x} - \boldsymbol{x}_n)\right) ,$$

where α_n is chosen so $\|g_n\|_2 = 1$ and

$$\psi_n(\boldsymbol{x}) := e^{i\lambda_n(\boldsymbol{x})} e^{i\boldsymbol{k}\cdot\boldsymbol{x}} g_n(\boldsymbol{x}) \ .$$

Then $\{\psi_n\}_{n \in \mathbb{N}}$ is obviously an orthonormal sequence in $L^2(\mathbb{R}^v)$ which goes weakly to 0. Thus, we have only to show (6.7). Note that

$$\|[H(\boldsymbol{a}) + 1]^{-1}[H(\boldsymbol{a}) - \boldsymbol{k}^2]\psi_n\| = \|[H(\boldsymbol{a}_n) + 1]^{-1}[H(\boldsymbol{a}_n) - \boldsymbol{k}^2]\phi_n\| \ ,$$

where $\phi_n(\boldsymbol{x}) := \exp(i\boldsymbol{k} \cdot \boldsymbol{x}) g_n(\boldsymbol{x})$. Now we insert the identity

$$H(\boldsymbol{a}_n) = H_0 + 2\boldsymbol{a}_n^2 + (i\nabla - \boldsymbol{a}_n) \cdot \boldsymbol{a}_n + i\boldsymbol{a}_n \cdot \nabla$$

and use

$$(H_0 - \boldsymbol{k}^2)\phi_n(\boldsymbol{x}) = e^{i\boldsymbol{k}\cdot\boldsymbol{x}} H_0 g_n(\boldsymbol{x}) + 2i e^{i\boldsymbol{k}\cdot\boldsymbol{x}} \boldsymbol{k} \cdot \nabla g_n(\boldsymbol{x})$$

to get

$$\|[H(\boldsymbol{a}_n)+1]^{-1}[H(\boldsymbol{a}_n)-\boldsymbol{k}^2]\phi_n\| \le \|H_0 g_n\| + 2|\boldsymbol{k}|\,\|\nabla g_n\| + \|[H(\boldsymbol{a}_n)+1]^{-1}(-i\nabla-\boldsymbol{a}_n)\|\,\|\boldsymbol{a}_n g_n\| + 2\|(\boldsymbol{a}_n^2+\boldsymbol{k}\boldsymbol{a}_n)g_n\| + \|\boldsymbol{a}_n\cdot\nabla g_n\| \ .$$

The first two terms on the R.H.S. above go to 0 as $n\to\infty$, since $\|\nabla g_n\|\le c1/n$ and $\|\Delta g_n\|\le c1/n^2$ for suitable $c>0$, and the last three terms go to 0 because of (7.6) in Step 4. Note that $[H(\boldsymbol{a}_n)+1]^{-1}(-i\nabla-\boldsymbol{a}_n)$ is a vector of bounded operators.

Thus, we have shown that, for any $\boldsymbol{k}^2\in\mathbb{R}^+$, we have $\boldsymbol{k}^2\in\sigma_{\text{ess}}(H(\boldsymbol{a},0))$ by Step 3, i.e., $[0,\infty)\subseteq\sigma_{\text{ess}}(H(\boldsymbol{a},0))$, which concludes the proof of the theorem. □

6.2 A Schrödinger Operator with Dense Point Spectrum

We will give next an example of a magnetic Schrödinger operator which has rather surprising spectral properties. Depending on the decay of the magnetic field B we can show that the operator has purely absolutely continuous spectrum, dense point spectrum in $[0,\infty)$, or it has a "mobility edge", i.e., there is a $d>0$ such that the spectrum is a dense point spectrum in $[0,d]$ and absolutely continuous in $[d,\infty)$.

Theorem 6.2 (Miller and Simon [244]). Let $v=2$, and consider the Hamiltonian

$$H := \left(p_x - c\frac{y}{(1+\rho)^\gamma}\right)^2 + \left(p_y + c\frac{x}{(1+\rho)^\gamma}\right)^2 ,$$

where $c>0$, $\rho := |\boldsymbol{x}| = (x^2+y^2)^{1/2}$ and $\gamma\in(0,\infty)$. Then we have the following cases:

(a) if $\gamma>1$, then $\sigma(H)=\sigma_{ac}(H)=[0,\infty)$, $\sigma_{pp}=\sigma_s=\phi$,
(b) if $\gamma<1$, then $\sigma(H)=\overline{\sigma_{pp}(H)}=[0,\infty)$, $\sigma_{ac}=\sigma_s=\phi$,
(c) if $\gamma=1$, then $\overline{\sigma_{pp}(H)}=[0,c^2]$, $\sigma_{ac}(H)=[c^2,\infty)$, and $\sigma_{sc}(H)=\phi$, i.e. c^2 is a "mobility edge."

Proof. First of all, we remark that

$$B(\boldsymbol{x}) = \frac{\partial}{\partial x}a_y - \frac{\partial}{\partial y}a_x = \frac{2c}{(1+\rho)^\gamma} + \frac{\rho c}{(1+\rho)^{\gamma+1}} ,$$

where

$$\boldsymbol{a} = (a_x, a_y) = \left(-\frac{cy}{(1+\rho)^\gamma}, \frac{cx}{(1+\rho)^\gamma}\right) .$$

Thus, by Theorem 6.1, we have

$$\sigma(H) = \sigma_{\text{ess}}(H) = [0, \infty) \ . \tag{6.8}$$

Now, if we expand H, we get

$$H = -\Delta + \frac{c^2\rho^2}{(1+\rho)^{2\gamma}} + \frac{2c}{(1+\rho)^\gamma} L_z \ ,$$

where $L_z := -\mathrm{i}(x(\partial/\partial y) - y(\partial/\partial x))$ is the operator of the angular momentum (pointing into the "z-direction"). This means that H commutes with rotations (in the plane $\mathbb{R}^2$), since all terms are rotational invariant. We can express this by

$$[L_z, H] = 0 \ .$$

Therefore, we can write the Hilbert space $\mathrm{H} = L^2(\mathbb{R}^2)$ as a direct sum of eigenspaces of L_z. L_z has the eigenvalues $m \in \{0, \pm 1, \pm 2, \ldots\}$ (see [182, p. 231]), and if we restrict H to the eigenspace with eigenvalue m, i.e., $\{\varphi \in \mathrm{H} \,|\, L_z\varphi = m\}$, we get

$$H_m := H \restriction (L_z = m) = -\Delta + \frac{c^2\rho^2}{(1+\rho)^{2\gamma}} + \frac{2mc}{(1+\rho)^\gamma} \ .$$

But this is just a Schrödinger operator without magnetic field with the potential

$$V_m := \frac{c^2\rho^2}{(1+\rho)^{2\gamma}} + \frac{2mc}{(1+\rho)^\gamma} \ .$$

Now we consider

Case (a). If $\gamma > 1$, then V_m is a short-range potential, and it is well known [305, 226] that H_m only has a.c. spectrum, and by (6.8), we have the assertion of case (a).

Case (b). If $\gamma < 1$, then $V_m(\boldsymbol{x}) \to \infty$ as $|\boldsymbol{x}| \to \infty$. So H_m, and therefore also H, has pure point spectrum [295, Theorem XIII.16]. But by (6.8), this must be necessarily dense in $[0, \infty)$.

Case (c). If $\gamma = 1$, then $V_m - c^2$ is a long-range potential going to 0 as $|\boldsymbol{x}| \to \infty$. Thus, H_m (and therefore H) has absolutely continuous spectrum in $[c^2, \infty)$ and pure point spectrum in $[0, c^2]$, which must be dense because of (6.8). □

Remark. We note that in the case $\gamma = 0$, we have a constant magnetic field (orthogonal to the (x, y)-plane). It is a classical result that one has point spectrum in this case. The eigenfunctions correspond to closed orbits, the so-called Landau orbits (see [224, Sect. 111]).

6.3 Supersymmetry (in 0-Space Dimensions)

We discuss now a simple abstract set-up which has, at first sight, nothing to do with Schrödinger operators (it is actually more related to high energy physics).

But we will see that there are quite interesting Schrödinger operator examples having this structure.

The structure is a specialization of supersymmetry from field theory contexts. There one introduces operators $Q_{\mu a \alpha}$, where μ is a "vector index", a a "spin index", and α an internal index. The Q's transform under Lorentz transformation by

$$U(\Lambda)Q_{\mu a\alpha}U(\Lambda)^{-1} = \sum_{\nu,b} \Lambda_{\nu\mu}A(\Lambda)_{ab}Q_{\nu b\alpha} ,$$

where $A(\Lambda)$ is a spinor representation of the Lorentz group, and obey "commutation" relations

$$\{Q_{\mu a\alpha}, Q_{\nu b\beta}\} = 2g_{\mu\nu}\Delta_{ab}\delta_{\alpha\beta}P_\nu ,$$

where $\{A, B\} := AB + BA$, g is the metric tensor, Δ a bi-linear form in spinor indices transforming as a scalar, and P_ν is the four momentum.

This extension of Lorentz invariance is of especial interest since the Q's link Bose and Fermi states. There is an especially attractive supersymmetry version of gravity called $N = 8$ supergravity. These are areas of intense current interest, although they have no experimental verification. See [369] for further discussion.

Below we specialize to zero-space dimensions. In terms of the operators P, Q discussed below, the operators $Q_1 = Q$, $Q_2 = iQP$ obey

$$Q_\alpha^* = Q_\alpha, \quad \{Q_\alpha, Q_\beta\} = 2\delta_{\alpha\beta}H .$$

This picture of zero-dimensional supersymmetry was emphasized especially by *Witten* [370]; see Chap. 11.

Consider the Hilbert space H and let H and Q be self-adjoint, and P be a bounded self-adjoint operator in H such that

$$H = Q^2 \geq 0 , \tag{6.9a}$$

$$P^2 = 1 \tag{6.9b}$$

$$\{Q, P\} := QP + PQ = 0 . \tag{6.9c}$$

Then we say the system (H, P, Q) has *supersymmetry*. Since P is self-adjoint and $P^2 = 1$, it only has the eigenvalues $1, -1$. We denote the associated eigenspaces by

$$\mathrm{H}_\mathrm{f} := \{\varphi \in \mathrm{H} | P\varphi = -\varphi\}$$

$$\mathrm{H}_\mathrm{b} := \{\varphi \in \mathrm{H} | P\varphi = \varphi\}$$

and we have the decomposition

$$\mathrm{H} = \mathrm{H}_\mathrm{f} \oplus \mathrm{H}_\mathrm{b} .$$

We call the vectors in H_f the *fermionic* states, and the vectors in H_b the *bosonic* states. Using this decomposition, we can write

$$P = \begin{pmatrix} \mathbb{1}_b & 0 \\ 0 & -\mathbb{1}_f \end{pmatrix} ,$$

where $\mathbb{1}_b$ and $\mathbb{1}_f$ are the unit operators in H_b and H_f, respectively, but in the following, we drop $\mathbb{1}_f$ and $\mathbb{1}_b$ and write

$$P = \begin{pmatrix} 1 & 0 \\ 0 & -1 \end{pmatrix} .$$

Since P and Q anti-commute and Q is self-adjoint, Q has always the form

$$Q = \begin{pmatrix} 0 & A^* \\ A & 0 \end{pmatrix} , \tag{6.10}$$

where A is an operator which maps H_b into H_f, and its adjoint A^* maps H_f into H_b. This implies that Q "flips fermionic and bosonic states." i.e.,

$$Q: H_f \to H_b \quad \text{and} \quad Q: H_b \to H_f \ .$$

Remark. As we will see in Example 1a below, we might call A the "annihilation operator" and A^* the "creation operator," concepts which are well-known for the harmonic oscillator [292, p. 142; 182, p. 211]. From (6.9) and the above, we have the representation

$$H = \begin{pmatrix} A^*A & 0 \\ 0 & AA^* \end{pmatrix} . \tag{6.11}$$

Thus, P commutes with H, and H_f and H_b are invariant under H.

We now define a "supersymmetric" index of H which has some remarkable stability properties.

Definition:

$$\operatorname{ind}_s(H) := \dim(\operatorname{Ker}(H \upharpoonright H_b)) - \dim(\operatorname{Ker}(H \upharpoonright H_f)) \ .$$

Remark. ind_s should not be mixed up with the usual index one defines for a semi-Fredholm operator F, i.e.,

$$\operatorname{ind}(F) := \dim(\operatorname{Ker} F) - \dim(\operatorname{Ker} F^*) \ ,$$

(see [196, p. 230]). But there is a connection between these two concepts, i.e.,

$$\operatorname{ind}_s(H) = \operatorname{ind}(A) \ ,$$

where A is the "annihilation" operator defined by (6.10). Now we show a fundamental property of supersymmetric systems, which says that non-zero eigenvalues have the same number of bosonic and fermionic eigenstates.

Theorem 6.3. If the system (H, P, Q) has supersymmetry, then for any bounded open set $\Omega \subseteq (0, \infty)$ we have

$$\dim(E_\Omega(H) \restriction \mathrm{H}_\mathrm{b}) = \dim(E_\Omega(H) \restriction \mathrm{H}_\mathrm{f}) ,$$

where $E_\Omega(H)$ is the spectral projector of H on Ω.

Proof. Denote by $P_\pm$, the projectors on H_b and H_f respectively, and by

$$E_\Omega^\pm := E_\Omega(H) P_\pm .$$

Note by the above discussion that we know that P, and therefore also $P_\pm$, commutes with $E_\Omega(H)$. Since Q anti-commutes with P, and Q is bounded on $E_\Omega(H)\mathrm{H}$, we get

$$Q E_\Omega^\pm = E_\Omega^\mp Q . \tag{6.12}$$

Now, because $0 \notin \Omega$, Q is invertible on $E_\Omega(H)\mathrm{H}$, (6.12) implies that

$$\dim(E_\Omega^+) = \dim(E_\Omega^-) .$$

But $\dim(E_\Omega(H) \restriction \mathrm{H}_\mathrm{b}) = \dim(E_\Omega^+)$, and $\dim(E_\Omega(H) \restriction \mathrm{H}_\mathrm{f}) = \dim(E_\Omega^-)$, and therefore we have the theorem. □

Now we discuss some examples.

Example 0. Laplace-Beltrami operator on forms on compact manifolds. This will be discussed in Chap. 11.

Example 1 (*Deift* [77]). Let $\mathrm{H} = L^2(\mathbb{R}) \otimes \mathbb{C}^2$, and q be a polynomial in x.

Set $A := (d/dx) + q(x)$, and thus (on the same domain), $A^* = -d/dx + q(x)$. Then we have, with (6.10),

$$Q = \begin{pmatrix} 0 & -\dfrac{d}{dx} + q \\ \dfrac{d}{dx} + q & 0 \end{pmatrix} \quad \text{and} \quad P = \begin{pmatrix} 1 & 0 \\ 0 & -1 \end{pmatrix} .$$

Note that (on suitable domains)

$$A^* A = -\frac{d^2}{dx^2} + q^2(x) - q'(x) \quad \text{and}$$

$$A A^* = -\frac{d^2}{dx^2} + q^2(x) + q'(x) .$$

Thus,

$$H := Q^2 = \begin{pmatrix} -\dfrac{d^2}{dx^2} + q^2 - q' & 0 \\ 0 & -\dfrac{d^2}{dx^2} + q^2 + q' \end{pmatrix} ,$$

and the system (H, P, Q) is supersymmetric.

Example 1a (Harmonic Oscillator). Set $q(x) = x$ in Example 1. Then $A^*A = -(d^2/dx^2) + q^2(x) - 1$ is the (shifted) harmonic oscillator. A is known as the "annihilation" operator and A^* as the "creation" operator [182, p. 211]. We have

$$AA^* = A^*A + 2 \tag{6.13}$$

and, therefore

$$\dim(E_\Delta(A^*A)) = \dim(E_{\Delta-2}(AA^*)) \quad \text{for } \Delta \subseteq (0, \infty) . \tag{6.14}$$

Using the fact that $\sigma(H) = \sigma(A^*A) \cup \sigma(AA^*)$, we can almost read off the spectrum H: Since $A^*A \geq 0$, we know by (6.13) that $AA^* \geq 2$, thus $\sigma(AA^*) \geq 2$. By (6.14), we know there is no spectrum of H in $(0, 2)$, and therefore no spectrum in $(2n, 2(n+1))$, $n \in \mathbb{N}$. Thus, H can only have spectrum in the set $\{2, 4, 6, \ldots\}$.

Example 1b (*Herbst* and *Simon* [158]). Set $q(x) = x + gx^2$ in Example 1, where $g \in [0, \infty)$ is a coupling constant. Then

$$q^2(x) \pm q'(x) = x^2(1 + gx)^2 \mp (1 + 2gx) ,$$

which means that A^*A and AA^* have almost the same potentials. They are actually equivalent by "parity," i.e.,

$$A^*A = U(AA^*)U^{-1} , \quad \text{where}$$

$$U\varphi(x) := \varphi\left(-x - \frac{1}{g}\right), \quad \varphi \in L^2(\mathbb{R}) .$$

If $g \neq 0$, then $\mathrm{ind}_s(H) = 0$, since $H \restriction \mathrm{H}_f$ and $H \restriction \mathrm{H}_b$ are unitarily equivalent. This means, in physical terms, that the only way to get new eigenvalues at 0 is for a pair of eigenvalues, a fermion one and a boson one to come down. But if $g = 0$, $\mathrm{ind}_s(H) = 1$, since $H \restriction \mathrm{H}_f = AA^* \geq 2$, $H \restriction \mathrm{H}_b = A^*A$ and $\dim(\mathrm{Ker}(A^*A)) = 1$. Physically this can be understood as the fermionic eigenstate being localized farther and farther out, and vanishing eventually if $g \to 0$. This example was invented in [158], because the ground state eigenvalue goes to zero exponentially in g^{-2} as $g \to 0$.

Our last example states a supersymmetry result for magnetic fields in two dimensions. We state it as a theorem.

Theorem 6.4. Let $v = 2$ and $\boldsymbol{a} \in (C^1(\mathbb{R}^2))^2$, such that $B = \nabla \times \boldsymbol{a}$. Then $(\boldsymbol{p} - \boldsymbol{a})^2 + B$ and $(\boldsymbol{p} - \boldsymbol{a})^2 - B$ have the same spectrum except perhaps at 0.

Proof. Let $\boldsymbol{a} = (a_1, a_2)$, $\boldsymbol{\sigma} = (\boldsymbol{\sigma}_1, \boldsymbol{\sigma}_2)$, and choose

$$Q = \not{p} - \not{a} = \sum_{i=1}^{2} (p_i - a_i)\sigma_i$$

$$P = \begin{pmatrix} 1 & 0 \\ 0 & -1 \end{pmatrix}.$$

Then, by (6.3′)

$$H = \tilde{H}(\boldsymbol{a}, 0) = Q^2 = \begin{pmatrix} (\boldsymbol{p} - \boldsymbol{a})^2 + B & 0 \\ 0 & (\boldsymbol{p} - \boldsymbol{a})^2 - B \end{pmatrix}.$$

Since $P = \boldsymbol{\sigma}_3$, we have $\{Q, P\} = 0$. Thus, the system $(\tilde{H}(\boldsymbol{a}, 0), P, Q)$ is supersymmetric, which implies, by Theorem 6.3, that $H \restriction \mathrm{H_b} = (\boldsymbol{p} - \boldsymbol{a})^2 + B$ and $H \restriction \mathrm{H_f} = (p - \boldsymbol{a})^2 - B$ have the same spectrum except at 0. □

We note that this is a result which is not true in 3 dimensions with non-constant $\boldsymbol{B}$.

6.4 The Aharonov-Casher Result on Zero Energy Eigenstates

It is an almost classical piece of folklore that the Hamiltonians of constant magnetic fields, restricted to a finite region in the plane orthogonal to the field, have eigenvalues with finite degeneracy (see, for example, [224, Sect. 111]). If the field is extended to the whole plane, there is an infinite degeneracy. This is also true for some non-constant fields (see [26] for rigorous arguments in the case of polynomial fields).

In this section, we will discuss two-dimensional Pauli-Hamiltonians with non-constant B's. We know from Theorem 6.4 that the two components of $\tilde{H}(\boldsymbol{a}, 0)$, i.e. $(p - \boldsymbol{a})^2 + B$ and $(p - \boldsymbol{a})^2 - B$ have the same spectrum except at 0. Here we will discuss a result due to *Aharonov* and *Casher* [6], which states that the number of zero-eigenstates is equal to the integral part of the magnetic flux. It says also that there is no supersymmetry at zero energy, i.e., depending on the sign of the flux, there are only bosonic or only fermionic zero-eigenstates. It is essentially a physical example of the Atiyah-Singer index theorem saying that $\mathrm{ind}_s(\tilde{H}(\boldsymbol{a}, 0))$ depends only on the flux. We will prove the theorem only for bounded B's with compact support which are just convenient conditions. There are more general results, however (see [243]).

Theorem 6.5. Let $v = 2$, B be bounded with compact support, and $\boldsymbol{a} = (a_x, a_y)$ be a suitable vector field associated with B. Define $\{y\}$ for $y > 0$, to be the largest integer *strictly* less than y and $\{0\} = 0$. Let $\phi_0 := \int B(\boldsymbol{x})\, d\boldsymbol{x} \geq 0$, the flux of B. Then the operator $\tilde{H}(\boldsymbol{a}, 0) = (\not{p} - \not{a})^2$ has exactly $\{\phi_0/2\pi\}$ eigenvectors with eigenvalue 0, all with $\sigma_3 = -1$. The index is only dependent on the flux, i.e., $\operatorname{ind}_s(\tilde{H}) = (\operatorname{sign}\phi_0)\{|\phi_0|/2\pi\}$. If $\phi_0 \leq 0$, there are the same number of zero-energy eigenstates, but with $\sigma_3 = 1$.

Proof. Consider the auxiliary potential

$$\phi(\boldsymbol{x}) := \frac{1}{2\pi}\int \ln(|\boldsymbol{x} - \boldsymbol{x}'|)B(\boldsymbol{x}')\,d^2\boldsymbol{x}' \ , \quad \text{then}$$

$$\phi(\boldsymbol{x}) = \frac{\phi_0}{2\pi}\ln|\boldsymbol{x}| + \mathrm{O}(|\boldsymbol{x}|^{-1}) \quad \text{as } |\boldsymbol{x}| \to \infty \ . \tag{6.15}$$

The reason for this choice is that $-\Delta\phi = B$. So if we choose

$$\boldsymbol{a} = (a_x, a_y) := \left(\frac{\partial\phi}{\partial y}, -\frac{\partial\phi}{\partial x}\right) , \quad \text{we have}$$

$$\nabla \times \boldsymbol{a} = -\Delta\phi = B \ ,$$

i.e., $\boldsymbol{a}$ is a vector potential associated with B. We are interested in solutions of

$$(\not{p} - \not{a})^2\psi = 0, \quad \psi = \begin{pmatrix}\psi_+ \\ \psi_-\end{pmatrix} \in L^2(\mathbb{R}^2) \otimes \mathbb{C}^2 \ . \tag{6.16}$$

Since $(\not{p} - \not{a})^2$ is a square of a single self-adjoint operator, (6.16) is equivalent to

$$(\not{p} - \not{a})\psi = 0 \ .$$

But this means (see the introduction at the beginning of this chapter)

$$\begin{pmatrix} 0 & (p_x - a_x) - \mathrm{i}(p_y - a_y) \\ (p_x - a_x) + \mathrm{i}(p_y - a_y) & 0 \end{pmatrix}\begin{pmatrix}\psi_+ \\ \psi_-\end{pmatrix} = 0 \ ,$$

and this is equivalent to the equations

$$\left[\frac{\partial}{\partial x} - \mathrm{i}\frac{\partial}{\partial y}\right]\mathrm{e}^{\phi(x,y)}\psi_-(x,y) = 0 \tag{6.17a}$$

$$\left[\frac{\partial}{\partial x} + \mathrm{i}\frac{\partial}{\partial y}\right]\mathrm{e}^{-\phi(x,y)}\psi_+(x,y) = 0 \ . \tag{6.17b}$$

Equations (6.17a, b) mean that $f_- := \mathrm{e}^{\phi}\psi_-$ is analytic in $\bar{z} := x - \mathrm{i}y$, and $f_+ := \mathrm{e}^{-\phi}\psi_+$ is analytic in $z := x + \mathrm{i}y$, respectively, since they are equivalent to the Cauchy-Riemann equations.

Now assume $\phi_0 \geq 0$. Then (6.15) implies that

$$e^{-\phi(z)} = |z|^{-\phi_0/2\pi}\left[1 + \mathrm{O}\left(\frac{1}{|z|}\right)\right] \quad \text{for } |z| \to \infty \ .$$

Since ψ_+ is in $L^2(\mathbb{R}^2)$, f_+ is also in $L^2(\mathbb{R}^2)$ and it is analytic. But there are no analytic functions in L^2, so ψ_+ must be 0. For $\psi_- = \mathrm{e}^{-\phi} f_-$ to be in $L^2(\mathbb{R}^2)$, f_- must increase no faster than a polynomial. But since it is analytic, it must be a polynomial in $\bar{z}$ of degree $u < (\phi_0/2\pi) - 1$. Since there are just $\{\phi_0/2\pi\}$ linearly independent polynomials of this type, we have exactly $\{\phi_0/2\pi\}$ "spin-down" eigenstates (i.e. with $\sigma_3 = -1$) with zero energy. It is obvious from the above that if $\phi_0 \leq 0$, then there are exactly $\{|\phi_0/2\pi\}$ "spin-up" (i.e. with $\sigma_3 = 1$) eigenstates and no spin-down ones. Thus, in both cases, the supersymmetric index of $\tilde{H}$ is $(\operatorname{sign} \phi_0)\{|\phi_0|/2\pi\}$. □

We want to give a second proof, due to *Avron-Tomaras* [32], of the Casher-Aharonov theorem, or more precisely, its analog when R^2 is replaced by a compact manifold like S^2, the two-dimensional sphere. We present the proof in part because it foreshadows our proof of the Gauss-Bonnet-Chern theorem in Chap. 12. We will not present the proofs of the technicalities here, but leave it as an exercise to the reader to use the machinery in Sect. 12.6 to provide the steps which we skip.

We must begin by explaining what one means by a Hamiltonian with magnetic field on S^2 and why the total flux is quantized. We begin with $H(\boldsymbol{a})$. One might decide that $\boldsymbol{a}$ should be defined globally, i.e., on all of S^2. If one demands that, then $B = \mathrm{d}a$ (a is a one-form, B a two form) has zero integral (i.e. zero flux) by Stokes' theorem. Instead, we use the idea that one will need distinct gauges in distinct coordinate patches, and imagine that a vector potential is actually a pair a_+, a_- of one-forms with a_+ defined on $S^2_+ = S^2 \backslash \{s\}$ and a_- on $S^2_- = S^2 \backslash \{n\}$ where s (resp. n) are the south (resp. north) pole. We want $da_+ = da_- = B$ on the cylinder, $C := S^2_+ \cap S^2_- = \{(\theta, \phi) | 0 < \theta < \pi, \phi \in S^1$, the circle$\}$ in poles coordinate where B is globally defined. Similarly, we want our wave functions to be a pair (φ_+, φ_-) of smooth functions on $S^2_\pm$. The two φ's must be related by a gauge transformation

$$\varphi_+ = \mathrm{e}^{\mathrm{i}\lambda} \varphi_- \;. \tag{6.18}$$

Because C is a cylinder, only $\exp(\mathrm{i}\lambda)$ is smooth, and λ may not be smooth on C, but we can extend λ to $(0, \pi) \times R$ so that

$$\lambda(\theta, \varphi + 2\pi) = \lambda(\theta, \varphi) + 2\pi n \tag{6.19}$$

for some *integer n*. We will set $H = \pi_\pm^2$ on $\varphi_\pm$ where $\pi_\pm = (-i\nabla - a_\pm)$, so we want

$$\pi_+ \varphi_+ = \mathrm{e}^{\mathrm{i}\lambda} \pi_- \varphi_- \;, \tag{6.20}$$

if $\varphi_\pm$ are related by (6.18). A simple calculation shows that (6.20) is equivalent to

$$a_+ - a_- = d\lambda \tag{6.21}$$

on C. Then quantization condition (6.19) then reads:

$$\frac{1}{2\pi}\int_0^{2\pi} d\phi\, \hat{e}_\phi \cdot (a_+ - a_-) = n \ . \tag{6.22}$$

But by Stokes' theorem

$$\int_0^{2\pi} d\phi\, \hat{e}_\phi \cdot a_\pm(\theta, \phi) = \pm \int_{S^\pm(\theta)} B \ ,$$

where $S^\pm(\theta_0) = \xi(\theta, \phi) | \phi \in S^1, \pm\theta_0 \geq 0\}$ and, in particular,

$$\frac{1}{2\pi}\int_{S^2} B = n \tag{6.23}$$

by (6.22). Thus, by this formalism, we succeed in doing quantum theory on S^2 in a magnetic field so long as the total flux is $2\pi n$.

The above formalism has a natural geometric meaning: $\varphi_\pm$ being related by (6.18) really says that φ is a section of a complex line bundle over S^2; $\exp(i\lambda)$ acts as a transfer function for going from one trivial bundle obtained by restricting to S^2_+ and the other obtained by restricting to S^2_-. The symbol $a_\pm$ defines a connection in the bundle, B is its curvature and n is the Chern integer of the bundle. For additional information on this point of view, see *Grümm* [144].

Now let $\{a_\pm\}$ be a vector potential with flux $2\pi n$, and define $H_0(a)(\varphi_+, \varphi_-) = (\pi_+^2 \varphi_+, \pi_-^2 \varphi_-)$. Now view φ_+, φ_- as having values in $\mathbb{C}^2$ rather than $\mathbb{C}$ [still related by (6.18)] and σ_3 acts on (φ_+, φ_-) by multiplying by $\begin{pmatrix} 1 & 0 \\ 0 & -1 \end{pmatrix}$. (Note: Do not confuse φ_+, φ_- with the two components of each of φ_+ and φ_-; σ_3 acts on the components of φ_+ and φ_-.) Then σ_3 commutes with $\tilde{H} = H - \sigma_3 B$. Moreover, one can write $\tilde{H} = \not{\pi}^2$ with $\{\not{\pi}, \sigma_3\} = 0$ so, as usual, nonzero eigenvalues of $\tilde{H}$ have equal multiplicities on $\{\varphi | \sigma_3 \varphi = \varphi\}$ and $\{\varphi | \sigma_3 \varphi = -\varphi\}$. Let $\mathrm{ind}_s(\tilde{H})$ be $\dim\{\varphi | \tilde{H}\varphi = 0, \sigma_3\varphi = \varphi\} - \dim\{\varphi | \tilde{H}\varphi = 0, \sigma_3\varphi = -\varphi\}$. The analog of the Aharonov-Casher theorem is

$$\mathrm{ind}_s(\tilde{H}) = \frac{1}{2\pi}\int_{S^2} B \ . \tag{6.24}$$

Here is a sketch of the Avron-Tomares proof of (6.24): We first claim that

$$\mathrm{ind}_s(\tilde{H}) = \mathrm{Tr}(\sigma_3 e^{-\beta\tilde{H}}) \tag{6.25}$$

for any β, since the supersymmetry provides a cancellation of the contribution of nonzero eigenvalues. The integral kernel $\exp(-\beta\tilde{H})(x, x)$ defines a map on $\mathbb{C}^2$ (actually on the fiber of a two-dimensional vector bundle), and we let tr_x denote the trace over this space. Then (6.25) becomes

$$\mathrm{ind}(\tilde{H}) = \int dx\, \mathrm{tr}(\sigma_3 e^{-\beta\tilde{H}}(x, x)) \ .$$

If B were constant, $\exp(-\beta\tilde{H}) = \exp(-\beta H)\exp(+\beta B\sigma_3)$; but for small β, path integral intuition suggests that the leading order should be the same—the proof of this is precisely where one needs the machinery of Sect. 12.6. Thus, one expects

$$\operatorname{ind}(\tilde{H}) = \int dx \lim_{\beta\downarrow 0} [\operatorname{tr}(\sigma_3 e^{+\beta B(x)\sigma_3}) e^{-\beta H}(x,x)] \ . \tag{6.26}$$

But $\operatorname{tr}\{\sigma_3 \exp[+\beta B(x)\sigma_3]\} = \beta B(x) + O(\beta^3)$ since $\operatorname{tr}(\sigma_3) = 0$ and $\exp(-\beta H)\cdot(x,x) = (2\pi\beta)^{-1}[1 + O(\beta)]$, so (6.26) becomes

$$\operatorname{ind}(\hat{H}) = \int dx (2\pi\beta)^{-1} \beta B(x) = (2\pi)^{-1} \int B(x)\,dx \ .$$

6.5 A Theorem of Iwatsuka

Two-dimensional purely magnetic Hamiltonians can have a wide variety of spectral properties. We saw an example with dense point spectrum in Sect. 6.2, the case $B(\boldsymbol{x}) = B_0$ has isolated point spectrum of infinite multiplicity, and if $B(\boldsymbol{x}) \to \infty$ as $|\boldsymbol{x}| \to \infty$, one can prove that $H(\boldsymbol{a})$ has a compact resolvent. Recently, *Iwatsuka* [181] has proven:

Theorem 6.6 (Iwatsuka). Suppose that $B(\boldsymbol{x}) \equiv B(x, y)$ is a function $b(x)$ of x alone, and that $\lim_{x\to\pm\infty} b(x) = b_\pm$ exist with both $b_\pm$ nonzero and unequal. Then $H(\boldsymbol{a})$ has purely absolutely continuous spectrum.

Remarks. (1) Iwatsuka deals with more general b's. It is a reasonable conjecture that, so long as b is not a.e. constant, then $H(\boldsymbol{a})$ has purely absolutely continuous spectrum.

(2) To get some feel for why this is true, consider the case $b(x) = b_\pm$ if $\pm x > 0$. Then classically there are orbits which are a succession of semicircles with diameters along $x = 0$; these wander off to infinity if $b_+ \neq b_-$; see Fig. 6.1.

Proof. Work in a gauge $\boldsymbol{a} = (0, a(x))$ where $a(x) = \int_0^x b(\zeta)\,d\zeta$. By passing to a Fourier transform in the y variable, $H(\boldsymbol{a})$ is a direct integral of the one-dimensional operators

$$h(k) = \frac{-d^2}{dx^2} + [k - a(x)]^2 \ .$$

See Sect. XIII.16 of *Reed* and *Simon* [295] for a discussion of such direct integrals. Since $b_\pm \neq 0$, $a(x)^2 \to \infty$ as $x \to \pm\infty$, so each $h(k)$ has discrete spectrum. Moreover $h(k)$ is analytic in k. Thus, by Theorem XIII.16 of [295], it suffices to show that, for each n, the nth eigenvalue $E_n(k)$ of $h(k)$ is not constant.

If b_+ and b_- have opposite signs, then either $a(x) \to +\infty$ at both $x = \pm\infty$,

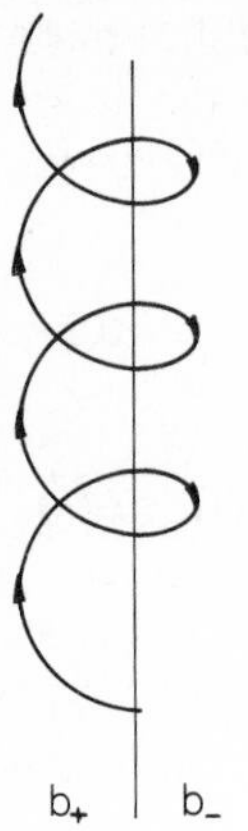

Fig. 6.1. A classical orbit

in which case $\lim_{k\to-\infty} E_n(k) = \infty$, or at $a(x) \to -\infty$ at both $x = \pm\infty$, in which case $\lim_{k\to\infty} E_n(k) = \infty$. In either case, no $E_n(k)$ can be constant.

If b_+ and b_- have the same sign, then suppose $0 < b_- < b_+$. Then $a(x)/x \to b_\pm$ as $x \to \pm\infty$ and $[a(x) - k]^2$ as $k \to \pm\infty$ looks like a translated harmonic well. From this, one easily proves that

$$\lim_{k\to\pm\infty} E_n(k) = (2n + 1)b_\pm \ ,$$

so again, $E_n(k)$ is not constant. □

6.6 An Introduction to Other Phenomena in Magnetic Fields

We have just touched the surface concerning the many subtle aspects of Schrödinger operators in magnetic field. Here we want to briefly indicate some other interesting phenomena: See [22–25] for further discussion.

(a) *Enhanced Binding.* Recall that in three dimensions, $-\Delta + V$ may not have any bound states for a compact support, $V \leq 0$, but small, while in one dimension, $-d^2/dx^2 + V$ always has a negative eigenvalue if $V \leq 0$ has compact support ($V \neq 0$). Constant magnetic fields bind in Landau orbits in two directions. Thus, if $\boldsymbol{a}$ corresponds to constant B, and $V \leq 0$ is in C_0^∞ with $V \neq 0$, then one expects that $H(\boldsymbol{a}, V)$ has an eigenvalues below the essential spectrum. A similar idea suggests that once negatively charged ions always exist in non-zero constant field. See [22, 25] where these expectations are verified.

(b) *Translational Symmetries.* Let $v = 2$, and let H_0 denote the Hamiltonian of a particle in a constant magnetic field B (with $V = 0$). The physics is invariant under translations, so we expect there will be operators $U(\boldsymbol{b})(\boldsymbol{b} \in R^2)$ with

$$U(\boldsymbol{b})H_0 U(\boldsymbol{b})^{-1} = H_0; \quad U(\boldsymbol{b})\boldsymbol{x}U(\boldsymbol{b})^{-1} = \boldsymbol{x} + \boldsymbol{b} \ .$$

This is correct, but $U(\boldsymbol{b})$ cannot just translate the wave function. The vector potential $\boldsymbol{a}$ is not translation invariant, so $U(\boldsymbol{b})$ must also have a phase factor which provides the gauge transformation from $a(\boldsymbol{x} + \boldsymbol{b})$ back to $\boldsymbol{a}(\boldsymbol{x})$. A calculation shows that

$$U(b_1,0)U(0,b_2) = e^{2\pi i\Phi} U(0,b_2)U(b_1,0) \ , \tag{6.27}$$

where Φ is the flux $Bb_1 b_2$ through the rectangle formed by the two translations. In the first place, (6.27) provides some subtleties in the removal of center of mass motion; see [23, 165]. Moreover, (6.27) implies some subtleties, still not resolved, in the analysis of $H_0 + V$, where V is periodic. If the flux through a unit cell is integral, one can make a Bloch analysis similar to that done if $B = 0$ (see [295, Sect. XIII.16]). By taking a larger unit cell, one can analyze similarly if the flux is rational (the net result is that $H_0 + V$ has only a.c. spectrum with the possibility, presumably non-existent if $V \neq 0$, of eigenvalues of infinite multiplicity). But if the flux is irrational, such an analysis is not possible. Indeed, *Grossman* [143] has shown that the U's commuting with $H_0 + V$ in that case generate a type II von Neumann algebra. It is believed that in that case, $H_0 + V$ will look like an almost periodic Schrödinger operator (see Chap. 10).

(c) *Paramagnetism.* The diamagnetic inequalities imply that $\inf \operatorname{spec}(H(\boldsymbol{a}, V)) \geq \inf \operatorname{spec}(H(0, V))$. There is a tendency for the opposite inequality to be true for $\tilde{H}$. This is illustrated by the following theorem of Lieb (see [22]):

Theorem 6.7. Let $\boldsymbol{a}$ be the vector potential of a constant magnetic field. Then

$$\inf \operatorname{spec}(\tilde{H}(\boldsymbol{a}, V)) \leq \inf \operatorname{spec}(\tilde{H}(0, V)) \ . \tag{6.28}$$

Since this is equivalent, in this constant field case, to

$$\inf \operatorname{spec}(H(\boldsymbol{a}, V)) \leq \inf \operatorname{spec}(H(0, V)) + B$$

and, if $V \to 0$ at infinity, we have

$$\inf \operatorname{ess\,spec} H(\boldsymbol{a}, V) = \inf \operatorname{spec}(H(0, V)) + B$$

(6.28) is an assertion about binding energies going up, and is connected to enhanced binding.

It was originally thought that (6.28) might hold for all $\boldsymbol{a}$, V, but *Avron* and *Simon* [27] found a counterexample. Perhaps (6.28) still holds for general $\boldsymbol{a}$ and selected sets of V, including those which, in a suitable limit, will trap particles in a convex box.

(d) *Zeeman Effect Perturbation Theory.* Much is known about the perturbation theory (in B^2) for $H_0(\boldsymbol{a}(B)) - 1/r$ where $\boldsymbol{a}(B)$ is the vector potential in a constant field $\boldsymbol{B}$. This series is Borel summable [25]. Due to symmetries, many terms in the Rayleigh-Schrödinger series can be computed: Indeed, over 100 have been [20]. *Avron* [19] has a not quite rigorous approach to large orders which matches the numerical values beautifully. Various summability methods seem to work very well; see *LeGuillou* and *Zinn-Justin* [146], *Silverman* [316]. Recently, *B. Helffer* and *J. Sjostrand* (Nantes preprint) have obtained a rigorous proof of Avron's formula.

(e) *Strong Coupling.* If one looks at atoms in magnetic fields, there are two natural distance scales, r_{at}, the Bohr radiis, and r_{cyc}, the cyclotron radius, which is the radius of the classical orbit in such a constant field (when $V = 0$). One has that $r_{\text{cyc}} \sim B^{-1}$. For typical laboratory fields, r_{cyc} is many orders of magnitude larger than r_{at}, but it is believed that in certain astrophysical situations, r_{cyc} will be a small fraction of r_{at}. The mathematical physics in this strong field regime is quite interesting: See [22, 24, 25] and references therein.

7. Electric Fields

Stark Hamiltonians play an exceptional role in the theory of Schrödinger operators since the particular singularity of the electric potential (in one coordinate multiplication by x extending over the whole space) yields a Schrödinger operator which is not semi-bounded. This needs mathematical methods differing from the usual treatment, and leads to unusual, sometimes surprising spectral properties (see Sect. 8.5). We concentrate in this chapter mainly on representations of the time evolution operator or propagator solving the Schrödinger equation. Thus, our discussion of the Stark effect is far from complete. See, for example, [140, 160, 377, 378] and the references quoted there, for a different approach.

In Sect. 7.1, we give a very useful, explicit formula for the time evolution for the free Stark Hamiltonian with constant electric field, due to Avron and Herbst. This will then be applied to Sect. 7.2 in order to show that multiplication operators with some averaging properties, when "localized" on the spectrum of the Stark Hamiltonian, are compact. This result was used in Chap. 4 in the Mourre theory of the Stark Hamiltonian, where we showed absence of singular continuous spectrum for certain Stark Hamiltonians with periodic potentials.

Section 7.3 deals with time-dependent Stark fields, and we give a representation of the free propagator due to Kitada and Yajima, analogous to the Avron-Herbst formula. The general case of time-dependent Hamiltonians can be treated by extending the configuration space by the time variable and then by solving the new "time-independent" evolution problem. This idea of Howland is discussed in Sect. 7.4. There is an application of this in the special case of Stark fields periodic in time, due to Yajima, and a method due to Tip in Sect. 7.5.

7.1 The Two-Body Stark Effect

Formally, the Hamiltonian describing a quantum mechanical particle in a constant electric field in the $-x_1$ direction is given by

$$-\Delta + Ex_1 + V(x) \quad \text{on } S(\mathbb{R}^\nu),\ x = (x_1, x_2, \ldots)\ ,$$

where $E > 0$ is the strength of the electric field (not to be confused with energy!), and V is a local decaying potential. Since Ex_1 is not a small perturbation of $(-\Delta)$ in any obvious mathematical sense, we consider the closure of

$$-\Delta + Ex_1 \quad \text{on } S(\mathbb{R}^\nu)$$

denoted by K_0, as the unperturbed, the "free" Hamiltonian, and treat V as perturbation.

We will give here a representation of the time evolution operator describing the free dynamics due to *Avron* and *Herbst* [21]. It plays a central role in the following sections (see also Chap. 8).

We set $E = 1$ in the following, for the sake of convenience. Then we have the theorem

Theorem 7.1 [21] (The Avron-Herbst Formula). Let K_0 be the closure of $(-\Delta + x_1)$ on $S(\mathbb{R}^\nu)$. Then K_0 is self-adjoint, and the time evolution is

$$\exp(-itK_0) = \exp(-it^3/3)\exp(-itx_1)\exp(-itp^2)\exp(ip_1t^2) \tag{7.1}$$

for $t \in \mathbb{R}$ and $p = (p_1, p_\perp)$, $x = (x_1, x_\perp)$.

Proof. Consider the decomposition $L^2(\mathbb{R}^\nu) = L^2(\mathbb{R}) \otimes L^2(\mathbb{R}^{\nu-1})$ according to the coordinate decomposition $x = (x_1, x_\perp)$ and $p = (p_1, p_\perp)$ in position space as well as in momentum space. One easily checks that, on $S(\mathbb{R}^\nu)$,

$$K_0 = \exp(ip_1^3/3)(p_\perp^2 + x_1)\exp(-ip_1^3/3) \ ,$$

since in momentum space x_1 acts as $-i(\partial/\partial p_1)$. But $p_\perp^2 + x_1$ is unitarily equivalent to a real-valued multiplication operator just by the Fourier transform in the $x_\perp$-variable. Thus, the self-adjointness of K_0 follows.

Now let $t \in \mathbb{R}$. Using

$$\exp(itx_1)p_1\exp(-itx_1) = p_1 - t \ ,$$

we can write, as operators on $S(\mathbb{R}^\nu)$,

$$\begin{aligned}
\exp(-itK_0) &= \exp(ip_1^3/3)\exp[-it(p_\perp^2 + x_1)]\exp(-ip_1^3/3) \\
&= \exp(-itp_\perp^2)\exp(-itx_1)\exp(itx_1)\exp(ip_1^3/3)\exp(-itx_1) \\
&\quad \times \exp(-ip_1^3/3) \\
&= \exp(-itp_\perp^2)\exp(-itx_1)\exp\left(\frac{i(p_1 - t)^3}{3}\right)\exp(-ip_1^3/3) \\
&= \exp(-itp_\perp^2)\exp(-itx_1)\exp(ip_1t^2)\exp(-ip_1^2t)\exp(-it^3/3) \\
&= \exp(-it^3/3)\exp(-itx_1)\exp[-it(p_1^2 + p_\perp^2)]\exp(it^2p_1) \ ,
\end{aligned}$$

and we arrive at (7.1) by taking closures on both sides. □

This formula has a classical interpretation. Look at the Hamiltonian function of the classical problem

$$H(p, x) := p^2 + x_1 \ .$$

Solve the equations of motion

$$\dot{x} = 2p \ ,$$

$$\dot{p} = (-1, 0, \ldots, 0) =: -\hat{x}_1 \qquad \text{and get}$$

$$x(t) = x_0 + p_0 t - \hat{x}_1 t^2 \ ,$$

where $x_0 \in \mathbb{R}^\nu$ and $p_0 \in \mathbb{R}^\nu$ are suitable initial values. This is just the "free" motion $x_0 + p_0 t$ plus a translation in the x_1 direction by the constant amount $-t^2$. Looking at (7.1), one sees that in the quantum mechanical case this is the same, except for a multiplicative phase factor. This phase factor will play a crucial role in Sect. 8.5. This shift of the free motion by $-t^2$ suggests that if we perturb K_0 with a potential V, set $K := K_0 + V$ and consider problems like existence or completeness of the wave operators $\Omega(K, K_0)$, then the borderline case should be a decay of V in the x_1-direction like $|x_1|^{-1/2}$, since it is the interaction at time t which should be integrable. This has actually been proven by *Avron* and *Herbst* [21] (existence) and *Herbst* [154] (completeness); see also [43, 377, 378].

7.2 A Theorem Needed for the Mourre Theory of the One-Dimensional Electric Field

In Chap. 4, we discussed an example which applies the Mourre theory to the Stark Hamiltonian with a periodic potential. There we used a theorem which we prove now by applying the Avron-Herbst formula.

We first state a lemma.

Lemma 7.2. Let K_0 and $K := K_0 + V$ as in Sect. 7.1. Assume that $D(K_0) = D(K)$. Then for any bounded operator C, $E_\Delta(K) C E_\Delta(K)$ is compact for all bounded intervals, $\Delta \subseteq \mathbb{R}$ if and only if $E_\Delta(K_0) C E_\Delta(K_0)$ is compact for all bounded intervals, $\Delta \subseteq \mathbb{R}$.

Proof. The first assertion is equivalent to the compactness of $(K + \mathrm{i})^{-1} C (K + \mathrm{i})^{-1}$ {for $(K + \mathrm{i}) E_\Delta(K)$ is bounded and $\lim_{a\to\infty} \|(K + \mathrm{i})^{-1}[1 - E_{(-a,a)}(K)]\| = 0$}, and the same holds for K replaced by K_0. Then, use that by the assumption, $(K_0 + \mathrm{i})^{-1}(K + \mathrm{i})$ and $(K + \mathrm{i})^{-1}(K_0 + \mathrm{i})$ are bounded. □

Now we prove the theorem.

Theorem 7.3 [45]. Let $\nu = 1$ and $K = K_0 + V$, and assume that $D(K) = D(K_0)$. Let F be a function in $\mathbb{R}$ which is bounded and uniformly continuous and has the property

$$\lim_{r\to\infty} \left(\sup_{x\in\mathbb{R}} \left| \frac{1}{2r} \int_{x-r}^{x+r} F(y)\, dy \right| \right) = 0 \ . \tag{7.2}$$

Then for any bounded interval $\Delta \subseteq \mathbb{R}$,

$$E_\Delta(K) F E_\Delta(K) \text{ is compact } .$$

Proof. Let $\Delta \subseteq \mathbb{R}$ be a bounded interval, and F as above. By the preceding lemma, we have only to show that

$$E_\Delta(K_0)FE_\Delta(K_0) \text{ is compact} .$$

Consider the unitary operator

$$U := \exp(\mathrm{i}p^3/3) .$$

Then we have (see Sect. 7.1)

$$UK_0U^{-1} = x \quad \text{and}$$
$$UxU^{-1} = x - p^2 .$$

This implies, by the spectral theorem for any function, G in $\mathbb{R}$

$$\begin{aligned} UE_\Delta(K_0)G(x)E_\Delta(K_0)U^{-1} &= E_\Delta(x)UG(x)U^{-1}E_\Delta(x) \\ &= E_\Delta(x)G(x-p^2)E_\Delta(x) . \end{aligned} \tag{7.3}$$

Note that the spectral projections $E_\Delta(x)$ are simply the characteristic functions of Δ. Hence the operator in (7.3) is Hilbert Schmidt if we can show that

$$UG(x)U^{-1} = G(x-p^2)$$

has a locally bounded kernel. It will be sufficient to show this for functions $\{G_r\}_{r\in\mathbb{R}}$ with

$$\|F - G_r\|_\infty \to 0 \quad \text{for } r \to \infty ,$$

since then the associated multiplication operators converge in the norm sense, and the compactness of $E_\Delta(K_0)FE_\Delta(K_0)$ follows by unitary equivalence.

We first show it for the function

$$x \mapsto \mathrm{e}^{\mathrm{i}sx} \quad \text{for } s \in \mathbb{R}, s \neq 0 .$$

We know by the Avron-Herbst formula (7.1) that

$$\begin{aligned} U\exp(\mathrm{i}sx)U^{-1} &= \exp[\mathrm{i}s(x-p^2)] \\ &= \exp(-\mathrm{i}s^3/3)\exp(-\mathrm{i}xs)\exp(-\mathrm{i}sp^2)\exp(\mathrm{i}ps^2) . \end{aligned}$$

This operator has an explicit kernel (note that the last term is just a translation by s^2), i.e.

$$\begin{aligned} &[U\exp(\mathrm{i}sx)U^{-1}](x,y) \\ &\quad = \exp(-\mathrm{i}s^3/3)\exp(-\mathrm{i}sx)(4\pi s)^{-1/2}\exp\left(\frac{\mathrm{i}[(x-y)+s^2]^2}{4s}\right) , \end{aligned} \tag{7.4}$$

which is obviously locally bounded.

Note next that if we convolute F with any L^1-function, h, then $F * h$ also satisfies (7.2) by Lebesgue's theorem. So if we consider the mollified function $F_\varepsilon := F * j_\varepsilon$ (j_ε being a mollifier family, see Chap. 1), then we know that F_ε satisfies (7.2), and since F is bounded and uniformly continuous, we have $\|F - F_\varepsilon\|_\infty \to 0$ as $\varepsilon \to 0$.

Thus, we can sssume without loss that F is in $C^\infty(\mathbb{R})$, has bounded derivatives, and satisfies (7.2). Now let

$$h_r := \frac{1}{2r}\chi_{[-r,r]}, \quad r > 0$$

the (normalized) characteristic function on $[-r, r]$. Then the assumption (7.2) can be restated as

$$\lim_{r\to\infty} \|F * h_r\|_\infty = 0 \ . \tag{7.5}$$

Let $k \in \mathbb{N}$, and set

$$G_r := F * \underbrace{(\delta - h_r) * (\delta - h_r) * \cdots * (\delta - h_r)}_{k\text{-times}} \ ,$$

where $f * (\delta - h_r)$ stands for $f - f * h_r$. Then, by (7.5), we know that

$$\|F - G_r\|_\infty \to 0 \quad \text{as } r \to \infty \ .$$

We will show that UG_rU^{-1} has a locally bounded kernel, which will complete the proof. Note that the advantage of G_r is that its Fourier transform has a kth order zero at the origin. Now we decompose G_r into "pieces" which have Fourier transforms decaying arbitrarily fast at infinity.

Consider $\chi_0 \in C_0^\infty(\mathbb{R})$ such that $\operatorname{supp}\chi_0 \subseteq [-1, 1]$, and such that the translates

$$\chi_n(\cdot) := \chi_0(\cdot + n), \quad (n \in \mathbb{Z})$$

satisfy $\sum_{n\in\mathbb{Z}} \chi_n = 1$. Then let

$$G_r^n := (\chi_n F) * (\delta - h_r)^{*k} \ ,$$

where $(\delta - h_r)^{*k}$ stands for

$$(\delta - h_r) * \cdots * (\delta - h_r) \quad k\text{-times} \ .$$

Note that obviously

$$G_r = \sum_{n\in\mathbb{Z}} G_r^n \ .$$

The Fourier transforms of these G_r^n's satisfy the following key estimates: For any α, l, $m \in \mathbb{N}$

$$\left|\left(\frac{d}{ds}\right)^\alpha e^{-ins}\hat{G}_r^n(s)\right| \leq c_{\alpha lm}|s|^l(1+|s|)^{-m} , \tag{7.6}$$

where $c_{\alpha lm}$ is a suitable constant independent of n (note that k has to be chosen suitably). These estimates can be seen by an explicit calculation (see [45, Lemma 1]).

Now we consider the operator

$$UG_r^n U^{-1} = (2\pi)^{-1/2} \int_{\mathbb{R}} ds\, \hat{G}_r^n(s)(Ue^{isx}U^{-1}) ,$$

where we used the Fourier representation of G_r^n. By (7.4), we know that the integrand has an explicit integral kernel. Thus, we get the kernel

$$\begin{aligned}(2\pi)^{1/2}(UG_r^n U^{-1})(x,y) &= \int_{\mathbb{R}} ds\, \hat{G}_r^n(s)(4\pi s)^{-1/2} e^{i\sigma(s,x,y)} \\ &= \int_{\mathbb{R}} ds\, e^{ins}[e^{-ins}G_r^n(s)](4\pi s)^{-1/2} e^{i\sigma(s,x,y)} ,\end{aligned}$$

where

$$\sigma(s,x,y) := \frac{(x-y+s^2)^2}{4s} - \frac{s^3}{3} - sx .$$

Now

$$e^{ins} \frac{1}{in}\frac{d}{ds} e^{ins} .$$

Integrating by parts, the estimate (7.6), and the fact that $\hat{G}_r^n$ has high-order zero at zero imply that, on bounded intervals $\Delta \subseteq \mathbb{R}$, we have

$$\sup_{x,y\in\Delta} |(UG_r^n U^{-1})(x,y)| \leq \frac{c}{1+n^2}$$

($c > 0$ suitably). Thus, we can sum up in n, and conclude that UG_rU^{-1} has a locally bounded kernel, which was what was left to show. □

7.3 Propagators for Time-Dependent Electric Fields

In this section, we discuss solutions of the Schrödinger equation with time-dependent Stark fields

$$i\frac{d}{dt}\varphi = K(t)\varphi , \tag{7.7}$$

where $K(t) = K_0(t) + V$ and $K_0(t)$ is the closure of $-\Delta + E(t)\cdot x$ on $S(\mathbb{R}^\nu)(t \in \mathbb{R})$, $E(\cdot)$ being an $\mathbb{R}^\nu$-valued function representing the time-dependent electric field. We assume, for the sake of convenience, that $E(\cdot)$ is bounded and piecewise continuous.

Let us first introduce the analog of unitary time evolution groups.

Definition 7.4. A two-parameter family of unitary operators, $U(s,t)$; $s, t \in \mathbb{R}$ is called a *propagator* if, for $r, s, t \in \mathbb{R}$

(i) $U(r,t) = U(r,s)U(s,t)$,
(ii) $U(t,t) = \mathbb{1}$,
(iii) $U(t,s)$ is jointly strongly continuous in s and t.

Note that it is enough to know $U(t,0)$ for all $t \in \mathbb{R}$ in order to know U.

We will now discuss a representation of the propagator $U_0(\cdot,\cdot)$, solving the "free" equation

$$\mathrm{i}\frac{d}{dt}U_0(t,0) = K_0(t)U_0(t,0) \tag{7.8}$$

generalizing one due to *Kitada* and *Yajima* [212, 213]. Motivated by the Avron-Herbst formula (7.1), we try the following ansatz

$$U_0(t,0) := T(t)\exp(-\mathrm{i}tp^2)\ , \quad \text{where}$$
$$T(t) := \exp[-\mathrm{i}a(t)]\exp[-\mathrm{i}b(t)\cdot x]\exp[-\mathrm{i}c(t)\cdot p]$$

with suitable real-valued and $\mathbb{R}^\nu$-valued functions $a(\cdot)$, $c(\cdot)$ and $b(\cdot)$ and initial conditions $a(0) = 0$, $b(0) = c(0) = 0$. Putting this into equation (7.8), we get [by formal calculations as operators on $S(\mathbb{R}^\nu)$]

$$\begin{aligned}
&\mathrm{i}\frac{d}{dt}U_0(t,0)\\
&\quad = \exp[-\mathrm{i}a(t)]\exp[-\mathrm{i}b(t)\cdot x](\dot a(t) + \dot b(t)\cdot x + \dot c(t)\cdot p + p^2)\\
&\qquad \times \exp[-\mathrm{i}c(t)\cdot p]\exp(-\mathrm{i}tp^2)\ .
\end{aligned}$$

We complete the square in the middle bracket by setting $\dot a = \frac{1}{4}(\dot c)^2$, commute [using $\exp(-\mathrm{i}b\cdot x)p\exp(\mathrm{i}b\cdot x) = p + b$] and get $\mathrm{i}(d/dt)U_0(t,0) =$

$$\begin{aligned}
&\left\{\dot b(t)\cdot x + \left[p + \frac{\dot c(t)}{2} + b(t)\right]^2\right\}\exp[-\mathrm{i}a(t)]\exp[-\mathrm{i}b(t)\cdot x]\\
&\qquad \times \exp[-\mathrm{i}c(t)\cdot p]\exp(-\mathrm{i}tp^2)\\
&\quad = \left\{\dot b(t)\cdot x + \left[p + \frac{\dot c(t)}{2} + b(t)\right]^2\right\}U_0(t,0)\ .
\end{aligned}$$

Now set $\dot{b}(t) = E(t)$, and $\dot{c}(t) = -2b(t)$; then the expression in the curly bracket above is $p^2 + E(t)\cdot x$. We can summarize this in

Theorem 7.5. Let $t \in \mathbb{R}$ and $K_0(t)$ be the closure of $-\Delta + E(t)\cdot x$ on $S(\mathbb{R}^\nu)$ where $E\colon \mathbb{R} \to \mathbb{R}^\nu$ is bounded and piecewise continuous. Denote $b(t) := \int_0^t E(s)\,ds$, $a(t) := \int_0^t b^2(s)\,ds$, $c(t) := -\int_0^t 2b(s)\,ds$, then the propagator

$$U_0(t,0) := T(t)\exp(-itp^2)\ , \quad \text{where} \tag{7.9a}$$

$$T(t) := \exp[-\mathrm{i}a(t)]\exp[-ib(t)\cdot x]\exp[-\mathrm{i}c(t)\cdot p] \tag{7.9b}$$

solves the "free" equation (7.8) with initial condition $U_0(0,0) = 1$.

Remark 1. Kitada and *Yajima* [212] give this formula for the case $E(t) = E_0 \cos \omega t$; the general case here does not seem to have appeared in print before, although it may well be known to workers in the field.

Remark 2. There is a classical physical interpretation of (7.9). Since

$$\ddot{c} = -2E\ ,$$

which is Newton's equation of motion, $c(\cdot)$ can be understood as the classical path. Since $\exp[-\mathrm{i}c(t)\cdot p]$ is a shift by the amount $c(t)$, the formula (7.9) for the quantum mechanical motion derived above can be written as

$$U(t,0)\varphi(t) = \exp[-ia(t)]\exp[-ib(t)\cdot x]\exp(-itp^2)\varphi(x - c(t))$$

for $\varphi \in L^2(\mathbb{R}^\nu)$. This can be interpreted, as in the time-independent case, as a traveling wave along the classical path.

Remark 3. It was noted by *Hunziker* [173] that the electric field can be "gauged away" by a time-dependent gauge transformation. By Maxwell's equations in electrodynamics, time-dependent electric fields can be realized via a time-dependent (magnetic) vector potential A, which obeys

$$E(t) = \frac{\partial}{\partial t}A(t) \quad \text{with} \quad \nabla \times A(t) = 0, \quad t \in \mathbb{R}$$

(disregarding normalization). Now look at the gauge transformation $M(t) := \exp[\mathrm{i}\phi(x,t)]$, where $\nabla\phi(x,t) = A(t)$, $x \in \mathbb{R}^\nu$, $t \in \mathbb{R}$ with a suitable real-valued time-dependent scalar potential ϕ. Then we get

$$M(t)[p^2 + E(t)\cdot x]M^*(t) = [p - A(t)]^2 + E(t)\cdot x\ .$$

Now if $U_0(t) := U_0(t,0)$ obeys (7.8), then

$$U(t) := M(t)U_0(t)M^*(0) \tag{7.10}$$

obeys

$$i\frac{d}{dt}U(t) = \{-\dot{\phi}(x,t) + [p - A(t)]^2 + E(t)\cdot x\}U(t) .$$

Now set $\dot{\phi}(x,t) = E(t)\cdot x$. Then $A(t) = b(t) = \int_0^t E(s)\,ds$ and the solution

$$U(t) = \exp\left\{-i\int_0^t [p - A(s)]^2\,ds\right\}$$

of the magnetic field Schrödinger equation

$$i\frac{d}{dt}U(t) = [p - A(t)]^2 U(t)$$

is related to the solution $U_0(t)$ of the electric field Schrödinger equation (7.8) by the gauge transformation (7.10). So writing (7.10) explicitly, one gets another way of understanding the formula (7.9):

$$U_0(t) = \exp[-ib(t)\cdot x]\exp\left[-i\int_0^t (p - b(s))^2\,ds\right]\exp[ib(0)\cdot x]$$

which is just the gauge transform of Hunziker's solution.

Now we give some examples:

Example 1 (The Constant Electric Field). If K_0 is the closure of $-\Delta + Ex_1$ on $S(\mathbb{R}^\nu)$, then $b(t) = E\hat{x}_1 t$, $c(t) = -E\hat{x}_1 t^2$, $a(t) = E^2 t^3/3$, and we get, for the propagator

$$\begin{aligned} U_0(t,0) = \exp(-itK_0) &= \exp(-iE^2t^3/3)\exp(-iEtx_1) \\ &\quad\times \exp(iEt^2p_1)\exp(-itp^2) , \end{aligned}$$

which is the Avron-Herbst formula (7.1) when $E = 1$.

Example 2 (Circular Polarized Photon Field). Let $E(t) := (\cos\omega t)\hat{x}_1 + (\sin\omega t)\hat{x}_2$; $\hat{x}_1$, $\hat{x}_2$ being unit vectors in the x_1 and x_2 coordinates. If we choose the initial conditions such that $b(0) = -(1/\omega)\hat{x}_2$, then

$$b(t) = \frac{1}{\omega}[(\sin\omega t)\hat{x}_1 - (\cos\omega t)\hat{x}_2]$$

$$c(t) = \frac{2}{\omega^2}E(t) \qquad \text{and}$$

$$a(t) = \frac{t}{\omega^2} .$$

Alternatively, we can look at the problem from the point of view of Hunziker (Remark 3 above), in which case we see that the unitary propagator for this problem is trivially related to that further problem

$$\tilde{H}(t) = [p - A(t)]^2 , \quad \text{where}$$

$$A(t) = \left(\frac{1}{\omega}\sin \omega t, -\frac{1}{\omega}\cos \omega t, 0\right) .$$

Later (Sect. 7.5) when we describe some work of Tip, we will start from this Hamiltonian (shifting t by $\pi/2\omega$).

Example 3 (*Kitada, Yajima*) [212] (The AC-Stark field). Let $E(t) := -\mu(\cos \omega t)\hat{x}_1$, for some $\mu > 0$, $\omega > 0$, then

$$b(t) = -\frac{\mu}{\omega}(\sin \omega t)\hat{x}_1$$

$$c(t) = -\frac{2\mu}{\omega^2}(\cos \omega t)\hat{x}_1 \quad \text{and}$$

$$a(t) = \frac{1}{2}\frac{t\mu^2}{\omega^2} - \frac{\mu^2}{4\omega^3}\sin 2\omega t ,$$

so we have an explicit formula [i.e. (7.9)] for solving the free AC-Stark problem.

We will finish this section with a short discussion of the perturbed case, i.e. of what happens if one adds an external potential V. We consider, for $t \in \mathbb{R}$

$$K(t) := K_0(t) + V ,$$

V being a suitable potential such that $K(t)$ is self-adjoint, and that

$$\mathrm{i}\frac{d}{dt}U(t,0) = K(t)U(t,0)$$

has a unitary propagator as solution.

As in the Hunziker remark (Remark 3) above, we try

$$U(t) = T(t)\tilde{U}(t)T^*(0), \quad t \in \mathbb{R} ,$$

where

$$U(t) := U(t,0), \quad T(t) \text{ as in (7.9)}$$

and $\tilde{U}(t) := \tilde{U}(t,0)$ solves the Schrödinger equation

$$\mathrm{i}\frac{d}{dt}\tilde{U}(t) = [p^2 + W(x,t)]\tilde{U}(t)$$

and where $W(x, t)$ is a suitable time-dependent potential fixed below. Then a direct formal calculation using $T^{-1}(t)p^2 T(t) = (p - b)^2$,

$$T^{-1}(t)E(t)\cdot xT(t) = E(t)\cdot(x + c(t)) \quad \text{and}$$
$$T^{-1}(t)V(x)T(t) = V(x + c(t))$$

suggests that

$$W(x, t) = V(x + c(t)) \ .$$

One has to tackle domain problems to make these arguments precise. But instead of doing this, we quote a theorem due to Kitada and Yajima for the special case of the AC-Stark effect.

Theorem 7.6 [212]. Assume that V is short range in the sense that

(i) V is H_0-bounded with relative bound >1
(ii) $\| V(1 + |x|)^{\alpha}(H_0 + 1)^{-\delta} \| < \infty$ for suitable $\alpha \in (0, 2)$ and $\delta \in [0, \frac{1}{2})$
(iii) $\| (H_0 + 1)^{1/2} V(H_0 + 1)^{-1} \| < \infty$.

Then

(a) there exist unitary propagators $U(\cdot, \cdot)$ and $\tilde{U}(\cdot, \cdot)$, solving

$$\mathrm{i}\frac{d}{dt}U(t, 0) = [H_0 + V(x) - \mu(\cos\omega t)x_1]U(t, 0)$$

and

$$\mathrm{i}\frac{d}{dt}\tilde{U}(t, 0) = \left\{H_0 + V\left[x - \frac{2\mu}{\omega^2}(\cos\omega t)\hat{x}_1\right]\right\}\tilde{U}(t, 0)$$

respectively, ($\mu > 0, \omega > 0$ suitably), and

(b) $U(t, 0) = T(t)\tilde{U}(t, 0)T(0)*$

where, as in Example 3 above

$$[T(t)\varphi](x) = \mathrm{e}^{\mathrm{i}g(x,t)}\varphi\left(x + \frac{2\mu}{\omega^2}(\cos\omega t)\hat{x}_1\right)$$

and

$$g(x, t) := -[a(t) + b(t)\cdot x] \quad \text{and}$$

$$a(t) = \frac{1}{2}\frac{t\mu^2}{\omega^2} - \frac{\mu^2}{4\omega^3}\sin 2\omega t$$

$$b(t) = -\frac{\mu}{\omega}(\sin\omega t)\hat{x}_1 \ ,$$

The proof in [212] consists essentially in showing the existence of $\tilde{U}$ and U with suitable domain properties, and then verifying the calculations we mentioned above. We note that Kitada and Yajima prove this theorem for more general potentials (including a long-range part).

Remark. If we look at the operators,

$$F := -\mathrm{i}\frac{\partial}{\partial t} + p^2 - \mu(\cos\omega t)x_1 + V(x)$$

and

$$\tilde{F} := -\mathrm{i}\frac{\partial}{\partial t} + p^2 + V\left(x - \frac{2\mu}{\omega^2}(\cos\omega t)\hat{x}_1\right)$$

[both defined on $S(\mathbb{R}^{\nu+1})$; note the $\nu+1$ for a time variable], then a direct formal calculation shows that they are unitarily equivalent except for a constant shift, i.e.

$$\tilde{T}(t)^{-1}F\tilde{T}(t) = \tilde{F} + \frac{3}{2}\frac{\mu^2}{\omega}$$

with the unitary operator

$$\tilde{T}(t) := \exp\left\{-\mathrm{i}\left[\frac{\mu}{\omega}(\sin\omega t)x_1 + \frac{\mu^2}{4\omega^3}\sin 2\omega t + \frac{2\mu}{\omega^2}(\cos\omega t)p_1\right]\right\} .$$

Thus, their self-adjoint extensions (if any) have the same spectrum (except for a shift).

Finally, with regard to adding potentials, we note that if one uses Hunziker's formalism, the gauge transformation commutes with potentials, so (7.10) holds if $U_0(t)$ is the propagator for $p^2 + E(t)\cdot x + V(x)$ and $U(t)$ for $[p - A(t)]^2 + V(x)$.

7.4 Howland's Formalism and Floquet Operators

There is a well-known method in classical mechanics of treating time-dependent Hamiltonians. If we consider the time-dependent classical Hamilton function

$$H(p,q,t); \quad p,q\in\mathbb{R}^\nu, \quad t\in\mathbb{R} ,$$

then the associated Hamilton equations of motion are

$$\frac{dq_i}{dt} = \frac{\partial H}{\partial p_i}, \quad \frac{dp_i}{dt} = -\frac{\partial H}{\partial q_i}, \quad i = 1,\ldots,\nu . \tag{7.11}$$

Since the Hamilton function depends on t, the energy is not conserved. But there is a standard method to get an energy-conserving system. Consider t as an additional coordinate, and the energy of the external force E as its conjugate momentum. Then the new Hamilton function is

$$F(p,q,E,t) := H(p,q,t) + E \ ,$$

and if we denote the new "time" variable by τ, then we have the new Hamilton equations

$$\frac{dq_i}{d\tau} = \frac{\partial H}{\partial p_i}, \quad \frac{dp_i}{d\tau} = -\frac{\partial H}{\partial q_i}; \quad i = 1, \dots, \nu$$

$$\frac{dt}{d\tau} = \frac{\partial F}{\partial E} = 1, \quad \frac{dE}{d\tau} = -\frac{\partial H}{\partial t} \ ,$$

which are equivalent to (7.11). Now this Hamilton function is independent of τ, and the third equation implies a simple relation between the old and the new time variables, i.e. $t = \tau + \text{const}$. So one gets the "time-independent" formalism just by adding the conjugate momentum of the time t to the Hamiltonian.

We sketch the analogous procedure in quantum mechanics due to *Howland* [168]. Consider the (time-dependent) Schrödinger equation

$$\mathrm{i}\frac{d}{dt}\varphi = H(t)\varphi, \quad \varphi \in \mathrm{H} \ , \tag{7.12}$$

where $\{H(t)\}$ $(t \in \mathbb{R})$ is a self-adjoint family of Schrödinger operators with constant domain D, and we assume that (7.12) has a unitary propagator, $U(\cdot,\cdot)$ as solution. The "extended" Hamiltonian analogous to the classical one is now formally the operator

$$F := -\mathrm{i}\frac{\partial}{\partial t} + H(t)$$

on the Hilbert space $\mathrm{H}_1 := L^2(\mathbb{R},\mathrm{H}) := \{f \mid f$ is strongly measurable H-valued, $\int_{\mathbb{R}} \|f(s)\|^2\, ds < \infty\}$. We assume that F has a self-adjoint representation (which we also denote by F). Then there should be a correspondence between the solution $\exp(-\mathrm{i}\tau F)$ of the extended Schrödinger equation

$$\mathrm{i}\frac{d}{d\tau}f = Ff, \quad f \in \mathrm{H}_1$$

and the solution $U(\cdot,\cdot)$ of (7.12). In fact, if we define the map $V\colon \mathbb{R} \to \mathrm{L}(\mathrm{H}_1)$ by

$$\tau \mapsto V(\tau)f(t) := U(t, t-\tau)f(t-\tau)$$

for $f \in \mathrm{H}_1$, $t \in \mathbb{R}$, then one verifies by direct computation that $V(\tau)$, $\tau \in \mathbb{R}$ is a strongly continuous unitary group in $\tau \in \mathbb{R}$. Moreover, by differentiation on a suitable core of F, say $C^1(\mathbb{R}, D)$, one gets that the infinitesimal generator of $V(\cdot)$

is F, i.e.

$$e^{-i\tau F}f(t) = U(t, t-\tau)f(t-\tau) \quad f \in H_1; t, \tau \in \mathbb{R} .$$

If, in addition, $H(t)$ is periodic in time t (with period T, say), then the propagator $U(\cdot,\cdot)$ is also periodic, i.e.

$$U(t+T, s+T) = U(t,s), \quad t, s \in \mathbb{R} .$$

This follows from the periodicity of the Hamiltonian and the uniqueness of the propagator $U(\cdot,\cdot)$ solving (7.12). The group $\tau \mapsto V(\tau)$ acts naturally on periodic, H-valued functions with period T, and it is obviously a unitary strongly continuous group on the Hilbert space

$$\tilde{H} := L^2(\mathbb{T}_T) \otimes H = L^2(\mathbb{T}_T, H) ,$$

where $\mathbb{T}_T := \mathbb{R}/\mathbb{Z}T$ is the one-dimensional torus. Then as above, one sees by differentiation (on a suitable core) that the infinitesimal generator F_T of this group is formally the same operator as F, but with periodic boundary conditions, i.e. the closure of

$$F_T\varphi := \left[-i\frac{d}{dt} + H(t)\right]\varphi \quad \varphi \in C^1(\mathbb{T}_T, D)$$

(see [379, Lemma 2.5]).

As it is well known in the Floquet theory of ordinary differential equations with periodic coefficients (see [151]), the solutions can be described by a matrix depending only on the initial time and the period. This has its analog here in the *Floquet operator* $U(s+T, s)$, which takes the system through a complete period starting at s (see [169, 376, 379] for a detailed analysis).

The eigenvalues of $U(s+T, s)$, if any, define the bound states of the system, namely, if $\lambda \in \mathbb{R}$ such that

$$U(s+T, s)f = e^{i\lambda T}f$$

for a suitable $f \in H$, then one gets (by using the periodicity of U)

$$U(t,s)f = \exp\left[-i\lambda T\left(\frac{t-s}{T}\right)\right]g_s(t) , \quad \text{where}$$

$$g_s(t) := U\left(s, s - T\left(\frac{(t-s)}{T} - \frac{t-s}{T}\right)\right)f$$

is periodic in t, and thus localized in space.

We mention, without going into details, that there are theorems which also relate the scattering states of the system to the (absolutely) continuous subspace of the Floquet operator, $U(s+T, s)$. These are the man results in [169, 212, 376] (see also [213]).

The important fact (which was stressed by *Yajima* [376, 379]) in the special case of the AC-Stark Hamiltonian, is that the Floquet operator is spectrally equivalent to the infinitesimal generator F_T (which is sometimes also called the Floquet Hamiltonian [170]) in the following sense. If $F_T f(t) = \lambda f(t)$, then $t \mapsto f(t)$ is an H-valued continuous and periodic function, with

$$U(t,s)f(s) = \mathrm{e}^{-\mathrm{i}\lambda(t-s)} f(t) \ .$$

In particular,

$$U(s+T,s)f(s) = \mathrm{e}^{-\mathrm{i}\lambda T} f(s)$$

and conversely, if

$$U(s+T,s)\phi = \mathrm{e}^{-\mathrm{i}\lambda T}\phi$$

for a suitable $\phi \in \mathrm{H}$, then

$$f(t) := \mathrm{e}^{\mathrm{i}\lambda(t-s)} U(t,s)\phi \in D(F_T)$$

and $F_T f = \lambda f$ (see [379, Lemma 2.9]).

7.5 Potentials and Time-Dependent Problems

The abstract framework we discussed in Sect. 7.4 has been applied to the time periodic Stark Hamiltonian (the AC-Stark effect) by *Yajima* [379] and *Graffi* and *Yajima* [141] (see also [170]). Consider the closure of $-\Delta - \mu x_1 \cos\omega t + V(t)$ on $\mathrm{S}(\mathbb{R}^3)$, which we denote by $\tilde{K}(t,\mu)$, $(t \in \mathbb{R})$ $(\mu > 0,\ \omega > 0$ suitably).

If we assume suitable conditions for V, we know by Theorem 7.6 that this can be transformed into the closure of

$$-\Delta + V\left(x_1 - \frac{2\mu}{\omega^2}\cos\omega t, x_2, x_3\right) \quad \text{on } \mathrm{S}(\mathbb{R}^3) \ ,$$

which we denote by $K(t,\mu)$. Putting this together with the results in Sect. 7.4, we can conclude that the discussion of the solutions of the AC-Stark Schrödinger equation

$$\mathrm{i}\frac{d}{dt}\varphi = \tilde{K}(t,\mu)\varphi$$

can be reduced to the discussion of the spectrum of the Floquet Hamiltonian

$$-\mathrm{i}\frac{d}{dt} + K(t,\mu) \quad \text{in } L^2(\mathbb{T}_T) \otimes \mathrm{H}$$

with domain $C^1(\mathbb{T}_T, \mathrm{H})$, where $T = 2\pi/\omega$, $\mathrm{H} = L^2(\mathbb{R}^3)$. We will come back to this in Chap. 8 (Sect. 8.6), where we discuss resonances.

There is a similar formalism due to *Tip* [356] which handles a different but related case, i.e. Hamiltonians with time-dependent fields, describing atoms in circularly polarized fields. Tip's formalism does not apply to linearly polarized fields (Example 3), while the Howland-Yajima formalism can handle his polarization. Following Tip, we will derive an explicit representation of the propagator of the system. In the simplest case (one atom and one electron) we can, by the reduction in Example 2, study the Hamiltonia $H(t)$ defined as the closure of $[p - A(t)]^2 + V$ on $\mathrm{S}(\mathbb{R}^3)$, for $t \in \mathbb{R}$. $A(t) := (A\cos\omega t, A\sin\omega t, 0)$, $A > 0$, $\omega > 0$ and V being rotationally invariant in $\mathbb{R}^3$. We assume, in addition, that V is H_0-bounded with bound smaller than one, and essentially bounded outside some ball. Since $A(t)$ is bounded, we have $D([p - A(t)]^2) = \mathrm{H}_{+2}$ the second Sobolev space, and by the resolvent version of the diamagnetic inequality (6.5), we can conclude that $H(t)$ is self-adjoint with constant domain $D(H(t)) = \mathrm{H}_{+2}$.

One can now transform this operator unitarily into a time-independent one by a time-dependent rotation. Consider the unitary operators (describing rotations on the z-axis in configuration space)

$$R(t) := \mathrm{e}^{\mathrm{i}\omega t L_z}, \quad t \in \mathbb{R}$$

where L_z is the operator of the angular momentum pointing in the z direction, i.e. the closure of $xp_y - yp_x$ on $(\mathrm{S}(\mathbb{R}^3)$. Thus, we get in momentum space a rotation through the angle ωt, i.e.

$$R(t)pR(t)^{-1} = (p_x\cos\omega t + p_y\sin\omega t, -p_x\sin\omega t + p_y\cos\omega t, p_z) \ ,$$

where we denote $p = (p_x, p_y, p_z)$. Then $R(t)$ commutes with V (because of the rotational invariance of V), and $R(t)A(t)R(t)^{-1} = A(0)$. Therefore, we get, for $\varphi \in \mathrm{S}(\mathbb{R}^3)$, and therefore also for $\varphi \in D(H(t)) = \mathrm{H}_2$ by direct calculation

$$R(t)H(t)R^{-1}(t)\varphi = [(p - a)^2 + V]\varphi$$

where $a = (A, 0, 0)$. Now consider the operator

$$H := (p - a)^2 + V - \omega L_z$$

and note without proof (see *Tip* [356, Theorem 2.1]) that H is self-adjoint and has $\mathrm{S}(\mathbb{R}^3)$ as a core. Then the following theorem gives a representation of the unitary propagator $U(\cdot,\cdot)$, which solves

$$\mathrm{i}\frac{\partial}{\partial t}U(t,s) = H(t)U(t,s); \quad U(0,0) = 1 \ . \tag{7.13}$$

Theorem 7.7 [Tip]. Let $U(\cdot,\cdot)$ be the unitary propagator solving (7.13). Then

$$U(t,s) = \mathrm{e}^{-\mathrm{i}\omega L_z t}\mathrm{e}^{-\mathrm{i}(t-s)H}\mathrm{e}^{\mathrm{i}\omega L_z s}, \quad t, s \in \mathbb{R} \ .$$

Proof. We calculate $U^*(t,0) = U(0,t)$, which is also unique, determines $U(t,s)$ and satisfies the adjoint Schrödinger equation

$$-\mathrm{i}\frac{\partial}{\partial t}U^*(t,0) = U^*(t,0)H(t) \ . \tag{7.13'}$$

Now consider the unitary group

$$\hat{U}(t) := \mathrm{e}^{\mathrm{i}tH}\mathrm{e}^{\mathrm{i}\omega tL_z} \ .$$

Then we get, by differentiation, for $\varphi \in \mathrm{S}(\mathbb{R}^3)$

$$\begin{aligned}\frac{\partial}{\partial t}\hat{U}(t)\varphi &= \mathrm{i}\mathrm{e}^{\mathrm{i}tH}(H + \omega L_z)\mathrm{e}^{\mathrm{i}\omega tL_z}\varphi \\ &= \mathrm{i}\hat{U}(t)\{\mathrm{e}^{-\mathrm{i}\omega tL_z}[(p-a)^2 + V]\mathrm{e}^{\mathrm{i}\omega tL_z}\}\varphi \\ &= \mathrm{i}\hat{U}(t)H(t)\varphi \ .\end{aligned}$$

Since $R(t)\varphi \in \mathrm{S}(\mathbb{R}^3)$ for $\varphi \in \mathrm{S}(\mathbb{R}^3)$, all expressions above are well defined. Thus, $\hat{U}(t)$ solves (7.13′) on the common core $\mathrm{S}(\mathbb{R}^3)$ of $H(t)$, H and L_z. Now by a standard limiting argument, $\hat{U}(t)$ is differentiable on $D(H(t)) = \mathrm{H}_2$ [use that $H(s)(1 + H_0)^{-2}$ is bounded]. Thus, $\hat{U}(t) = U^*(t,0)$ and

$$U(t,0) = \hat{U}^*(t) = \mathrm{e}^{-\mathrm{i}\omega tL_z}\mathrm{e}^{-\mathrm{i}tH}$$

and by

$$U(t,s) = U(t,0)U(0,s)$$

we get the desired result. □

We mention, without going into details, that Tip uses this representation to describe resonances with dilation techniques, similar to those in Chap. 8.

For additional information, see [220].

8. Complex Scaling

Complex scaling, also known as the method of complex coordinates, coordinate rotation or dilation analyticity, has developed rapidly in the last fifteen years (see [8] and the references there), and has become a powerful tool in numerical studies of resonances.

The first ideas go back to the early sixties in connection with the study of Regge poles (for references, see [295]). But a rigorous mathematical theory was not developed until the work of *Aguilar* and *Combes* [5] and *Balslev* and *Combes* [39] in 1971. These authors applied their ideas to spectral analysis. Their usefulness in studying embedded eigenvalues and resonances was realized by *Simon* [319, 320].

We will not give a complete treatment, but rather pick out some aspects and recent developments.

In Sect. 8.1, we review the basic definitions and results of "ordinary" complex scaling, and discuss some technical details which arise in the N-body case. Section 8.2 deals with translation analyticity, which was the first attempt to describe resonances in electric fields. In some ways, Mourre theory (see Chap. 3) can be viewed as first-order complex scaling. Section 8.3 contains some discussions of higher order Mourre theory, a technique which can be understood as an nth order Taylor approximation to analytic complex scaling.

In Sect. 8.4, we make some remarks about computational aspects of complex scaling, which has recently attracted considerable attention from physicists and chemists.

In Sect. 8.5 and 6, we discuss the dilation analytic treatment of resonances in the DC and AC-Stark effect, and Sect. 8.7 closes this chapter with some remarks about extensions to larger classes of potentials and to molecular systems (Born-Oppenheimer approximation).

8.1 Review of "Ordinary" Complex Scaling

We just recall the basic definitions and results. We give very few details, and refer the reader to the original papers of *Aguilar* and *Combes* [5] and *Balslev* and *Combes* [39] or *Reed* and *Simon* [295] for complete proofs.

First we discuss the two-body case, i.e. the Hamiltonian

$$H := H_0 + V, \quad H_0 := \overline{(-\Delta)\restriction C_0^\infty}$$

for suitable V. Consider the one-parameter family of unitary dilations in $L^2(\mathbb{R}^\nu)$

$$U_\theta \varphi(x) := e^{\theta(\nu/2)} \varphi(e^\theta x), \quad \varphi \in L^2(\mathbb{R}^\nu) \tag{8.1}$$

for $\theta \in \mathbb{R}$ and $x \in \mathbb{R}^\nu$. A direct calculation shows that

$$H_0(\theta) := U_\theta H_0 U_\theta^{-1} = e^{-2\theta} H_0 \ .$$

Thus, $H_0(\theta)$ is well defined for real θ, and can be analytically continued into regions of complex θ. Note that if $\operatorname{Im}\theta > 0$, the spectrum of $H_0(\theta)$ is the "rotated" semi-axis, $\exp[-2i(\operatorname{Im}\theta)]\mathbb{R}_+$, $[\mathbb{R}_+ := [0, \infty)]$. We want to restrict the potentials to a class where the same is true for the essential spectrum when $H = H_0 + V$ replaces H_0. We will denote this class by C_α.

Definition 8.1. An operator V on $L^2(\mathbb{R}^\nu)$ is called *dilation analytic* (or in C_α) if there exists a strip

$$S_\alpha := \{\theta \in \mathbb{C} \mid |\operatorname{Im}\theta| < \alpha\}$$

for suitable $\alpha > 0$ such that

(i) $D(V) \supseteq D(H_0)$ and V is symmetric
(ii) V is H_0-compact
(iii) the family $V(\theta) := U_\theta V U_\theta^{-1}$, $\theta \in \mathbb{R}$ has an analytic continuation into the strip S_α, in the sense that $V(\theta)(H_0 + 1)^{-1}$ is a bounded operator-valued analytic function on S_α.

Note there is also a form-analog of this definition (see [295, p. 184]). If V is dilation analytic, then

$$H(\theta) := H_0(\theta) + V(\theta), \quad \theta \in S_\alpha \tag{8.2}$$

is an analytic family [of type (A)], and since (by analytic continuation) $V(\theta)$ is also $H_0(\theta)$-compact, $\sigma_{\mathrm{ess}}(H(\theta)) = \exp[-2i(\operatorname{Im}\theta)]\mathbb{R}_+$. Furthermore, discrete (i.e. isolated and finitely degenerated) eigenvalues move analytically in $\theta \in S_\alpha$ so long as the eigenvalues avoid the essential spectrum. But since U_θ is unitary, the group property of U_θ for real θ implies that they do not move at all. For, if ϕ is real, $H(\theta + \phi)$ is unitarily equivalent to $H(\theta)$, and so $E(\theta + \phi) = E(\theta)$ by analyticity, so E is constant. Thus, eigenvalues persist, except that they can be absorbed when continuous spectrum moves over them or "uncovered" as continuous spectrum moves past.

We will call the non-real eigenvalues of $H(\theta)$ the *resonances of* H. Thus, we have the following picture for the spectrum of $H(\theta)$:

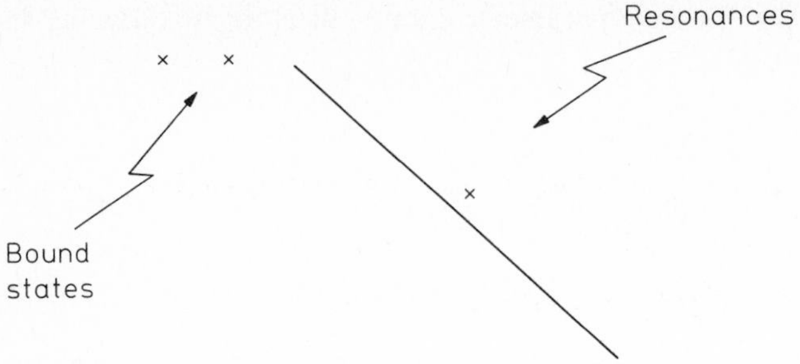

consider now the matrix element of the resolvent

$$f(z) := \langle \varphi, (H - z)^{-1} \varphi \rangle$$

for $z \in \rho(H)$ and φ an analytic vector (in the sense that $U_\theta \varphi$ can be analytically continued into a strips S_α). Then, for $\operatorname{Im} \theta > 0$, $\operatorname{Im} z > 0$

$$f_\theta(z) := \langle U_\theta \varphi, (H(\theta) - z)^{-1} U_\theta \varphi \rangle$$

is equal to $f(z)$ since it is analytic in θ and constant for real θ. Thus, $f_\theta(\cdot)$ is a meromorphic continuation of $f(\cdot)$ (as a function of z) across the real line into the cone spanned by $\mathbb{R}_+$ and $\sigma_{\text{ess}}(H(\theta)) = \exp[-2i(\operatorname{Im} \theta)]\mathbb{R}_+$. Since eigenvalues of $H(\theta)$ are poles of $z \mapsto (H(\theta) - z)^{-1}$ and therefore of $f_\theta(\cdot)$, the poles of $f_\theta(\cdot)$ are the resonances of H (this assertion has to be taken with some care. For a more detailed discussion, see [324] (Howland's razor)).

This justifies the name resonances since one can show (at least in some cases) that the complex poles of $f_\theta(\cdot)$ coincide with the complex poles of the scattering matrix of the system (see [37, 147, 183, 320]), which are usually interpreted as resonances in the physics literature.

In the N-body case, i.e. if

$$H = H_0 + \sum_{i<j}^{N} V_{ij}$$

(H_0 being the free Hamiltonian with center of mass motion removed), the situation is basically the same if one assumes that the V_{ij}'s are dilation analytic in the associated space of two-particle sybsystems.

In the two-body case, we have seen that a crucial role is played by the fact that one can explicitly find $\sigma_{\text{ess}}(H(\theta))$. This is a more involved problem in the N-body case: Every subsystem contributes to $\sigma_{\text{ess}}(H(\theta))$ by a semi-axis rotated at the individual thresholds.

Denote the set of thresholds belonging to a cluster decomposition, $D := \{C_1, \ldots, C_k\}$, by

$$\sum_D(\theta) := \{E_1 + E_2 + \cdots + E_k | E_i \in \sigma_{\text{disc}}(H_{C_i}(\theta))\}$$

and let

$$\sum(\theta) := \bigcup_{D} \sum_{D}(\theta)$$

be the set of thresholds (see Chap. 3). We will show in the following that $\sigma_{\rm ess}(H(\theta))$ is contained in the set

$$S := \{\mu + \exp(-2{\rm i}\,{\rm Im}\,\theta)\lambda \mid \mu \in \sum(\theta), \lambda \in \mathbb{R}_+\} \ .$$

The dilation group in this (N-body) case is defined by

$$U_\theta \varphi(x) := \exp\left(\frac{\nu N - \nu}{2}\theta\right)\varphi[{\rm e}^\theta x], \quad \varphi \in L^2(\mathbb{R}^{\nu N-\nu})$$

$$(x \in \mathbb{R}^{\nu N - \nu}, \theta \in \mathbb{R}) \ .$$

Then $H(\theta) := U_\theta H U_\theta^{-1}$ has a well-defined type (A)-continuation into the strip S_α, if the V_{ij} are dilation analytic, and we have

Proposition 8.2. Let $H = H_0 + \sum_{i<j}^N V_{ij}$ in $L^2(\mathbb{R}^{3N-3})$, where H_0 is the free Hamiltonian with center of mass motion removed, and V_{ij} are dilation analytic. Then

$$\sigma_{\rm ess}(H(\theta)) \subseteq S \ . \tag{8.3}$$

Proof. The first step of the proof is as in [295, Proposition 2, p. 189], but we repeat it for the reader's convenience. If $N = 2$, then $\sum(\theta) = \{0\}$, and we know from the two-body case that the assertion (8.3) is true. So we assume (8.3) is true for all M-body systems with $M \leq N - 1$. We will prove it for N-body systems.

As a first step, we show

$$\sigma(H_D(\theta)) \subseteq S \tag{8.4}$$

for any $D := \{C_1, \ldots, C_k\}$ being a decomposition of the N-body system with at least two clusters. Recall that

$$H_D(\theta) = \sum_{l=1}^{k} H_{C_l}(\theta) + T_D(\theta)$$

(see Sect. 3.2) where $H_{C_l}(\theta)$ is the dilated cluster Hamiltonian and $T_D(\theta)$ is the dilated kinetic energy of the center of mass of the individual clusters in D. Since the V_{ij}'s are H_0-bounded with relative bound 0, each $H_{C_l}(\theta)$ is a strictly m-sectorial operator (see [196, p. 338]).

Moreover, in the natural decomposition

$$L^2(\mathbb{R}^{3N-3}) = \mathrm{H}_D \otimes \mathrm{H}_{C_1} \otimes \cdots \otimes \mathrm{H}_{C_k}$$

each summand of $H_D(\theta)$ acts in a different factor of the tensor product.

By Ichinose's Lemma [295, p. 183], we know that the spectrum of a tensor-sum of m-sectorial operators can be found just as in the self-adjoint case, i.e.

$$\sigma(H_D(\theta)) = \sum_{l=1}^{k} \sigma(H_{C_l}(\theta)) + \sigma(T_D(\theta)) \ .$$

Since

$$\sigma(T_D(\theta)) = \exp(-2\mathrm{i}\,\mathrm{Im}\,\theta)\mathbb{R}_+ \quad \text{and}$$

$$\sigma_{\mathrm{ess}}(H_{C_l}(\theta)) \subseteq S$$

by the induction hypothesis, we have (8.4) for any D with at least two clusters.

Now as the second step, following *Sigal* [310, 312], we mimic the proof of the HVZ-theorem (see Chap. 3), thereby avoiding the technicalities of the Weinberg-van Winter equations. Let $\{j_a\}$ be a Ruelle-Simon partition of unity in $\mathbb{R}^{3N-3}$, where a runs over all two-cluster decompositions (see Definition 3.4, Chap. 3). Then

$$H(\theta) = H_a(\theta) + I_a(\theta) \ , \quad \text{where}$$

$$I_a(\theta) := \sum_{(i,j)\nsubseteq a} V_{ij}(\theta) \quad \text{and}$$

$$H_a(\theta) := \sum_{i=1}^{2} H_{C_i}(\theta) + T_a(\theta) \ .$$

Then we have

$$\begin{aligned}(H(\theta) - z)^{-1} &= \sum_a j_a^2 (H(\theta) - z)^{-1} \\ &= -\sum_a j_a^2 \{(H_a(\theta) - z)^{-1} I_a(\theta)(H(\theta) - z)^{-1}\} \\ &\quad + \sum_a j_a^2 (H_a(\theta) - z)^{-1} \\ &= I(z)(H(\theta) - z)^{-1} + D(z) \ , \end{aligned} \tag{8.5}$$

where

$$I(z) := -\sum_a j_a^2 \{(H_a(\theta) - z)^{-1} I_a(\theta)\} \quad \text{and}$$

$$D(z) := \sum_a j_a^2 \{(H_a(\theta) - z)^{-1} \ .$$

We know by (8.4) that both $I(z)$ and $D(z)$ are analytic in $\mathbb{C}\backslash S$ (note this is a connected set). Since $I(z)$ is compact (Proposition 3.6, Chap. 3) and $\|I(z)\| \to 0$ as $\mathrm{Re}\, z \to \infty$, we know by the analytic Fredholm theorem [292, p. 201] that $[1 - I(z)]^{-1}$ is meromorphic in $\mathbb{C}\backslash S$, the poles are contained in a discrete set D_0, and the residues are finite rank operators. So D_0 does not belong to $\sigma_{\mathrm{ess}}(H(\theta))$.

Furthermore, by (8.5), we know that $[H(\theta) - z]^{-1}$ is bounded and analytic in $\mathbb{C}\backslash\{S \cup D_0\}$ and this implies the assertion (8.3). □

Remark. One can actually show that $\sigma_{\text{ess}}(H(\theta)) = S$. The missing inclusion corresponds to the easy part of the HVZ theorem, and can be shown by an appropriate construction of a Weyl-sequence.

Having located the essential spectrum, all the arguments can be carried over from the two-body case to obtain stability of the discrete eigenvalues, etc. Suppose that $\overline{\sum(\theta = 0)}$ is countable for each subsystem. Then we can control boundary values of $(f, (H - z)^{-1} f)$ for a dense set of f and z away from a closed countable set. One concludes that $\sigma_{\text{sing}}(H) = \phi$ and eigenvalues can only accumulate at $\overline{\sum(0)}$. In this way, one inductively shows that $\overline{\sum(0)}$, for the whole N-body system, is countable. Thus, we have

Theorem 8.3 (Balslev and Combes [39]). If $H = H_0 + \sum_{i<j}^{N} V_{ij}$ is the Hamiltonian of an N-body system with C.M. motion removed and the V_{ij} are dilation analytic, then H has empty singular continuous spectrum, and its set of thresholds union eigenvalues is a closed countable set.

This theorem can be proven also by Mourre's method, which is an infinitesimal version of complex scaling (see Sect. 8.3).

8.2 Translation Analyticity

We consider the Stark (-effect) Hamiltonian H, which is defined as the closure of

$$h := -\Delta + Ex_1 + V \quad \text{on } S(\mathbb{R}^\nu) ,$$

where $E > 0$ and V is a suitable multiplication operator (see Chap. 7).

At first sight, since h has no threshold, the method of complex scaling does not seem to be an appropriate way to study spectral properties such as resonance phenomena of H. We return to this in Sect. 8.5. *Avron* and *Herbst* [21] therefore introduced a class of potentials which have same special smoothness properties.

Definition 8.4. Let $x = (x_1, x_\perp)$ with $x_\perp = (x_2, \ldots, x_\nu)$. Suppose V is a multiplication operator such that for a.e. $x \in \mathbb{R}^\nu$, the map $\lambda \mapsto V_\lambda := V(x_1 + \lambda, x_\perp)$ is analytic in the strip $S_\alpha := \{\lambda \in \mathbb{C} \,|\, |\text{Im}\,\lambda| < \alpha\}$, ($\alpha > 0$ suitable). Then V is called K_0-*translation analytic* if $\lambda \mapsto V_\lambda(K_0 - \mathrm{i})^{-1}$ is a compact analytic operator-valued function in S_α, where K_0 denotes the closure of $-\Delta + Ex_1$ on $S(\mathbb{R}^\nu)$.

We remark that while the Coulomb potential $c/|x|$ is not translation analytic, the "smeared out" Coulomb potential $V_\rho(x) := c|x|^{-1} * \rho(x)$ (ρ being a suitable Gaussian), which is arbitrarily close to it, is in this class.

Let V be translation analytic, then H_λ, defined as the closure of $h_\lambda := -\Delta + E(x_1 + \lambda) + V_\lambda$ on $S(\mathbb{R}^\nu)$, is an analytic family [of type (A)] for $\lambda \in S_\alpha$. Furthermore, since V_λ is K_0-compact, $K_0 + E\lambda$ and H_λ have the same essential spectrum, i.e.

$$\sigma_{\text{ess}}(H_\lambda) = \sigma_{\text{ess}}(K_0 + \lambda) + E\lambda = \mathbb{R} + \mathrm{i}E \operatorname{Im} \lambda$$

for $\lambda \in S_\alpha$. This means that the essential spectrum of the "translated" operator H_λ is the shifted real axis.

If we look at the map $U_\alpha \varphi(x_1, x_\perp) := \varphi(x_1 + \alpha, x_\perp)$, which is unitary if α is real, the identity $U_\alpha H_\lambda U_\alpha^{-1} = H_{\lambda+\alpha}$ shows that, as in the ordinary dilation analytic case, the eigenvalues of H_λ are independent of λ as long as the line $\mathbb{R} + \mathrm{i}E$ does not intersect them. These eigenvalues, which are discrete and occur only in the strip

$$\{z \mid \operatorname{Im} z \in (0, \operatorname{Im} \lambda)\}$$

are called *resonances* of H.

It is shown in [156] that this has a physical justification in the sense that those resonances, say $E_1, E_2, \ldots$ are due to the exponential decay of the expectation value of the time evolution of certain (translation analytic) states, ψ, i.e.

$$\langle \psi, \mathrm{e}^{\mathrm{i}tH} \psi \rangle = \sum_{\operatorname{Im} E_n > -\alpha} C_n(\psi, E) \mathrm{e}^{-\mathrm{i}tE_n} + O(\mathrm{e}^{-\alpha t})$$

where the constants C_n depend on ψ and E.

In addition, translation analyticity implies certain operators have no singular continuous spectrum. The application of Mourre's method to Stark Hamiltonians using $A = p$, described in Example 3 of Sect. 4.1, is an infinitesimal version of translation analyticity.

8.3 Higher Order Mourre Theory

Let $A := \frac{1}{2}(xp + px)$, $[p := \overline{-\mathrm{i}\nabla_x \restriction S(\mathbb{R}^\nu)}]$. Then A is the generator of the dilation group U_θ, defined in (8.1), which means that $U_\theta = \exp(\mathrm{i}\theta A)$ (see Chap. 4), and for the dilation analytic V, the map

$$\theta \mapsto \mathrm{e}^{\mathrm{i}\theta A} H \mathrm{e}^{-\mathrm{i}\theta A} =: H(\theta)$$

can be analytically continued into a suitable strip C_α. Looking at the Mourre method (Chap. 4), which deals essentially with the family

$$H_1(\theta) := H + \theta[H, \mathrm{i}A], \quad \theta \in C_\alpha$$

one sees that $H_1(\theta)$ is just the first-order Taylor approximation of $H(\theta)$. The advantage of dealing with $H_1(\theta)$ instead of $H(\theta)$ is that one can ease the conditions on V considerably. (But of course one then gets less explicit information about H.)

It seems to be natural to try to use nth order Taylor approximations on $H(\theta)$, i.e.

$$H_n(\theta) := H + \sum_{k=1}^{n} \frac{\theta^k}{k!}[H, \mathrm{i}A]^{(k)}$$

for some $n \in \mathbb{N}$, where

$$[H, \mathrm{i}A]^{(k)} := [\ldots[[H, \mathrm{i}A], \mathrm{i}A], \ldots]$$

is the kth commutator on a suitable domain. This was used to get more detailed estimates of the resolvent $R(z) := (H - z)^{-1}$ of H near the real axis by *Cycon* and *Perry* [74], and by *Jensen, Mourre* and *Perry* [184]. The main results of these techniques are local decay estimates for the time evolution of scattering states. *Cycon* and *Perry* [74] showed a "semi-global" estimate

$$\|\mathrm{e}^{-\mathrm{i}tH} g(H)\varphi\|_{-s} \le c(1 + |t|)^{-s+\varepsilon} \|\varphi\|_s$$

for $t \in \mathbb{R}$, $\varphi \in \mathrm{H}$, $\|\cdot\|_s$ being the weighted norm in $L^2(\mathbb{R}^\nu, \langle x \rangle^{2s}\,dx)$ and g is a smooth cut-off function which cuts off the lower part of the spectrum. (This is a slightly improved version of a result of *Kitada* [210]). *Jensen, Mourre* and *Perry* [184] proved similar but "local" resolvent estimates [i.e. g is replaced by a function in $C_0^\infty(\mathbb{R} \backslash \sigma_p(H))$]. These estimates were used to get some results in two-body scattering theory.

8.4 Computational Aspects of Complex Scaling

Since the numerical computation of eigenvalues has proved quite successful, one would like to have a similarly well-developed machinery for calculating resonances. There is a large literature which has appeared in the last ten years, dealing with numerical calculations of resonances (see, for example, [166, 237, 297] and the references there). We will describe in a non-mathematical way some of the ideas involved here.

The "naive" way to compute resonances would be to calculate the matrix elements (with respect to a suitable basis of trial functions) of the "rotated", i.e. dilated Hamiltonian, and then try to diagonalize. Then one would follow the trajectories in θ and look for "stationary" points. This needs a lot of computing and a careful handling of the data to avoid unreasonably large errors.

For Coulombic systems (where V is a linear combination of Coulomb potentials) there is an easier way to implement this strategy (see [297]). Since

$$H(\theta) = \mathrm{e}^{-2\theta} H_0 + \mathrm{e}^{-\theta} V \ ,$$

one can start by computing the matrix elements of the "unrotated" H_0 and V, multiply the first by $\exp(-2\theta)$ and the second by $\exp(-\theta)$ and then diagonalize. The big advantage is that, for different θ's, one does not have to calculate new matrix elements. This method works well in the 2-electron systems, but not so well for N-electron systems because, for the inner electrons, simple Gaussian are not appropriate trial functions (see [297, p. 239]).

There is another cookbook procedure for computing resonances, which looks puzzling at first sight. Take the Hamiltonian and a set of (orthonormalized) trial functions $\{\varphi_n\}$ which are not real, and calculate the matrix elements

$$\langle \overline{\varphi_n}, H\varphi_m \rangle \ , \tag{8.6}$$

and then compute the eigenvalues associated with this matrix. If these numbers are stable under variation of some nonlinear parameter, there is a good chance that they are resonances!

This can be "justified" by complex scaling. Suppose the φ_n's are dilation analytic (usually they are built out of polynomials, Gaussians and exponentials). Then (8.6) is equal to

$$\langle \overline{\varphi_n}, H\varphi_m \rangle = \langle \overline{U(\theta)\varphi_n}, H(\theta)U(\theta)\varphi_m \rangle \tag{8.7}$$

by the Schwarz reflection principle and the fact that $U_\theta^* = (U_{\bar{\theta}})^{-1}$ [U_θ and $H(\theta)$ being as in (8.1) and (8.2)]. Now because of $H(\overline{\theta}) = H(\theta)^*$, (8.7) obeys a certain variational principle. Suppose that

$$H(\theta)\varphi = E\varphi, \quad \|\varphi\| = 1, \quad E \in \mathbb{C} \ .$$

Then a short calculation shows that, for any η

$$\langle \bar{\eta}, H(\theta)\eta \rangle = E\langle \bar{\eta}, \eta \rangle + \langle \overline{(\varphi - \eta)}, [H(\theta) - E](\varphi - \eta) \rangle \ .$$

So the error is quadratic in $(\varphi - \eta)$.

This "explains" why the matrix elements "near" eigenvalues of $H(\theta)$ should show some kind of stability. This method has its advantage especially in the cases where the potential is dilation analytic only outside a certain ball (see Sect. 8.7—exterior scaling), since the complex rotation of the Hamiltonian by exterior scaling leads to very complicated expressions.

There are still mathematical problems, however, since there is no theorem which says that, for non-selfadjoint operators, the eigenvalues of the matrix (calculated for a set of trial functions) tend to the eigenvalues of the operator as the number of trial-functions is increased, as is the case for selfadjoint operators (see [295, p. 83]). Moreover, no effective error bounds are known analogous to Temple's bound for the selfadjoint case.

8.5 Complex Scaling and the DC-Stark Effect

Translation analyticity as described in Sect. 8.2 suffers from the defect that it does not include Coulomb perturbations.

The "ordinary" complex scaling does not seem to be appropriate for the Stark

Hamiltonian since there is no threshold on which the spectrum can "turn" [$\sigma(K_0) = \mathbb{R}$!]. But if one, nevertheless, turns on the dilation, i.e. if one considers $K_0(\theta)$, the closure of $-\Delta \exp(-2\theta) + x_1 \exp\theta$ on $S(\mathbb{R}^\nu)$ for $\operatorname{Im}\theta > 0$, the surprising fact is that $K_0(\theta)$ has empty spectrum! (So one might think that the spectrum has been turned on the point $-\infty$.)

This was shown by *Herbst* [155], who was motivated by the success of numerical calculations. If we perturb $K_0(\theta)$ by a dilation analytic potential, the spectrum becomes purely discrete and one can then interpret the (complex) eigenvalues as resonances, as in Sect. 8.1. We will now show the emptiness of the spectrum of the complex-scaled free Stark Hamiltonian.

Theorem 8.5 [155]. Let $k_0(\theta) := -\Delta \exp(-2\theta) + x_1 \exp(\theta)$ defined on $S(\mathbb{R}^\nu)$, $\theta \in \mathbb{C}$. Then

(i) $k_0(\theta)$ is closable. We call the closure $K_0(\theta)$.
(ii) $\sigma(K_0(\theta)) = \phi$, if $\operatorname{Im}\theta \in (0, \pi/3)$.

Proof. (We give only an outline; see [155] for details). Let $\operatorname{Im}\theta \in (0, \pi/3)$. Consider the operator

$$l_0 := \mathrm{i}\mathrm{e}^{-\theta} k_0(\theta) = \mathrm{i}(\mathrm{e}^{-3\theta} p^2 + x_1)$$

defined on $S(\mathbb{R}^\nu)$, where as usual p_j is the closure of $-\mathrm{i}(\partial/\partial x_j)$ on $S(\mathbb{R}^\nu)$ and $p^2 := -\Delta = \sum_j p_j^2$. It is easy to see that the numerical range of l_0 is contained in the half plane

$$\{z \in \mathbb{C} \,|\, \operatorname{Re} z \geq 0\} \ ,$$

which suggests that l_0 may generate a contraction semigroup.

In fact, a formal computation (analytic continuation of the Avron-Herbst formula in Chap. 7) suggests that

$$P_s := \exp(-\mathrm{i}sx_1)\exp\left\{-\mathrm{i}s\mathrm{e}^{-3\theta}\left[p_\perp^2 + \left(p_1 - \frac{s}{2}\right)^2 + \frac{s^2}{12}\right]\right\}, \quad s \in \mathbb{R}_+ \tag{8.8}$$

defined on $S(\mathbb{R}^\nu)$ is a reasonable candidate for this semigroup. Since $\operatorname{Re}[\mathrm{i}\exp(-3\theta)] > 0$, we can estimate

$$\|P_s\| \leq \exp\left[-\operatorname{Re}\{\mathrm{i}\exp(-3\theta)\} \cdot \frac{s^3}{12}\right] \leq 1, \quad s \in \mathbb{R}_+ \ . \tag{8.9}$$

Thus, P_s can be extended to all $L^2(\mathbb{R}^\nu)$ (we denote this extension also by P_s), and some straightforward calculations show that $\{P_s\}_{s\in\mathbb{R}_+}$ is, in fact, a contraction semigroup.

This semigroup has a closed and densely defined infinitesimal generator, L_0, [293, p. 237], and some simple arguments show that, in fact, $\bar{l}_0 = L_0$, which implies that $k_0(\theta)$ is also closable. Now we calculate the resolvent of L_0

$$(L_0 - z)^{-1} = \int_0^\infty P_s e^{zs}\, ds \ .$$

But from the estimate (8.9), it follows that

$$\|(L_0 - z)^{-1}\| \le c < \infty \quad \text{for all } z \in \mathbb{C} \ .$$

This implies that $\sigma(L_0) = \sigma(K_0(\theta)) = \phi$. □

Now let V be a dilation analytic potential, i.e. $V \in C_\alpha$ for $\alpha = \pi/3$. Then $H(\theta) := K_0(\theta) + V(\theta)$ is well defined, and one can show [see [155]] that $H(\theta)$ is closed analytic of type (A) in the region $0 < \operatorname{Im} \theta < \pi/3$ and has only discrete spectrum. However, there are some subtleties as compared to the usual complex scaling theory. In the usual theory, as $\operatorname{Im} \theta \searrow 0$, the resolvents of $H(\theta)$ converage in *norm*; now they only converge strongly.

The Herbst theory of complex scaling has been extended to N-body systems by *Herbst* and *Simon* [160] (see also *Herbst* [157]). Applications to the summability of the Rayleigh-Schrödinger series for the Stark effect in hydrogen and the width of the resonanaces there can be found in *Herbst* and *Simon* [159, 160] (see also *Graffi* and *Grecchi* [140] and *Harrell* and *Simon* [150]).

8.6 Complex Scaling and the AC-Stark Effect

Without going into details, we mention another application of the method of dilation analyticity to time-dependent Stark fields.

In Chap. 7 (Theorem 7.6), we saw that the Hamiltonian with a time-periodic Stark field, i.e. the closure of

$$-\Delta + V - \mu x_1 \cos \omega t \quad \text{defined on } \mathrm{S}(\mathbb{R}^3) \ ,$$

is equivalent to the operator with a "twiggling" field, i.e. to the closure $K(t, \mu)$ of

$$-\Delta + V\left(x_1 - \frac{2\mu}{\omega^2} \cos \omega t, x_2, x_3\right) \quad \text{on } \mathrm{S}(\mathbb{R}^3) \ ,$$

provided V has some suitable smoothness properties.

Following *Yajima* [379], we discussed the associated Floquet operator and the self-adjoint realization of its "generator"

$$F(\mu) := -\mathrm{i} \frac{\partial}{\partial t} + K(t, \mu)$$

with periodic boundary conditions on the Hilbert space

$$\tilde{\mathrm{H}} := L^2(\mathbb{T}_\omega) \otimes L^2(\mathbb{R}^3) \ ,$$

where $\mathbb{T}_\omega$ is the torus

$$\mathbb{T}_\omega := \mathbb{R} \Big/ \frac{2\pi}{\omega} \cdot \mathbb{Z} \ .$$

For $\mu = 0$, it is obvious that the spectrum of $F(0)$ is an infinite overlapping of the spectrum of $H = \overline{(-\Delta + V)\restriction S}$, i.e.

$$\sigma(F(0)) = \bigcup_{n=-\infty}^{\infty} \{n\omega + \sigma(H)\}$$
$$= \bigcup_{n=-\infty}^{\infty} (\{n\omega + \sigma_{\text{ess}}(H)\} \cup \{n\omega + \sigma_p(H)\}) \ .$$

Thus, all eigenvalues of H appear as embedded eigenvalues in $\sigma(F(0))$.

Yajima [379] has shown if $F(0)$ is perturbed by the Stark field (i.e. if $\mu > 0$), these eigenvalues turn into resonances of $K(t, \mu)$ in the sense of dilation analyticity. More precisely, assume that V is dilation analytic as well as translation analytic (see Sect. 8.1 and 8.2), and let

$$F(\theta, \mu) := -\mathrm{i}\frac{\partial}{\partial t} + K(t, \theta, \mu) \ ,$$

where $K(t, \theta, \mu)$ is the closure of

$$(-\Delta)\mathrm{e}^{-2\theta} + V\left(\mathrm{e}^{\theta}x_1 - \frac{2\mu}{\omega^2}\cos\omega t, \mathrm{e}^{\theta}x_2, \mathrm{e}^{\theta}x_3\right) \text{ on } S(\mathbb{R}^3) \ .$$

Then

(i) for each $\mu > 0$ is $\{F(\theta, \mu)\}_{\theta \in S_\alpha}$ a holomorphic family of type (A), ($S_\alpha := \{\theta \in \mathbb{C} \,|\, |\mathrm{Im}\,\theta| < \alpha\}$, $\alpha > 0$ suitable)

(ii) $\sigma_{\text{ess}}(F(\theta, \mu)) = \bigcup_{n=-\infty}^{\infty} \{n\omega + \exp(-2\,\mathrm{Im}\,\theta)\mathbb{R}_+\}$

(iii) the isolated eigenvalues of $F(\theta, \mu)$ are independent of θ and stable in μ (in the sense that they converge to suitable eigenvalues of $F(0)$ if $\mu \to 0$). Furthermore, they have strictly negative imaginary part if $\mathrm{Im}\,\theta > 0$, $\mu > 0$, small for almost all $\omega \in \mathbb{R}_+$.

We give no proofs. Yajima also discusses the fact that the leading order for $\mathrm{Im}\,E$ depends on how many photons it takes to excite the state to the continuum.

As in Sect. 8.2, this theory has the disadvantage that the Coulomb potential is not included because of the translation analyticity needed. This difficulty has been overcome by *Graffi* and *Yajima* [141], where complex scaling is replaced by exterior scaling (see Sect. 8.7) which needs analyticity of the potential only outside a certain region.

8.7 Extensions and Generalizations

There are several generalizations of complex scaling. They were all developed to discuss potentials and physical situations where one expects resonances but the usual complex scaling technique does not apply.

The most important case which cannot be treated by usual complex scaling is the Born-Oppenheimer approximation. In the simplest case, this is a molecule consisting of two nuclei with fixed (clamped) distance and an electron which is moving in the electric field of the two nuclei.

Then one has the Hamiltonian

$$H_{BO} := (-\Delta) - \frac{z_A}{|x - R_A|} - \frac{z_B}{|x - R_B|} \quad \text{on } C_0^\infty(\mathbb{R}^3) ,$$

where R_A, R_B and z_A, z_B are the positions and charges of the nuclei, and x is the coordinate for the (electron) wave functions. The problem is that one wants to scale only the electronic coordinate but not the nuclear distance. Thus, one gets, for the potential

$$V(e^\theta x) := \frac{-z_A}{|e^\theta x - R_A|} + \frac{-z_B}{|e^\theta x - R_B|}, \quad x \in \mathbb{R}^3, \theta \in \mathbb{C} ,$$

which has a circle of square root branch point singularity at

$$\{x \in \mathbb{R}^3 | (e^\theta x - R_A)^2 = 0\}$$

and at a similar set around R_B. It is obvious that $\theta \mapsto V[\exp(\theta)x]$ has no analytic continuation, and V is therefore not dilation analytic (see [33, Theorem 4.1]).

To overcome this difficulty, *Simon* [328] introduced a method which scales only outside a ball containing the singularities. It is called, therefore, *exterior scaling.* More precisely, in this method the unitary scaling transformation (8.1) is replaced by the exterior scaling map

$$U(\theta, R)\varphi(x) := \left[\det \frac{\partial S(\theta, R)}{\partial x}\right]^{1/2} \varphi(S(\theta, R)x)$$

for $\theta \in \mathbb{R}$, $R > 0$, $\varphi \in L^2(\mathbb{R}^\nu)$ and

$$S(\theta, R)x := \begin{cases} x & \text{for } |x| \leq R \\ (R + e^\theta(|x| - R))\dfrac{x}{|x|} & \text{for } |x| \geq R . \end{cases}$$

Then one can show that for

$$H_0(\theta, R) := U(\theta, R) H_0 U(\theta, R)^{-1}$$

$$\sigma(H_0(\theta, R)) = \sigma_{\text{ess}}(H_0(\theta, R)) = \exp(-2i \operatorname{Im} \theta)\mathbb{R}_+$$

(see [141, Proposition 2.6]), and it is obvious that any shifted Coulomb potential gives rise to an analytic family of operators for suitable R. Thus, many of the arguments used in the usual complex scaling can be used here also.

Other extensions have been made by *Babbitt* and *Balslev* [34], *Balslev* [38], *Sigal* [311], *Cycon* [72] and *Hunziker* [174] to include potentials with compact support into the theory of complex scaling. The papers [311] and [72] are based on the idea that a complex valued map in the momentum space with bounded imaginary part gives rise to a unitary map of L^2, and this generates an analytic continuation for multiplication operators with compact support. Then all arguments are similar, as in Sect. 8.1.

For additional information, see [36].

9. Random Jacobi Matrices

In this chapter and the next one, we discuss two rather new subjects of mathematical research: random and almost periodic operators. Those operators serve in solid state physics as models of disordered systems, such as alloys, glasses and amorphous materials. The disorder of the system is reflected by the dependence of the potential on some random parameters. Let us discuss an example. Suppose we are given an alloy, that is, a mixture of (say two) crystalline materials. Suppose furthermore that the atoms (or ions) of the two materials generate potentials of the type $\lambda_1 f(x - x_0)$ and $\lambda_2 f(x - x_0)$, respectively, where x_0 is the position of the atom. If atoms of the two kinds are spread randomly on the lattice $\mathbb{Z}^\nu$ with exactly one atom at each site, then the resulting potential should be given by

$$V_\omega(x) = \sum_{i \in \mathbb{Z}^\nu} q_i(\omega) f(x - i) \ ,$$

where q_i are random variables assuming the values λ_1 and λ_2 with certain probabilities.

Schrödinger operators with stochastic (i.e. random or almost periodic) potentials show quite "unusual" spectral behavior. We will give examples, in this and the next chapter, of dense point spectrum, singular continuous spectrum and Cantor spectrum.

Despite intensive research by many mathematicians since the seventies, the theory of stochastic Schrödinger operators is far from being complete. In fact, most of the basic problems are unsolved in dimension $\nu > 1$.

We will not attempt to give a complete treatment of the subject, but rather introduce some of the basic problems, techniques and fascinating results in the field. We will not discuss stochastic Schrödinger operators, i.e. operators of the form $H_\omega = H_0 + V_\omega$ on $L^2(\mathbb{R}^\nu)$, but a discretized version of those operators acting on the sequence space $l^2(\mathbb{Z}^\nu)$, namely Jacobi matrices. The operator H_0 is replaced by a finite difference operator, and the potential becomes a function on $\mathbb{Z}^\nu$ rather than on $\mathbb{R}^\nu$. This model is known in solid state physics as the "tight binding approximation." We refer to Schrödinger operators on $L^2(\mathbb{R}^\nu)$ as the continuous case, and to the tight binding model as the discrete case.

The advantage of this procedure is twofold: Some technically difficult but unessential problems of the continuous case disappear, and our knowledge (especially for the almost periodic case) is larger for the discrete case.

Two recent reviews of random Schrödinger operators from distinct points of view from each other and from our discussion here are *Carmona* [61] and *Spencer* [347].

In this chapter, we assume some knowledge of basic concepts of probability theory. The required preliminaries can be found in any textbook on probability theory (for example, *Breiman* [53]).

9.1 Basic Definitions and Results

Let $u = \{u(n)\}_{n \in \mathbb{Z}^\nu}$ denote an element of $l^2(\mathbb{Z}^\nu)$; i.e. $\|u\| := [\sum_{n \in \mathbb{Z}} |u(n)|^2]^{1/2} < \infty$. Set $|n| = \max |n_j|$ and $|n|_+ := \sum_{j=1}^{\nu} |n_j|$ for $n \in \mathbb{Z}^\nu$. We define a discrete analog Δ_d of the Laplacian on $L^2(\mathbb{R}^d)$ by:

$$(\Delta_d u)(i) = \sum_{j; |j-i|_+ = 1} [u(j) - u(i)] = \left[\sum_{j; |j-i|_+ = 1} u(j) \right] - 2\nu u(i) \ . \tag{9.1}$$

Here Δ_d is a bounded operator on $l^2(\mathbb{Z}^\nu)$, a fact that makes life easier than in the continuous case. The spectrum of Δ_d is purely absolutely continuous, and $\sigma(\Delta_d) = \sigma_{ac}(\Delta_d) = [-4\nu, 0]$. This can be seen by Fourier transformation.

Note that

$$\langle u, -\Delta_d u \rangle = \sum_{\substack{i,j \\ |i-j|_+ = 1}} |u(i) - u(j)|^2$$

(each pair occurring once in the sum), explaining why we can regard (9.1) as an analog of the Laplacian.

If V is a function on $\mathbb{Z}^\nu$ playing the role of a potential, a natural analog of the Schrödinger operator is the operator

$$\bar{H} = -\Delta_d + V \ . \tag{9.2}$$

However, it is common to consider $+\Delta_d$ instead of $-\Delta_d$, and furthermore, to subsume the diagonal terms of Δ_d into the potential. Since Δ_d is bounded, this procedure has no "essential" effect on the properties of $\bar{H}$. Indeed, since the operator $(-1)^N$ defined by $[(-1)^N u](n) = (-1)^{|n|_+} u(n)$ obeys $(-1)^N \bar{H} [(-1)^N]^{-1} = 4\nu + \Delta_d + V$, $\bar{H}$ and the operator H below are unitarily equivalent up to a constant. Thus, we consider the operators

$$(H_0 u)(n) = \sum_{j; |j-n|_+ = 1} u(j) \quad \text{and} \tag{9.3}$$

$$(Hu)(n) = (H_0 u)(n) + V(n) u(n) \ . \tag{9.4}$$

The potentials V we are interested in form a random field, i.e. for any $n \in \mathbb{Z}^\nu$, the potential $V(n)$ evaluated at n is a random variable (= measurable function) on a probability space (Ω, F, P). F is a σ-algebra on Ω, and P a probability measure on (Ω, F). We adopt the common use in probability theory to denote the integral with respect to P by $\mathbb{E}$ (for "expectation"), i.e. $\int f(\omega)\, dP(\omega) =: \mathbb{E}(f)$. Without loss of generality, we may (and will) assume that

$$\Omega = S^{\mathbb{Z}^\nu} = \bigtimes_{\mathbb{Z}^\nu} S \ , \tag{9.5}$$

where S is a (Borel-) subset of $\mathbb{R}$, and F is the σ-algebra generated by the cylinder sets, i.e. by sets of the form $\{\omega | \omega_{i_1} \in A_1, \ldots, \omega_{i_n} \in A_n\}$ for $i_1, \ldots, i_n \in \mathbb{Z}^\nu$ and $A_1, \ldots, A_n$ Borel set in $\mathbb{R}$. We define the *shift operators* T_i on Ω by

$$T_i \omega(j) = \omega(j - i) \ . \tag{9.6}$$

A probability measure P on Ω is called *stationary* if $P(T_i^{-1} A) = A$ for any $A \in$ F. A stationary probability measure is called *ergodic*, if any shift invariant set A, i.e. a set A with $T_i^{-1} A = A$ for all $i \in \mathbb{Z}^\nu$, has probability, $P(A)$, zero or one.

If $V_\omega(n)$ is a real-valued random field on $\mathbb{Z}^\nu$, it can always be realized on the above probability space in such a way that $V_\omega(n) = \omega(n)$. V is called stationary (ergodic), if the corresponding probability measure P is stationary (ergodic).

An important example of an ergodic random field is a family of independent, identically distributed random variables. In this case, the measure P is just the product measure

$$\bigtimes_{i \in \mathbb{Z}^\nu} dP_0$$

of the common distribution P_0 of the random variables $V_\omega(i)$, i.e. $P_0(A) = P(V_\omega(i) \in A)$ for any $A \in B(\mathbb{R})$ and $i \in \mathbb{Z}^\nu$. We have, for example:

$$\begin{aligned} &\int f(\omega_{i_1}, \ldots, \omega_{i_n}) \, dP(\omega) [= \mathbb{E}(f)] \\ &\quad = \int f(x_1, \ldots, x_n) \, dP_0(x_1) \, dP_0(x_2) \ldots dP_0(x_n) \ . \end{aligned}$$

The Hamiltonian H_ω with V_ω i.i.d. is referred to as the *Anderson model*.

Another important class of ergodic potentials are almost periodic potentials. We introduce and investigate those potentials in Chap. 10.

For a fixed ω, the operator H_ω is nothing but a discretized Schrödinger operator with a certain potential. Therefore, it may seem to the reader that the introduction of a probability space is useless since we could as well consider each V_ω as a deterministic potential. The point of random potentials is that we are no longer interested in properties of H_ω for a fixed ω, but only in properties for *typical* ω. More precisely, we are interested in theorems of the form: H_ω has a property, p, for all ω in a set $\Omega_1 \subset \Omega$ with $P(\Omega_1) = 1$. This will be abbreviated by: H_ω has the property, p, P-almost surely (or P-a.s. or a.s.).

If not stated otherwise, V_ω is assumed to be a stationary ergodic random field satisfying $|V_\omega(n)| \leq C < \infty$ for all n and ω. However, the boundedness assumption can be omitted (or replaced by a moment condition) for many purposes.

We state and prove the following proposition for later use, as well as to demonstrate typical techniques concerning ergodicity. A random variable f is called *invariant* under T_i if $f(T_i \omega) = f(\omega)$ for all i.

Proposition 9.1. Suppose the family of measure preserving transformation, T_i, is ergodic. If a random variable, $f\colon \Omega \to \mathbb{R}$, is invariant under $\{T_i\}$, then f is constant P-a.s.

Remark. The proof extends easily to $f\colon \Omega \to \mathbb{R} \cup \{\infty\}$.

Proof. Define $\Omega_M = \{\omega | f(\omega) \leq M\}$. Since f is invariant, the set Ω_M is invariant under T_i and consequently has probability zero or 1. For $M \leq M'$ we have $\Omega_M \subset \Omega_{M'}$. Moreover,

$$\bigcup_{M \in \mathbb{R}} \Omega_M = \bigcup_{M \in \mathbb{Z}} \Omega_M = \Omega \quad \text{and}$$

$$\bigcap_{M \in \mathbb{R}} \Omega_M = \bigcap_{M \in \mathbb{Z}} \Omega_M$$

has probability zero. Thus,

$$M_0 = \inf_{P(\Omega_M)=1} M$$

is finite. Since

$$\Omega_{M_0} = \bigcap_{n \in \mathbb{N}} \Omega_{M_0+(1/n)} \quad \text{and}$$

$$\tilde{\Omega}_{M_0} := \{\omega | f(\omega) < M_0\} = \bigcup_{n \in \mathbb{N}} \Omega_{M_0-(1/n)}$$

we have $P(\Omega_{M_0}) = 1$, $P(\tilde{\Omega}_{M_0}) = 0$, and consequently

$$P(\{\omega | f(\omega) = M_0\}) = P(\Omega_{M_0} \backslash \tilde{\Omega}_{M_0}) = 1 \ . \quad \square$$

Let us define, for $u \in l^2(\mathbb{Z}^\nu)$

$$(U_i u)(n) = u(n - i) \ . \tag{9.7}$$

If V_ω is an ergodic potential with corresponding measure preserving transformations $\{T_i\}_{i \in \mathbb{Z}^\nu}$, then

$$H_{T_i \omega} = U_i H_\omega U_i^* \ , \tag{9.8}$$

a relation basic to some elementary properties of H_ω. Stochastic operators H_ω satisfying (9.8) are sometimes called *ergodic operators.*

The following theorem is a basic observation of *Pastur* [271] (with some technical supplements for the unbounded and the continuous case in [221] and [205]):

Theorem 9.2 (Pastur). Let V_ω be an ergodic potential. Then there exists a set $\Sigma \subset \mathbb{R}$ such that

$$\sigma(H_\omega) = \Sigma \quad P\text{-a.s.} \ .$$

Moreover,

$$\sigma_{\text{dis}}(H_\omega) = \phi \quad P\text{-a.s.} \ .$$

To prove Theorem 9.2, we need a preparatory lemma. We say that a family $\{A_\omega\}_{\omega\in\Omega}$ of bounded operators on a Hilbert space H is *weakly measurable* if the mapping $\omega \mapsto \langle\varphi, A_\omega\psi\rangle$ is measurable for all $\varphi, \psi \in$ H.

Lemma 9.3. Suppose that $\{P_\omega\}_{\omega\in\Omega}$ is a weakly measurable family of orthogonal projections satisfying (9.8). Then

$$\dim \operatorname{Ran}(P_\omega) \text{ is zero } P\text{-a.s.} \quad \text{or}$$

$$\dim \operatorname{Ran}(P_\omega) \text{ is infinite } P\text{-a.s.} \ .$$

Proof. The P_ω are positive operators, hence the trace $\operatorname{tr} P_\omega$ is uniquely defined (possibly $+\infty$). Fixing ω and choosing an orthonormal basis $a_1(\omega), a_2(\omega) \ldots$ of $\operatorname{Ran}(P_\omega)$ and an orthonormal basis $b_1(\omega), b_2(\omega) \ldots$ of $\operatorname{Ran}(P_\omega)^\perp$ we see that

$$\begin{aligned} \operatorname{tr} P_\omega &= \sum \langle a_i(\omega), P_\omega a_i(\omega)\rangle + \sum \langle b_i(\omega), P_\omega b_i(\omega)\rangle \\ &= \sum \langle a_i(\omega), a_i(\omega)\rangle = \dim \operatorname{Ran}(P_\omega). \end{aligned}$$

Now let $\{e_i, i \in \mathbb{Z}^\nu\}$ be the standard orthonormal basis in $l^2(\mathbb{Z}^\nu)$, i.e. $e_i(n) = \delta_{in}$. Then $\operatorname{tr} P_\omega = \sum \langle e_i, P_\omega e_i\rangle$ is a random variable (= measurable). Moreover,

$$\begin{aligned} \operatorname{tr} P_{T_j\omega} &= \sum \langle P_{T_j\omega} e_i, P_{T_j\omega} e_i\rangle \\ &= \sum \langle P_\omega e_{i-j}, P_\omega e_{i-j}\rangle = \operatorname{tr} P_\omega \ . \end{aligned}$$

By Proposition 9.1, $\dim \operatorname{ran} P_\omega = \operatorname{tr} P_\omega$ is thus a.s. constant. Hence P-a.s.:

$$\begin{aligned} \operatorname{tr} P_\omega &= \mathbb{E}(\operatorname{tr} P_\omega) \geq \sum_{|i|\leq N} \mathbb{E}(\langle e_i, P_\omega e_i\rangle) \\ &= \sum_{|i|\leq N} \mathbb{E}(\langle e_0, P_{T_i\omega} e_0\rangle) = \sum_{|i|\leq N} \mathbb{E}(\langle e_0, P_\omega e_0\rangle) \ , \end{aligned}$$

(where we use that T_i are measure preserving)

$$= (2N+1)^\nu \mathbb{E}(\langle e_0, P_\omega e_0\rangle)$$

Since N was arbitrary, $\operatorname{tr} P_\omega = 0$ or $\operatorname{tr} P_\omega = \infty$ according to $\mathbb{E}(\langle e_0, P_\omega e_0\rangle) = 0$ or not. □

Proof of Theorem 9.2. Denote the spectral projections of H_ω by $E_\Delta(\omega)$. Equation (9.8) implies that

$$E_\Delta(T_i\omega) = U_i E_\Delta(\omega) U_i^* \ . \tag{9.9}$$

This follows from the fact that the right-hand side of (9.9) is the spectral resolution for the operator $U_i H_\omega U_i^*$.

We now prove that for a fixed Borel set Δ the function $\omega \mapsto E_\Delta(\omega)$ is weakly measurable. It is not difficult to see that products of bounded, weakly measurable functions are weakly measurable. So, in particular, H_ω^n is weakly measurable for any $n \in \mathbb{N}$. We can approximate $E_\Delta(\omega)$ in the strong topology by ω-independent polynomials in H_ω. Thus, $E_\Delta(\omega)$ is weakly measurable. Therefore, by Lemma 9.3, for fixed Δ dim Ran(E_Δ) is either zero a.s. or infinite a.s.

For any pair $\langle p, q \rangle$ of rational numbers, we set $\eta(p,q) := 0$ if $\dim \operatorname{Ran}(E_{(p,q)}(\omega)) = 0$ P-a.s. and $\eta(p,q) := \infty$ if $\dim \operatorname{Ran}(E_{(p,q)}(\omega)) = \infty$ P-a.s. According to Lemma 9.3, $\eta(p,q)$ is well defined. Define $\Omega_{p,q} := \{\omega \mid \dim \operatorname{Ran} E_{(p,q)}(\omega) = \eta(p,q)\}$ and

$$\Omega_0 = \bigcap_{p,q \in \mathbb{Q}} \Omega_{p,q} .$$

Since $\Omega_{p,q}$ has probability 1 and the intersection over p, $q \in \mathbb{Q}$ is countable, we have $P(\Omega_0) = 1$.

We claim that for $\omega_1, \omega_2 \in \Omega_0$ the spectra $\sigma(H_{\omega_1})$ and $\sigma(H_{\omega_2})$ coincide. Indeed, if $\lambda \notin \sigma(H_{\omega_1})$, then

$$\dim \operatorname{Ran} E_{(\lambda_1, \lambda_2)}(\omega_1) = 0$$

for all λ_1, λ_2 with $\lambda_1 < \lambda < \lambda_2$ sufficiently near to λ. Since $\omega_1, \omega_2 \in \Omega_0$ we have

$$\dim \operatorname{Ran} E_{(p,q)}(\omega_1) = \dim \mathbb{E}_{(p,q)}(\omega_2)$$

for p, $q \in \mathbb{Q}$, so

$$\dim \operatorname{Ran} E_{(\lambda_1, \lambda_2)}(\omega_2) = 0$$

for $\lambda_1, \lambda_2 \in \mathbb{Q}$ with $\lambda_1 < \lambda < \lambda_2$ sufficiently near to λ. This implies $\lambda \notin \sigma(H_{\omega_2})$. The claim follows by interchanging the roles of ω_1 and ω_2.

Now, suppose that $\lambda \in \sigma_{\text{dis}}(H_\omega)$ for an $\omega \in \Omega_0$. Then

$$0 < \dim \operatorname{Ran} E_{(\lambda_1, \lambda_2)}(\omega) < \infty$$

for some $\lambda_1 < \lambda < \lambda_2$, λ_1, $\lambda_2 \in \mathbb{Q}$. But this contradicts the choice of $\omega \in \Omega_0$. So $\sigma_{\text{dis}}(H_\omega) = \phi$ P-a.s. $\square$

Remark. (1) The use of the countable set of pairs $\mathbb{Q} \times \mathbb{Q}$ in the above proof is essential, since an uncountable intersection of sets of full probability may have probability strictly less than 1.

(2) To prove the result for unbounded operators needs a bit more technique to prove the weak measurability of $E_\Delta(\omega)$ (see [205]).

The following theorem is due to *Kunz-Souillard* [221], and was extended to a more general context by *Kirsch* and *Martinelli* [205]:

Theorem 9.4 (Kunz-Souillard). Let V_ω be an ergodic potential. Then there exist sets $\sum_{\text{ac}}, \sum_{\text{sc}}, \sum_{\text{pp}} \subset \mathbb{R}$ such that

$$\sigma_{ac}(H_\omega) = \Sigma_{ac} \quad P\text{-a.s.}$$

$$\sigma_{sc}(H_\omega) = \Sigma_{sc} \quad P\text{-a.s.}$$

$$\sigma_{pp}(H_\omega) = \Sigma_{pp} \quad P\text{-a.s.}$$

Notational Warning. By $\sigma_{pp}(H)$ we denote the *closure* of the set $\varepsilon(H) := \{\lambda | \lambda$ is an eigenvalue of $H\}$. This notation disagrees with that of *Reed* and *Simon* I [292]. In fact, the above theorem would be wrong for $\varepsilon(H_\omega)$!

Remark. At first glance, given Theorem 9.2, Theorem 9.4 looks rather trivial. However, there is a pitfall: The necessary measurability of certain projections is nontrivial.

Proof. We define $E_\Delta^c(\omega) := E_\Delta(\omega)P_c(\omega)$ where $P_c(\omega)$ is the projector onto the continuous subspace w.r.t. H_ω. Analogously, we define $E_\Delta^{ac}(\omega) := E_\Delta(\omega)P_{ac}(\omega)$, etc. Then the proof of Theorem 9.2, with $E_\Delta(\omega)$ replaced by $E_\Delta^{ac}(\omega)$ etc., proves Theorem 9.4 except for one point: We have to prove the weak measurability of $E_\Delta^{ac}(\omega)$, $E_\Delta^{sc}(\omega)$ and $E_\Delta^{pp}(\omega)$. For this, it suffices to prove the weak measurability of $E_\Delta^c(\omega)$ and $E_\Delta^s(\omega)$.

It is not difficult to verify by the RAGE-theorem (Theorem 5.8) that

$$\langle \varphi, P_c(\omega)\psi \rangle = \lim_{J\to\infty} \lim_{T\to\infty} \frac{1}{2T} \int_{-T}^{T} \langle \varphi, e^{itH_\omega} F(|j| > J) e^{-itH_\omega}, \psi \rangle \, dt \tag{9.10}$$

where $F(A)$ is multiplication with the characteristic function of A. Since

$$e^{itH_\omega} = \sum \frac{(itH_\omega)^n}{n!} ,$$

the right-hand side of (9.10) is measurable, hence $P_c(\omega)$ and therefore $E_\Delta^c(\omega) := E_\Delta(\omega)P_c(\omega)$ is weakly measurable.

We prove the weak measurability of $E_\Delta^s(\omega)$ using an argument of *Carmona* [61]. We need a lemma:

Lemma. Let $\mathscr{I}$ be the family of finite unions of open intervals, each of which has rational endpoints. Then, for any Borel set A,

$$\mu_{sing}(A) = \lim_{n\to\infty} \sup_{I\in\mathscr{I}, |I|<n^{-1}} \mu(A \cap I) =: \nu(A) ,$$

where $|\cdot|$ = Lebesgue measure.

Proof of Lemma. Note first that the sup decreases as n increases, so the limit defining $\nu(A)$ exists.

Write $d\mu_{a.c.}(x) = f(x)\,dx$ and set $g(R) = \mu_{a.c.}(\{x | f(x) > R\})$ so $g(R) \searrow 0$ as

$R \to \infty$. Then $\mu_{\text{a.c.}}(I) \leq R|I| + g(R)$ so

$$\mu(A \cap I) \leq \mu_s(A) + R|I| + g(R) \ .$$

Thus, for any R, $v(A) \leq \mu_s(A) + g(R)$, and thus $v(A) \leq \mu_s(A)$.

Conversely, find B with $|B| = 0$ and $\mu_s(\mathbb{R}\backslash B) = 0$. Find open sets C_m so $A \cap B \subset C_m$ and $|C_m| \searrow 0$. Given n and ε, find m so $|C_m| < n^{-1}$ and then $I \in \mathscr{I}$, so $\mu(C_m \backslash I) \leq \varepsilon$. Then

$$\mu(A \cap I) \geq \mu(A \cap C_m) - \mu(C_m \backslash I) \geq \mu(A \cap B) - \varepsilon = \mu_s(A) - \varepsilon \ .$$

Thus, $v(A) \geq \mu_s(A) - \varepsilon$ so $v(A) \geq \mu_s(A)$. □

Conclusion of the Proof of Theorem 9.4. For any φ,

$$(\varphi, E^s_\Delta(\omega)\varphi) = \lim_{n\to\infty} \sup_{I \in \mathrm{I}, |I| < n^{-1}} (\varphi, E_{\Delta \cap I}(\omega)\varphi) \ ,$$

since I is countable and $(\varphi, E_{\Delta \cap I}(\omega)\varphi)$ is measurable, we conclude that $(\varphi, E^s_\Delta(\omega)\varphi)$ is measurable. By polarization, $E^s_\Delta(\omega)$ is weakly measurable. □

We close this section with the following observation due to *Pastur* [271] which is special to the one-dimensional case:

Theorem 9.5 (Pastur). If $v = 1$, then for any given λ, $P(\{\omega | \lambda$ is an eigenvalue of $H_\omega\}) = 0$.

Remarks (1). It does *not* follow from Theorem 9.5 that H_ω has no eigenvalues. An uncountable union of sets of probability zero may have positive probability (or may even be unmeasurable).

(2) While this observation of Pastur and the proof we give is one-dimensional, the result is true in any dimension. It follows from Theorem 9.9 below (see *Avron* and *Simon* [31]). This multidimensional result is more subtle, and is still not proven for the continuous case.

Proof. $\operatorname{tr} E_{\{\lambda\}} = 0$ a.s. or $\operatorname{tr} E_{\{\lambda\}} = \infty$ a.s. according to Lemma 9.3. But our one-dimensional finite difference equation has at most a two-dimensional space of solutions, hence $\operatorname{tr} E_{\{\lambda\}} = 0$ a.s. □

Corollary. If the point spectrum $\Sigma_{\text{pp}}[= \sigma_{\text{pp}}(H_\omega)\text{a.s.}]$ is non-empty, then it is locally uncountable.

9.2 The Density of States

In this section, we briefly discuss an important quantity for disordered systems, the density of states $k(E)$. For recent surveys on this subject, see [202, 342]. The quantity $k(E)$ measures, in some sense, "how many states" correspond to energies below the level E.

Recall that our Hamiltonians H_ω model the motion of a single particle (electron) in a solid with infinitely many centers of forces (nuclei, ions) located at some fixed positions. This is the so-called one-body approximation. However, in a solid with infinitely many nuclei, we also have infinitely many electrons. We cannot handle directly a problem with infinitely many particles, but we should at least take into account the fermionic nature of the electrons via the Pauli exclusion principle. This principle states that two fermions (e.g. electrons or protons) cannot occupy the same quantum mechanical state (see also Chap. 3). This leads to the well-known fact that electrons in an atom do not all have the "ground-state energy", but fill up the energy levels starting from the ground state energy up to a certain level. Such a phenomenon also should occur in our disordered solid. However, we are faced with the problem of having to distribute *infinitely many* fermions on a *continuum* of *energy levels*. To get rid of this problem, we will restrict the problem first to a finite domain. In such a domain, we should have only finitely many electrons. To do this, let, as usual, $E_\Delta(\omega)$ denote a spectral projection measure associated with H_ω and denote, by χ_L, the characteristic function of the "cube" $C_L = \{i \in \mathbb{Z}^\nu | -L \leq i_k \leq L; k = 1, \ldots, \nu\}$. The "number of electrons" in C_L should be a density times $\# C_L = (2L+1)^\nu$. We define a measure dk_L by

$$\int_A dk_L = \frac{1}{(2L+1)^\nu} \dim \operatorname{Ran}(\chi_L E_A(\omega) \chi_L) = \frac{1}{(2L+1)^\nu} \operatorname{tr}(E_A(\omega)\chi_L) \ , \qquad (9.11)$$

This measures "how many electrons per lattice site (i.e. per nucleus) can be put into energy levels in the set A without violating the Pauli principle if we restrict the whole problem to the cube C_L." We may hope that the measures dk_L converge (in some sense), if we send C_L to $\mathbb{Z}^d$ (i.e. $L \to \infty$).

The appropriate convergence of measures is the vague convergence, i.e. $d\mu_n \to d\mu$ if $\int f\, d\mu_n \to \int f\, d\mu$ for any continuous function f with compact support. We define the measure dk by

$$\int f(\lambda)\, dk(\lambda) := \mathbb{E}(f(H_\omega)(0,0)) \ ,$$

where $A(0,0)$ is a shorthand notation for $\langle \delta_0, A\delta_0 \rangle$.

Theorem 9.6. For any bounded measurable function f there is a set Ω_f of probability 1 such that

$$\int f(\lambda)\, dk_L(\lambda) \to \int f(\lambda)\, dk(\lambda) \quad \text{as } L \to \infty \ . \qquad (9.12)$$

Proof. Fix a bounded measurable function f. Define $\bar{f}(\omega) := f(H_\omega)(0,0)$. By stationarity,

$$f(H_\omega)(n,n) = \bar{f}(T_n\omega) \ .$$

We have

$$\int f(\lambda)\,dk_L(\lambda) = \frac{1}{(2L+1)^\nu}\operatorname{tr}(f(H_\omega)\chi_L)$$

$$= \frac{1}{(2L+1)^\nu}\sum_{|n|\le L} f(H_\omega)(n,n) = \frac{1}{(2L+1)^\nu}\sum_{|n|\le L} \bar{f}(T_n\omega) = (*) \ .$$

To the last expression we apply Birkhoff's ergodic theorem (see e.g. *Breiman* [53]) which states that

$$\frac{1}{(2L+1)^\nu}\sum_{|i|\le L} \varphi(T_n\omega) \text{ converges to } \mathbb{E}(\varphi) \ P\text{-a.s. if } \varphi\in L^1(P) \ .$$

Thus,

$$\lim(*) = \mathbb{E}(\bar{f}(\omega)) = \mathbb{E}(f(H_\omega)(0,0)) = \int f(\lambda)\,dk(\lambda) \ . \quad \square$$

Since the set Ω_f may depend on f, it is not clear that there is an $\omega\in\Omega$ such that (9.12) is true for all bounded measurable functions f. However, we have

Theorem 9.7. dk_L converges vaguely to dk for P-a.s.

Remark. Notice that the limit measure dk is non-random.

Proof. There exists a countable subset F, of C_0, the continuous functions with compact support, such that for any $f\in C_0$ there is a sequence $\{f_n\}$ in F with $f_n\to f$ uniformly, and $\bigcup_n \operatorname{supp} f_n$ is contained in a (f-dependent) compact set. Set

$$\Omega_0 := \bigcap_{g\in \mathrm{F}} \Omega_g \ .$$

We have $P(\Omega_0) = 1$. Moreover, one checks that (9.12) holds for any $\omega\in\Omega_0$ and any $f\in C_0$. $\square$

We define

$$k(E) := \int \chi_{(-\infty,E)}(\lambda)\,dk(\lambda)$$

and call this quantity the *integrated density of states.* (Note that it is sometimes this quantity that is called "density of states" in the literature.)

The following theorem states a connection between the spectrum and the density of states.

Theorem 9.8 (Avron and Simon [31]).

$$\operatorname{supp}(dk) = \Sigma[=\sigma(H_\omega)\text{a.s.}] \ .$$

Remark. From our intuition at the beginning of this section, the theorem certainly should hold.

Proof. If $\lambda_0\notin\Sigma$, there is a non-negative continuous function with $f(\lambda_0) = 1$ and $f = 0$ on Σ. Thus, $f(H_\omega)\equiv 0$ and so $\int f(\lambda)\,dk = (f(H_\omega)(0,0)) = 0$ so $\lambda_0\notin\operatorname{supp} dk$.

Conversely, if $\lambda_0 \notin \operatorname{supp}(dk)$, then there is a positive continuous function f with $f(\lambda_0) = 1$ and $\int f(\lambda)\, dk = 0$. Thus, for a.e. ω, $f(H_\omega)(0,0) = 0$, and so by $f(H_\omega)(n,n) = f(H_{T^n\omega})(0,0)$ we know that a.e. ω, $(\delta_n, f(H_\omega)\delta_n) = 0$ for all n. But since $f(H_\omega) \geq 0$, this implies that $f(H_\omega) = 0$. Since f is continuous and $f(\lambda_0) = 1$, this implies that $\lambda_0 \notin \Sigma$. □

It is easy to see that the measure dk is a continuous measure, i.e. the function $k(E)$ is continuous, in the one-dimensional case. It was proven by *Craig* and *Simon* [69] that in the multidimensional case, $k(E)$ is even log-Hölder continuous (see [69] for details). Those authors use a version of the Thouless formula (see Chap. 9.4) for a strip to establish this result. Recently *Delyon* and *Souillard* [86] found a very elementary proof for the continuity (not log-Hölder continuity) of k.

Theorem 9.9 (Craig-Simon, Delyon-Souillard). $k(E)$ is a continuous function.

Proof. We follow *Delyon* and *Souillard* [86]. Fix λ. Let f_n be a sequence of continuous functions with $f_n(\lambda) = 1$ and $f_n(x) \downarrow 0$ if $x \neq \lambda$. Then $f_n(H_\omega)(0,0) \downarrow E_{\{\lambda\}}(0,0)$ and $\int f_n(x)\, dk(x) \downarrow k(\lambda + 0) - k(\lambda - 0)$. Thus, by the definition of dk and Theorem 9.6, it is enough to prove that

$$\mathbb{E}(E_{\{\lambda\}}(0,0)) = \lim \frac{1}{(2L+1)^\nu} \operatorname{tr}(E_{\{\lambda\}}\chi_L) = 0 .$$

We remark that the set where the first equality holds may be λ-dependent, but that does not change the fact that for λ fixed it holds a.e., and we need only look at a typical point.

A solution ψ of $H_\omega \psi = E\psi$ is uniquely determined inside C_L by its values on

$$\tilde{C}_L = \bigcup_{j=1}^{\nu} \tilde{C}_{L,j} ,$$

where $\tilde{C}_{L,1} = \{i \in C_L | i_1 = -L \text{ or } -L+1\}$ and $\tilde{C}_{L,k} = \{i \in C_L | i_k = -L \text{ or } L\}$ if $k \geq 2$. For

$$\psi(\alpha) = [E - V(\alpha - \delta_1)]\psi(\alpha - \delta_1) - \psi(\alpha - 2\delta_1) - \sum_{j=2}^{\nu} [\psi(\alpha - \delta_1 + \delta_j) + \psi(\alpha - \delta_1 - \delta_j)]$$

allows us then to determine ψ inductively for $\alpha_1 = -L+2, \ldots, L$. Thus,

$$\dim \chi_L[\operatorname{Ran} E_{\{\lambda\}}] \leq \#(\tilde{C}_L) \leq 2\nu(2L+1)^{\nu-1} .$$

But

$$\operatorname{tr}(\chi_L E_{\{\lambda\}}) \leq \dim \chi_L(\operatorname{Ran} E_{\{\lambda\}}) \|\chi_L E_{\{\lambda\}}\| \leq \dim \chi_L(\operatorname{Ran} E_{\{\lambda\}})$$

so $(2L+1)^{-\nu} \operatorname{tr}(\chi_L E_{\{\lambda\}}) \to 0$. □

There are some more results on the regularity of the density of states. We mention Wegner's proof of the existence of a density for dk for the Anderson model when the common distribution has a density (see *Wegner* [366] or *Fröhlich* and *Spencer* [119] Appendix C). *Constantinescu Fröhlich* and *Spencer* [67] proved the analyticity of $k(E)$ if $|E|$ is large for i.i.d. Gaussian $V_\omega(n)$.

However, for $\nu > 1$, there seems to be no regularity result for $k(E)$ in the continuous case, so far. For $\nu = 1$, it is not difficult to show that $k(E)$ is continuous.

Frequently the density of states is defined in a slightly different way than above. Instead of restricting $E_\Lambda(\omega)$ to cubes, one restricts H_ω itself. Let us set $C_{N,M} := \{k \in \mathbb{Z}^\nu | N \le k_i \le M;\ i = 1, \ldots, \nu\}$ for N, $M \in \mathbb{Z}$. We then define an operator $H_\omega^{(N,M)}$ on $l^2(C_{N,M}) \simeq \mathbb{C}^{(M-N+1)^\nu})$ by its matrix elements:

$$H_\omega^{(N,M)}{}_{i,j} = \langle \delta_i, H_\omega^{(N,M)} \delta_j \rangle := \langle \delta_i, H_\omega \delta_j \rangle \tag{9.13}$$

for i, $j \in C_{N,M}$. Equation (9.13) can be looked upon as imposing "boundary conditions" $u(k) = 0$ for $k \notin C_{N,M}$, k nearest neighbors to $C_{N,M}$.

We set

$$\rho_{N,M}(A) := \#\{\lambda \in A | \lambda \text{ is an eigenvalue of } H_\omega^{(N,M)}\} \ . \tag{9.14}$$

It can be shown (see *Avron* and *Simon* [31]) that the measures $(\# C_{N,M})^{-1}\, d\rho_{N,M}$ converge vaguely to dk as $|M - N| \to \infty$.

The rigorous investigation of the density of states goes back to *Benderskii-Pastur* [44], who proved the existence of k as the limit of $d\rho_{-L,L}/(2L+1)^\nu$. The existence of dk was proven in increasing generality and by different methods by *Pastur* [269], *Nakao* [262], *Kirsch* and *Martinelli* [206] and others. The way of defining dk through dk_L is due to *Avron* and *Simon* [31]. The definition of dk via the rotation number in the one-dimensional continuous case was introduced by *Johnson* and *Moser* [186] (see *Delyon* and *Souillard* [85] for the discrete case).

There is large interest in the asymptotic behavior of $k(E)$ for large and small values in E. In the continuous case, $k(E)$ behaves like $\tau_\nu E^{\nu/2}/(2\pi)^\nu$ as $E \to \infty$ (τ_ν is the volume of the unit ball in $\mathbb{R}^\nu$); see *Pastur* [269], *Nakao* [262], *Kirsch* and *Martinelli* [206]. This is the same behavior as for the free operator H_0. However, as E goes to $E_0 := \inf \sigma(H_\omega)$, the behavior of $k(E)$ differs heavily from the free case. As a rule, $k(E)$ decays for $E \searrow E_0$ much faster than $k_0(E)$—the density of states for H_0. For $E_0 > -\infty$, it was predicted by *Lifshitz* [234] on the basis of physical arguments that $k(E)$ should behave like $C_1 \exp[-c_2(E - E_0)^{-\nu/2}]$ as $E \searrow E_0$, which is now called the Lifshitz behavior. For rigorous treatment of the Lifshitz behavior for the discrete case, see *Fukushima* [123], *Romerio* and *Wreszinski* [298] and *Simon* [341]; for the continuous case, see *Nakao* [262], *Pastur* [270], *Kirsch* and *Martinelli* [207] and *Kirsch* and *Simon* [208]. The behavior of $k(E)$ as $E \searrow E_0 = -\infty$ is treated in *Pastur* [269] (see also *Fukushima, Nagai* and *Nakao* [124], *Nakao* [262], *Kirsch* and *Martinelli* [206]).

9.3 The Lyaponov Exponent and the Ishii-Pastur-Kotani Theorem

For most of the rest of this chapter, we suppose that $v = 1$, for it is the one-dimensional case which is best understood. What makes the one-dimensional case accessible is that, for fixed E, a solution of $(H_\omega - E)u = 0$ is determined by its values at two succeeding points (initial value problem for a second-order difference equation).

Fix E. We consider the one-dimensional difference equation of second order

$$u(n+1) + u(n-1) + [V_\omega(n) - E]u(n) = 0 \ . \tag{9.15}$$

and introduce the vector-valued function

$$\underline{u}(n) = \begin{pmatrix} u(n+1) \\ u(n) \end{pmatrix} .$$

Define

$$A_n(E) := A_n(E, \omega) := \begin{pmatrix} E - V_\omega(n) & -1 \\ 1 & 0 \end{pmatrix} .$$

A function $u(n)$ is a solution of (9.15) if and only if

$$\underline{u}(n+1) = A_{n+1}(E)\underline{u}(n) \ .$$

Set

$$\Phi_n(E) := \Phi_n(\omega, E) := A_n(E)A_{n-1}(E) \dots A_1(E) \ . \tag{9.16}$$

Then

$$\underline{u}(n) = \Phi_n(E)\underline{u}(0)$$

defines the solution of (9.15) "to the right" with initial condition

$$\underline{u}(0) = \begin{pmatrix} u(1) \\ u(0) \end{pmatrix} .$$

Similarly $\underline{u}(-n) = \Phi_{-n}(E)\underline{u}(0)$ with $\Phi_{-n}(E) := A_{-n+1}(E)^{-1} \dots A_0(E)^{-1}$ defines the solution to the left. Note that

$$A_n(E)^{-1} = \begin{pmatrix} 0 & 1 \\ -1 & E - V_\omega(n) \end{pmatrix} .$$

We now define

$$\overline{\gamma}^{\pm}(\omega, E) = \overline{\lim_{N\to\pm\infty}} \frac{1}{|N|} \ln \|\Phi_N(\omega, E)\| \tag{9.17a}$$

$$\underline{\gamma}^{\pm}(\omega, E) := \underline{\lim_{N\to\pm\infty}} \frac{1}{|N|} \ln \|\Phi_N(\omega, E)\| \ . \tag{9.17b}$$

Remark. For definiteness, let $\|A\|$ denote the operator norm of the matrix A. However, the limits (9.17a) and (9.17b) do not change, if we use another norm.

These quantities measure the growth of the matrix norm $\|\Phi_N(\omega, E)\|$. Since $\det A_i(E) = 1$ and hence $\det \Phi_N = 1$, it follows that $\|\Phi_N(\omega, E)\| \geq 1$, and consequently $0 \leq \underline{\gamma}^{\pm}(\omega, E) \leq \overline{\gamma}^{\pm}(\omega, E)$. Indeed, if $\det A = 1$, at least one eigenvalue of A satisfies $|\lambda| \geq 1$, thus, $\|A\| \geq 1$.

Theorem 9.10 (Furstenberg and Kesten [126]). For fixed E and P-almost all ω

$$\gamma^{\pm}(E) := \lim_{N\to\pm\infty} \frac{1}{|N|} \ln \|\Phi_N(\omega, E)\| \tag{9.18}$$

exists, is independent of ω and

$$\gamma^{+}(E) = \gamma^{-}(E) \ . \tag{9.19}$$

We call $\gamma(E) := \gamma^{\pm}(E)$ the *Lyaponov exponent* for H_ω. We will see in a moment that $\gamma(E)$ plays a central role in the investigation of one-dimensional stochastic Jacobi matrices.

To prove (9.18), we will exploit Kingman's subadditive ergodic theorem [200] which we state without proof. We remark that a multi-dimensional version of the subadditive ergodic theorem can be used to prove the existence of the (integrated) density of states by means of Dirichlet-Neumann bracketing (see [206, 325]).

Since now $\nu = 1$, we have $T_n = (T_1)^n$. We set $T := T_1$, so $T_n = T^n$. We call T ergodic if $\{T^n\}_{n\in\mathbb{Z}}$ is ergodic. A sequence $\{F_N\}_{N\in\mathbb{N}}$ of random variables is called a *subadditive* process if

$$F_{N+M}(\omega) \leq F_N(\omega) + F_M(T^N\omega)$$

where T is a measure preserving transformation.

Theorem 9.11 (Subadditive Ergodic Theorem, Kingman [200]). If F_N is a subadditive process satisfying $\mathbb{E}(|F_N|) < \infty$ for each N, and $\Gamma(F) := \inf \mathbb{E}(F_N)/N > -\infty$, then $F_N(\omega)/N$ converges almost surely. If, furthermore, T is ergodic, then

$$\lim_{N\to\infty} \frac{1}{N} F_N(\omega) = \Gamma(F)$$

almost surely.

Proof of Theorem 9.10. Define $F_N(\omega) = \ln\|\Phi_N(\omega, E)\|$. Since

$$\begin{aligned}
F_{N+M}(\omega) &= \ln\left\|\prod_{j=N+1}^{N+M} A_j(\omega, E)\cdot\prod_{i=1}^{N} A_i(\omega, E)\right\| \\
&= \ln\left\|\prod_{j=1}^{M} A_j(T_N\omega, E)\cdot\prod_{i=1}^{N} A_i(\omega, E)\right\| \\
&\le \ln(\|\Phi_M(T_N\omega, E)\|\,\|\Phi_N(\omega, E)\|) \\
&= F_M(T_N\omega) + F_N(\omega)
\end{aligned}$$

the process F_N is subadditive.

Moreover,

$$\begin{aligned}
\frac{1}{N}\mathbb{E}(|F_N|) &= \frac{1}{N}\mathbb{E}\left(\ln\left\|\prod_{j=1}^{N} A_j(\omega, E)\right\|\right) \\
&\le \frac{1}{N}\mathbb{E}\prod_{j=1}^{N}\ln\|A_j(\omega, E)\| \\
&= \frac{1}{N}\prod_{j=1}^{N}\mathbb{E}(\ln\|A_j(\omega, E)\|) \\
&= \mathbb{E}(\ln\|A_0(\omega, E)\|)\ ,
\end{aligned}$$

where we used the stationarity of $V_\omega(n)$ in the last step. Moreover, $\mathbb{E}(\ln^+|V_\omega(0)|) < \infty$ implies $\mathbb{E}(\ln\|A_0(\omega, E)\|) < \infty$. In addition, as noted above, $F_N \ge 0$ so $\inf[\mathbb{E}(F_N)/N] \ge 0 > -\infty$. Thus, Theorem 9.11 implies that

$$\lim_{N\to+\infty}\frac{1}{N}\ln\|\Phi_N(\omega, E)\| = \inf_{N>0}\frac{1}{N}\mathbb{E}(\ln\|\Phi_N(\omega, E)\|) \quad \text{a.s.}$$

and

$$\lim_{N\to-\infty}\frac{1}{|N|}\ln\|\Phi_N(\omega, E)\| = \inf_{N<0}\frac{1}{|N|}\mathbb{E}(\ln\|\Phi_N(\omega, E)\|) \quad \text{a.s.}$$

Now we prove (9.19). Since, for $N > 0$, $\Phi_{-N} = A_{-N}^{-1}\dots A_{-1}^{-1}A_0^{-1} = (A_0A_{-1}\dots A_{-N})^{-1}$, we have by stationarity

$$\mathbb{E}(\ln\|\Phi_{-N+1}\|) = \mathbb{E}(\ln\|\Phi_N^{-1}\|)\ . \tag{9.20}$$

Moreover, for

$$J = \begin{pmatrix} 0 & -1 \\ 1 & 0 \end{pmatrix},$$

we have

$$(J\Phi_N J^{-1})^t = \Phi_N^{-1} .$$

Thus, since $\|J\underline{u}\| = \|J^{-1}\underline{u}\| = \|\underline{u}\|$ and we have (9.20), it follows that $\gamma^+ = \gamma^-$. □

The following "multiplicative ergodic theorem" of *Osceledec* [267] connects the large N behavior of Φ_N with the behavior of solutions $\Phi_N\underline{u}$.

Theorem 9.12 (Multiplicative Ergodic Theorem, Osceledec). Suppose $\{A_n\}_{n\in\mathbb{N}}$ is a sequence of real 2×2 matrices satisfying (i) $\lim_{n\to\infty}(1/n)\ln\|A_n\| = 0$ and (ii) $\det A_n = 1$. If $\gamma := \lim_{n\to\infty}(1/n)\ln\|A_n\cdot\ldots\cdot A_1\| > 0$, then there exists a one-dimensional subspace $V^- \subset \mathbb{R}^2$ such that

$$\lim \frac{1}{n}\ln\|A_n\cdot\ldots\cdot A_1 v\| = -\gamma \quad \text{for } v\in V_-,\, v\neq 0$$

and

$$\lim \frac{1}{n}\ln\|A_n\cdot\ldots\cdot A_1 v\| = \gamma \quad \text{for } v\notin V_- .$$

Osceledec proved a probabilistic version of this theorem (see Osceledec [267], *Raghunathan* [291]). Ruelle realized it was a deterministic theorem; for a proof, see *Ruelle* [301].

Osceledec's theorem tells us that under the hypothesis $\gamma(E) > 0$, there exists P-a.s. only "exponentially growing" and "exponentially decaying" solution (to the right) of the equation $H_\omega u = Eu$. The exponentially decaying solution occurs only for a particular initial condition $\lambda\underline{u}_+$; any other initial condition leads to an increasing solution. The same is true for solutions to the left with a particular initial condition $\lambda\underline{u}_-$. An (l^2-) eigenvector can only occur if the lines $\lambda\underline{u}_+$, $\lambda\underline{u}_-$ happen to coincide.

However, we have to be very careful with assertions as above, since these considerations are justified only for fixed E. If we allow E to vary through an uncountable set, it may happen that the exceptional ω for which $\gamma^\pm(\omega, E) \neq \gamma(E) > 0$ may add up to a set of measure 1!

For example, we *cannot* conclude that for P-a.a. ω any solution of $H_\omega u = Eu$ (with arbitrary $E\in\mathbb{R}$) is either exponentially increasing or decreasing! However, the Lyaponov exponent $\gamma(E)$ characterizes the absolutely continuous spectrum completely.

Suppose μ is a measure on $\mathbb{R}$, $\mu_{\text{a.c.}}$ its absolutely continuous part. We call a set A an *essential support* of $\mu_{\text{a.c.}}$ if: (1) There is a set, B, of Lebesgue measure zero such that $\mu(\mathbb{R}\backslash(A\cup B)) = 0$. (2) If $\mu(C) = 0$, then the Lebesgue measure $|A\cap C|$ is zero. The essential support is defined uniquely up to sets of Lebesgue measure zero. We define the *essential closure* $\bar{A}^{\text{ess}} := \{\lambda \,|\, |A\cap(\lambda-\varepsilon, \lambda+\varepsilon)| > 0$ for all $\varepsilon > 0\}$.

Theorem 9.13 (Ishii-Pastur-Kotani). Suppose that V_ω is a bounded ergodic process. Then

$$\sigma_{\text{a.c.}}(H_\omega) = \overline{\{E|\gamma(E) = 0\}}^{\text{ess}} \ . \tag{9.21}$$

Moreover, the set $\{E|\gamma(E) = 0\}$ is the essential support of $E_\Delta^{\text{a.c.}}(H_\omega)$.

Ishii [176] and *Pastur* [271] proved that $\sigma_{\text{a.c.}}(H_\omega) \subset \overline{\{E|\gamma(E) = 0\}}^{\text{ess}}$. Kotani [215] proved the converse for the continuous case. His method was adopted to the discrete case by *Simon* [336]. See *Minami* [247] for further information. The Kotani part of the proof requires the use of theorems on H^2-functions on the unit disc. We will only give the Ishii-Pastur part.

Proof. We only prove $\sigma_{\text{a.c.}}(H_\omega) \subset \overline{\{E|\gamma(E) = 0\}}^{\text{ess}}$. Suppose $\gamma(E) > 0$ for Lebesgue-almost all E in the open interval (a, b). Thus, $A_E = \{\omega|\gamma(\omega, E) = 0\}$ has P-measure zero for almost any $E \in (a, b)$.

Set $A := \{(\omega, E)|\gamma^+(\omega, E) \neq \gamma^-(\omega, E)$ or limit does not exist or $\gamma(\omega, E) = 0;$ $E \in (a, b)\}$. Denote the Lebesgue measure by λ. Then

$$0 = \int_a^b P(A_E)\,dE = (\lambda \times P)(A) = (P \times \lambda)(A)$$

by Fubini's theorem. Therefore, for P-a.e. ω

$$\lambda(A_\omega) = \lambda(\{E|\gamma^+(\omega, E) \neq \gamma^-(\omega, E) \quad \text{or limit does not exist or}$$
$$\gamma(\omega, E) = 0;\ E \in (a, b)\}) = 0 \ ,$$

i.e. for P-a.e. ω, $\gamma(\omega, E) > 0$ for all E outside a set of Lebesgue measure zero.

We know (see Chap. 2) that

$S_\omega := \{E|H_\omega u = Eu$ has a polynomially bounded solution$\}$ satisfies

$$E_{\mathbb{R}\setminus S_\omega}(H_\omega) = 0 \ .$$

Moreover, since $\lambda(A_\omega) = 0$, it follows that

$$E_{A_\omega}^{\text{a.c.}}(H_\omega) = 0 \ .$$

But for $E \notin A_\omega$ the only polynomially bounded solutions are exponentially decreasing because of Theorem 9.12; hence, they are l^2-eigenvectors. There are only countably many of them, hence $\lambda(S_\omega \cap ((a, b)\setminus A_\omega)) = 0$. Thus,

$$E_{(a,b)}^{\text{a.c.}}(H_\omega) = E_{(a,b)\cap S_\omega}^{\text{a.c.}}(H_\omega) = E_{(a,b)\cap A_\omega}^{\text{a.c.}}(H_\omega) + E_{(a,b)\cap S_\omega\setminus A_\omega}^{\text{a.c.}}(H_\omega)$$
$$= 0 \ . \quad \square$$

Knowing whether $\gamma(E)$ is strictly positive or zero, we can answer the question for the measure theoretic nature of the spectrum of H_ω partially. However, in general, $\gamma(E)$ cannot distinguish between point spectrum and singularly continuous spectrum, as we shall see in Chap. 10.

Of course, to use Theorem 9.13 in concrete cases, we need a criterion to decide whether $\gamma(E) = 0$ or $\gamma(E) > 0$. The first such criterion was given by *Furstenberg* [125] for i.i.d. matrices $A_n(\omega, E)$. *Kotani* [215] proved that in a very general case, $\gamma(E) > 0$. (See *Simon* [336] for the discrete case.)

An ergodic potential, $V_\omega(n)$ is called *deterministic* if $V_\omega(0)$ is a measurable function of the random variables $\{V_\omega(n)\}_{n \le -L}$ for all L. It is called *nondeterministic* if it is not deterministic.

Thus, an ergodic process $V_\omega(n)$ is deterministic if the knowledge of $V_\omega(n)$ arbitrary far to the left allows us to compute $V_\omega(0)$, and hence the whole process $V_\omega(n)$.

Theorem 9.14 (Kotani [215]). If $V_\omega(n)$ is nondeterministic, then $\gamma(E) > 0$ for Lebesgue-almost all $E \in \mathbb{R}$. Thus, $\sigma_{\text{a.c.}}(H_\omega) = \varnothing$.

For the proof, see *Kotani* [215] and *Simon* [336].

Example 1. If the $V_\omega(u)$ are i.i.d., then $\sigma_{\text{a.c.}}(H_\omega) = \varnothing$.

One might believe that Theorem 9.14 covers all interesting cases of random potentials. However, as was pointed out in *Kirsch* [201] and *Kirsch, Kotani* and *Simon* [203], there are interesting examples of stochastic potentials that are really random in an intuitive sense, but deterministic in the above precise sense. In [203], it is shown that $V_\omega(x) = \sum q_i(\omega) f(x - i)$—our introductory example—is "typically" deterministic even for i.i.d. $\{q_i\}$ if f has noncompact support. Here is a discrete example.

Example 2. Let φ be a bijection from $\mathbb{Z}$ to $\mathbb{Z}^+ = \{n \in \mathbb{Z} \mid n \ge 0\}$. Set $f(n) = 3^{-\varphi(n)}$. Let $q_i(\omega)$ be i.i.d. random variables with $P(q_i(\omega) = 0) = p$; $P(q_i(\omega) = 1) = 1 - p$. Then for fixed $\lambda > 0$: $V_\omega(n) = \lambda \sum_m q_m(\omega) f(n - m)$ is an ergodic potential, which is random in an intuitive sense. However, q_m is essentially the decimal expansion of $\lambda^{-1} V_\omega(n)$ to the base 3, so the process V_ω is clearly deterministic. Especially for this example, the following theorem of *Kotani* [216] becomes useful.

To state Kotani's theorem, we regard our probability measures as measures on

$$\Omega = \mathop{\times}_{-\infty}^{\infty} [a, b] \quad \text{for some } a, b < \infty \ .$$

We can view Ω as a compact space under the topology of pointwise convergence. supp P can then be defined in the usual way.

Theorem 9.15 (Kotani). Suppose $V_\omega^{(1)}(n)$ and $V_\omega^{(2)}(n)$ are two bounded ergodic processes with corresponding probability measures P_1, P_2, corresponding spectra Σ_1, Σ_2 and absolutely continuous spectra $\Sigma_1^{\text{a.c.}}$ and $\Sigma_2^{\text{a.c.}}$. If $\operatorname{supp} P_1 \subset \operatorname{supp} P_2$, then

(i) $\Sigma_1 \subset \Sigma_2$ and

(ii) $\Sigma_2^{\text{a.c.}} \subset \Sigma_1^{\text{a.c.}}$.

Part (i) follows essentially from *Kirsch* and *Martinelli* [204]. The more interesting part (ii) uses heavily *Kotani*'s earlier paper [215], usng again H^2-function theory.

Kotani proves this theorem in the continuous case. Using *Simon* [336], it can be carried over to the discrete case without difficulties.

Kirsch, Kotani and *Simon* [203] use Theorem 9.15 to prove the absence of absolutely continuous spectrum for a large class of random, but deterministic potentials.

Example 2 (continued). Taking $q_n \equiv 0$ and $q_n \equiv 1$, we see that $W_0 \equiv 0$ and $W_1 \equiv 3\lambda/2$ are periodic potentials in $\operatorname{supp} P$. Hence, the point measure P_0 and P_1 on W_0 and W_1 respectively are ergodic measures with $\operatorname{supp} P_i \subset \operatorname{supp} P$, $i = 0, 1$. By Theorem 9.15, we have $\sigma_{\text{a.c.}}(H_0 + W_0) = \sigma_{\text{a.c.}}(H_0) = [-2, 2] \supset \sigma_{\text{a.c.}}(H_\omega)$ a.s. and $\sigma_{\text{a.c.}}(H_0 + W_1) = [-2 + 3\lambda/2, 2 + 3\lambda/2] \supset \sigma_{\text{a.c.}}(H_\omega)$ a.s. Thus, $\sigma_{\text{a.c.}}(H_\omega) = \emptyset$ a.s. if $\lambda \geq 8/3$. □

Deift and *Simon* [79] investigate those energies with $\gamma(E) = 0$. Interesting examples for $\gamma(E) = 0$ on a set of positive Lebesgue measure occur in the context of almost periodic potentials (see Chap. 10). Among other results, Deift and Simon [79] show:

Theorem 9.16 (Deift-Simon). For a.e. pair $(\omega, E_0) \in \Omega \times \{E | \gamma(E) = 0\}$ there are linearly independent solutions $u_\pm$ of $H_\omega u = Eu$ such that

(i) $u_+ = \bar{u}_-$

(ii) $0 < \varlimsup_{N \to \infty} \left(\dfrac{1}{2N + 1}\right) \sum_{n=-N}^{N} |u_\pm(n)|^2 < \infty$.

Moreover, $|u_\pm(n, \omega)| = |u_\pm(0, T^n \omega)|$.

For a proof, see [79].

9.4 Subharmonicity of the Lyaponov Exponent and the Thouless Formula

In this section, we establish an important connection of the Lyaponov exponent and the density of states: the Thouless formula. For the proof of this formula, as well as for other purposes, a certain regularity of the Lyaponov exponent—namely its subharmonicity—is useful.

Before proving this property of $\gamma(E)$, we recall some definitions and basic facts concerning subharmonic functions.

A function f on $\mathbb{C}$ with values in $\mathbb{R} \cup \{-\infty\}$ is called *submean* if

$$f(z_0) \leq \frac{1}{2\pi} \int_0^{2\pi} f(z_0 + re^{i\theta})\, d\theta \tag{9.22}$$

for $r > 0$ arbitrary. A function f is called *uppersemicontinuous* if, for any sequence $z_n \to z_0$, we have $\overline{\lim} f(z_n) \leq f(z_0)$. A function f is called *subharmonic* if it is both submean and uppersemicontinuous. It is an immediate consequence of the definitions that

$$f(z_0) \leq \underline{\lim}_{r \to 0} \frac{1}{\pi r^2} \int_{|z-z_0| \leq r} f(z)\, dz \ , \tag{9.23a}$$

if f is submean, and that

$$f(z_0) = \lim_{r \to 0} \frac{1}{\pi r^2} \int_{|z-z_0| \leq r} f(z)\, dz \ , \tag{9.23b}$$

if f is subharmonic.

Proposition 9.17. (i) If f_n are submean functions with

$$\sup_{|z|<R} |f_n(z)| < \infty \text{ for any } R, \quad \text{and}$$

$$f_0(z) = \overline{\lim_{n \to \infty}} f_n(z) \ ,$$

then f_0 is submean.

(ii) If $\{f_n\}$ is a decreasing sequence of subharmonic functions, then

$$f_0(z) = \inf_n f_n(z)$$

is subharmonic.

Proof. (i) For any n and any $N \leq n$

$$\begin{aligned} f_n(z_0) &\leq \frac{1}{2\pi} \int_0^{2\pi} f_n(z_0 + \mathrm{r e}^{\mathrm{i}\theta})\, d\theta \\ &\leq \frac{1}{2\pi} \int_0^{2\pi} \sup_{j \geq N} f_n(z_0 + \mathrm{r e}^{\mathrm{i}\theta})\, d\theta \ . \end{aligned}$$

Thus,

$$\begin{aligned} f_0(z_0) = \inf_N \sup_{j \geq N} f_j(z_0) &\leq \frac{1}{2\pi} \inf_N \int_0^{2\pi} \sup_{j \geq N} f_j(z_0 + \mathrm{r e}^{\mathrm{i}\theta})\, d\theta \\ &= \frac{1}{2\pi} \int_0^{2\pi} f_0(z_0 + \mathrm{r e}^{\mathrm{i}\theta})\, d\theta \end{aligned}$$

by the monotone convergence theorem.

(ii) follows from (i) since the inf of uppersemicontinuous functions is uppersemicontinuous. □

After these preliminaries, we come to the basic result of this paragraph, which is due to *Craig* and *Simon* [70]. Craig and Simon were motivated by *Herman* [162] who extensively used subharmonicity of γ in various auxiliary parameters.

Theorem 9.18 (Craig and Simon). (i) $\overline{\gamma}^{\pm}(\omega, E)$ is submean.
(ii) $\gamma(E)$ is subharmonic.

Proof. (i) The matrix-valued function $\Phi_N(\omega, E)$ is obviously analytic in E for any N. Thus, $\ln \|\Phi_N(\omega, E)\|$ is subharmonic [see e.g. *Katznelson* [199], Chapter III, Equation (3.2)]. Thus, by Proposition 9.17(i),

$$\overline{\gamma}^{\pm}(\omega, E) = \overline{\lim_{N\to\pm\infty}} \frac{1}{N} \ln \|\Phi_N(\omega, E)\|$$

is submean.

(ii) By the subadditive ergodic theorem (Theorem 9.11)

$$\gamma(E) = \inf \frac{1}{N} \mathbb{E}(\ln \|\Phi_N(\omega, E)\|) \ .$$

By Fubini's theorem and Fatou's lemma, $\mathbb{E}(\ln \|\Phi_N(\omega, E)\|)$ is subharmonic. By 9.17(ii), $\gamma(E)$ is subharmonic. □

We come to a first application of Theorem 9.18:

Theorem 9.19 (Craig and Simon [70]). For P-almost all ω and all E

$$\overline{\gamma}^{\pm}(\omega, E) \leq \gamma(E) \ .$$

Proof. From the very definition of $\overline{\gamma}^{\pm}(\omega, E)$ and $\gamma(E)$, it is obvious that for fixed E: $\overline{\gamma}^{\pm}(\omega, E) = \gamma(E)$ P-a.s. By use of Fubini's theorem, we conclude from this that $\overline{\gamma}(\omega, E) = \gamma(E)$ for P-almost all ω and Lebesgue-almost all E. Thus,

$$\int_{|E-E_0|\leq r} \overline{\gamma}^{\pm}(E, \omega)\, d^2E = \int_{|E-E_0|\leq r} \gamma(E)\, d^2E \quad P\text{-a.s.}$$

(d^2E indicates that we integrate over a complex domain). Thus, using Theorem 9.18 by (9.23), we know (P-a.s.)

$$\begin{aligned} \overline{\gamma}^{\pm}(E_0, \omega) &\leq \lim_{r\to 0} \frac{1}{\pi r^2} \int_{|E-E_0|\leq r} \overline{\gamma}^{\pm}(E, \omega)\, d^2E \\ &= \lim_{r\to 0} \frac{1}{\pi r^2} \int_{|E-E_0|\leq r} \gamma(E)\, d^2E = \gamma(E_0) \ . \quad \square \end{aligned}$$

Finally, we discuss an important connection between the Lyaponov exponent γ and the density of states k: The Thouless formula. It is named after Thouless, who gave a not completely rigorous proof of it [355]. The Thouless formula was discovered independently by *Herbert* and *Jones* [153]. Thouless' proof was made rigorous by *Avron* and *Simon* [31]. We follow *Craig* and *Simon* [70], who simplified the proof of [31] by using the subharmonicity of γ. In the continuous case, *Johnson* and *Moser* [186] have an alternative proof.

Theorem 9.20 (Thouless Formula).

$$\gamma(E) = \int \ln|E - E'|\, dk(E') \; . \tag{9.24}$$

Proof. We first prove (9.24) for $E \in \mathbb{C}\setminus\mathbb{R}$. For those E, the function $f(E') := \ln|E - E'|$ is continuous on $\mathrm{supp}(dk) \subset \mathbb{R}$. By the definition of $\Phi_L(E)$ [see (9.16)], it is easy to see that $\Phi_L(E)$ is of the form

$$\Phi_L(\omega, E) = \begin{pmatrix} P_L(\omega, E) & Q_{L-1}(\omega, E) \\ P_{L-1}(\omega, E) & Q_{L-2}(\omega, E) \end{pmatrix} ,$$

where P_l and Q_l are polynomials in E of degree l with leading coefficient 1. Moreover,

$$P_l(\omega, E) = 0 \; ,$$

if and only if $H_\omega u = Eu$ has a solution u satisfying $u(0) = 0$, $u(l+1) = 0$ and $Q_l(\omega, E) = 0$, if and only if there exists a solution with $u(1) = 0$ and $u(l+2) = 0$. Thus,

$$P_l(E) = \prod_{j=1}^{l} (E - E_j^{(l)}), \quad Q_l(E) = \prod_{j=1}^{l} (E - \tilde{E}_j^{(l)}) \; ,$$

where $E_j^{(l)}$ (resp. $\tilde{E}_j^{(l)}$) are the eigenvalues of H_ω restricted to $\{1, \ldots, l\}$ (resp. $\{2, \ldots, l+1\}$) with boundary condition $u(0) = u(l+1) = 0$ [resp. $u(1) = u(l+2) = 0$]. Thus, we conclude that [see (9.14)]

$$\ln|P_L(E)| = \int \ln|E - E'|\, d\rho_{1,L}(E') \quad \text{and}$$

$$\ln|Q_L(E)| = \int \ln|E - E'|\, d\rho_{2,L+1}(E') \; .$$

By (9.14), we conclude that

$$\frac{1}{L}\ln|P_L(E)| \to \int \ln|E - E'|\, dk(E')$$

and the same for $Q_L(E)$ if $E \in \mathbb{C}\setminus\mathbb{R}$. Thus, (9.24) follows for those values of E.

Now let E be arbitrary. We need

Lemma. The function $f(E) = \int \ln|E - E'|\, dk(E')$ (with $f(E) = -\infty$ if the integral diverges to $-\infty$) is subharmonic.

Before we prove this lemma, we continue the proof of Theorem 9.20: Since we know (9.24) for $E \in \mathbb{C} \backslash \mathbb{R}$, we have

$$\frac{1}{\pi r^2} \int_{|E-\tilde{E}| \leq r} \gamma(\tilde{E})\, d^2\tilde{E} = \frac{1}{\pi r^2} \int_{|E-\tilde{E}| \leq r} f(\tilde{E})\, d^2\tilde{E}$$

[$f(E)$ as in the lemma].

Taking the limit $r \to 0$ on both sides of the above equation, we obtain

$$\gamma(E) = f(E) = \int \ln|E - E'|\, dk(E') \ ,$$

since both γ and f are subharmonic. □

Proof of the Lemma. The function $\varphi(E) = \ln|E - E'|$ is subharmonic [see *Katznelson* [199] III, Equation (3.2)]. Thus, f is submean by Fubini. Define for $M > 0$

$$f_M(E) = \int \max\{\ln|E - E'|, -M\}\, dk(E') \ .$$

Here f_M is obviously continuous. By the monotone convergence theorem,

$$f(E) = \inf_{M>0} f_M(E) \ .$$

Thus, f is uppersemicontinuous. □

Craig and *Simon* [70] use the Thouless formula to prove that $k(E)$ is log-Hölder continuous in the one-dimensional case. In [69], these authors prove a version of the Thouless formula for strips in arbitrary dimension. From this result, they obtain the log-Hölder continuity of k in the multidimensional case.

9.5 Point Spectrum for the Anderson Model

In this section, we show that the Anderson model has pure point spectrum.

We first prove a criterion for point spectrum of H_ω that allows us to reduce the problem to uniform estimates for $H_\omega^{(N)} := H_\omega^{(0,N)}$ [see (9.13)].

We set

$$a(n, m) := \mathbb{E}\left(\sup_t |e^{-itH_\omega}(n, m)| \right) \ .$$

[If A is a bounded operator on $l^2(\mathbb{Z}^\nu)$, we denote by $A(n, m) = \langle \delta_n, A\delta_m \rangle$ with $\delta_n(i) = 0$ for $i \neq n$, $\delta_n(n) = 1$ the matrix elements of A.]

We say that *physical localization* holds if $\sum_{n \in \mathbb{Z}} |a(n, m)| < \infty$ for $m = 0$ and $m = 1$.

Theorem 9.21 (Kunz-Souillard [221]). Physical localization implies mathematical localization (i.e. $\sigma_c(H_\omega) = \phi$).

Proof. By the RAGE theorem (Sect. 5.4), in particular, formula (9.10) in Sect. 9.1, we conclude:

$$\|P^c(H_\omega)\varphi\|^2 = \lim_{J\to\infty}\lim_{T\to\infty}\frac{1}{2T}\int_{-T}^{T}\sum_{|j|\geq J}|\langle e^{-itH_\omega}\varphi,\delta_j\rangle|^2\,dt\ .$$

Thus, if $\sum_{n\in\mathbb{Z}}|a(n,0)|^2<\infty$, we have (since $|\exp(-itH_\omega)(m,n)|\leq 1$)

$$\mathbb{E}(\|P^c(H_\omega)\delta_0\|^2)\leq\lim_{J\to\infty}\sum_{|j|\geq J}|a(j,0)|=0\ .$$

Hence δ_0 is P-a.s. orthogonal to the continuous subspace. A similar argument shows that $P^c(H_\omega)\delta_1=0$ almost surely. It is easy to see that any δ_j can be written as

$$\delta_j=\sum_{n=0}^{N}\alpha_n(\omega)H_\omega^n\delta_0+\sum_{n=0}^{N}\beta_n(\omega)H_\omega^n\delta_1$$

(e.g. $\delta_2=H_\omega\delta_1-V_\omega\delta_1-\delta_0$). Hence, $P^c(H_\omega)\delta_j=0$ a.s. for any j. Thus, $\sigma_c(H_\omega)=\phi$ a.s. □

Remark. In terms of a direct physical interpretation, it would be better to define a with a square inside $\mathbb{E}$; the statement and proof of Theorem 9.21 still go through. Since it is easier to estimate a as we define it, we have used that definition.

The next result shows that $a(n,m)$ even determines the decay of the eigenfunctions:

Theorem 9.22. If, for $m=0$ and $m=1$,

$$|a(n,m)|\leq Ce^{-D|n|}\ , \tag{9.25}$$

then P-a.s. any eigenfunction φ_ω of H_ω satisfies

$$|\varphi_\omega(n)|\leq C_{\omega,\varepsilon}e^{-(D-\varepsilon)|n|}$$

for any $\varepsilon>0$.

Remark. The constant $C_{\omega,\varepsilon}$ may depend on the eigenvalue. We say that the eigenfunction φ_ω is *exponentially localized.*

Proof. Set

$$\beta(\omega,n,m):=\sup_t|e^{-itH_\omega}(n,m)|\ .$$

We first prove that (9.25) implies

$$\beta(\omega,n,m)\leq\tilde{C}_{\omega,\varepsilon}e^{-(D-\varepsilon)|n|} \tag{9.26}$$

for $m=0$ and $m=1$ P-a.s.

Equation (9.26) holds if we show that

$$P(\beta(\omega,n,m) > \mathrm{e}^{-(D-\varepsilon)|n|} \text{ for infinitely many } n) = 0 \ .$$

This, in turn, follows by the Borel-Cantelli Lemma (see any book on probability theory, e.g. *Breiman* [53]) if we show that

$$\sum_n P(\beta(\omega,n,m) > \mathrm{e}^{-(D-\varepsilon)|n|}) < \infty \tag{9.27}$$

for $m = 0$ and $m = 1$.

Since, by Tschebycheff's inequality

$$\begin{aligned} P(\beta(\omega,n,m) > \mathrm{e}^{-(D-\varepsilon)|n|}) &\le \mathrm{e}^{+(D-\varepsilon)|n|}\mathbb{E}(\beta(\omega,n,m)) \\ &= \mathrm{e}^{-\varepsilon|n|}[\mathrm{e}^{D|n|}a(n,m)] \le C\mathrm{e}^{-\varepsilon|n|} \ , \end{aligned}$$

(9.27) follows. Thus, we have proven (9.26).

Now we use the formula

$$P_{\{E\}}(H) = s - \lim_{T\to\infty} \frac{1}{T}\int_0^T \mathrm{e}^{\mathrm{i}sE}\mathrm{e}^{-\mathrm{i}sH}\,ds \ . \tag{9.28}$$

This follows from continuity of the functional calculus and the fact that functions

$$f_T(x) = \frac{1}{T}\int_0^T \mathrm{e}^{\mathrm{i}sE}\mathrm{e}^{-\mathrm{i}sx}\,ds$$

obey $|f_T(x)| \le 1$ and $f_T(x) \to 0$ (resp. 1) as $T \to \infty$ for $x \ne E$ (resp. $x = E$).

Suppose now that E is an eigenvalue of H_ω. Since $v = 1$, any eigenvalue is simple. Denote by $\varphi_{\omega,E}$ the normalized eigenfunction corresponding to E. Then (9.28) implies

$$\begin{aligned} |\varphi_{\omega,E}(0)|\,|\varphi_{\omega,E}(n)| &= |\langle\delta_0,\varphi_{\omega,E}\rangle\langle\varphi_{\omega,E},\delta_n\rangle| \\ &= |\langle\delta_0,P_{\{E\}}(H_\omega)\delta_n\rangle| \le \beta(\omega,n,0) \le \tilde{C}_{\omega,\varepsilon}\mathrm{e}^{-(D-\varepsilon)|n|} \end{aligned}$$

by (9.26). This proves the theorem if $\varphi_{\omega,E}(0) \ne 0$. If $\varphi_{\omega,E}(0) = 0$, we have $\varphi_{\omega,E}(1) \ne 0$, and obtain the above estimate for $m = 1$. □

Now we consider H_ω restricted to $l^2(-L,\dots,L)$. As usual, we denote the corresponding operator by $H_\omega^{(L)}$ [see (9.13)]. We define

$$a_L(n,m) := E\left(\sup_t |\exp[-\mathrm{i}tH_\omega^{(L)}](n,m)|\right) \ .$$

Proposition 9.23: $a(n,m) \le \overline{\lim}_{L\to\infty} a_L(n,m)$.

Remark. By Proposition 9.23 and Theorems 9.21 and 9.22, we can conclude that

H_ω has pure point spectrum with exponentially localized eigenfunctions if we have an estimate

$$a_L(n,m) \le Ce^{-D|n|} \quad \text{for } m = 0 \text{ and } m = 1$$

uniformly in L.

Proof. $H_\omega^{(L)}$ converges strongly to H_ω (with the understanding that $H_\omega^{(L)}(n,m) = 0$ for $|n| > L$ or $|m| > L$). Hence,

$$\exp[-itH_\omega^{(L)}](n,m) \to \exp(-itH_\omega)(n,m)$$

(cf. *Reed* and *Simon* I, VIII.20 [292]), and thus by Fatou's lemma

$$\mathbb{E}\left(\sup_t |\exp(-itH_\omega)(n,m)|\right) \le \overline{\lim}\, \mathbb{E}\left(\sup_t |\exp[-itH_\omega^{(L)}](n,m)|\right) . \quad \square$$

We denote by $\{E_\omega^{L,k}\}$ the eigenvalues of $H_\omega^{(L)}$ in increasing order. $\varphi_\omega^{L,k}$ denotes "the" normalized eigenfunction corresponding to $E_\omega^{L,k}$.

Finally, we define for any (Borel) set $A \subset \mathbb{R}$:

$$\rho_L(n,m,A) := \mathbb{E}\left(\sum_k |\varphi_\omega^{L,k}(n)||\varphi_\omega^{L,k}(m)|\chi_A(E_\omega^{L,k})\right) . \tag{9.29}$$

Note that the sum over k goes only over $2L+1$ terms since H_ω^L is a $(2L+1) \times (2L+1)$-matrix.

It is immediately clear that

$$a_L(n,m) \le \rho_L(n,m,\mathbb{R}) , \quad \text{since}$$

$$\exp(-itH_\omega^L)(n,m) = \sum_k \exp(-itE_\omega^{L,k})\overline{\varphi_\omega^{L,k}(n)}\varphi_\omega^{L,k}(m) .$$

Note that $\rho_L(n,m,\mathbb{R}) = \rho_L(n,m,[-M,M])$ for M large enough since the operators H_ω are uniformly bounded.

Theorem 9.24 (Kunz-Souillard). Suppose the $V_\omega(n)$ are independent random variables with a common distribution $r(x)\,dx$. If $r \in L^\infty$ and r has compact support, then

$$H_\omega u(n) = u(n+1) + u(n-1) + V_\omega(n)u(n)$$

has pure point spectrum (P-a.s.). The eigenfunctions of H_ω are P-a.s. exponentially localized.

Remark. The above theorem was conjectured by theoretical physicists since the early sixties. A continuous analog, where the potential $V_\omega(x)$ is a rather complicated diffusion process, was proven by *Goldsheid, Molchanov* and *Pastur* [138], and *Molchanov* [248] (see also *Carmona* [59]). The above theorem is due to *Kunz* and *Souillard* [221]. *Delyon Kunz* and *Souillard* [81] simplified this proof and extended the theorem to other types of disorder. We mainly follow their proof. Their proof has some elements in common with an earlier approach of *Wegner* [365].

Proof. We will give a uniform (in L) estimate on $\rho_L(n,m) := \rho_L(n,m,\mathbb{R})$ of the form

$$\rho_L(n,m) \leq C e^{-D|n|} \quad m = 0, 1 \ . \tag{9.30}$$

This implies the theorem by Proposition 9.23 and Theorems 9.21 and 9.22.

The proof of (9.30) is broken into three steps. First, we rewrite $\rho_L(n,m)$ as a multiple product of integral operators T_0 and T_1. This will be done by changing variables from $V(-L), \ldots, V(L)$ to $\lambda, x_{-L}, \ldots, x_{-1}, x_1, \ldots, x_L$ where λ is the eigenvalue and the x_i are simple expressions in terms of the eigenfunctions. Note that the expectation $\mathbb{E}$ in (9.29) is nothing but an integral in the variables $V(k)$; $|k| \leq L$.

In the second step (Proposition 9.25), we explore some properties of the operators T_0, T_1. This investigation allows us to estimate ρ_L in the last step.

We start with

$$\rho_L(n,m) = \rho_L(n,m,\mathbb{R})$$

$$= \int \sum_{k=0}^{2L+1} |\varphi_{\underline{V}}^{L,k}(n)||\varphi_{\underline{V}}^{L,k}(m)| \prod_{n=-L}^{L} r(V(n))\, d^{2L+1}\underline{V} \ ,$$

where $\underline{V} = (V(-L), \ldots, V(L))$ and the $\varphi_{\underline{V}}^{L,k}$ denote the eigenfunctions for the potential $\underline{V}$. For definiteness, we now assume $m = 0$, $n > 0$. The other cases are similar. After interchanging sum and integrals, we change variables from $\{V(n)\}_{n=-L}^{L}$ to $\{x_{-L}, \ldots, x_{-1}, \lambda, x_1, \ldots, x_L\}$ where

$$\lambda := E_{\underline{V}}^{L,k} \ , \tag{9.31a}$$

$$x_m := \frac{\varphi_{\underline{V}}^{L,k}(m-1)}{\varphi_{\underline{V}}^{L,k}(m)} \quad \text{for } m > 0 \ , \text{and} \tag{9.31b}$$

$$x_m := \frac{\varphi_{\underline{V}}^{L,k}(m+1)}{\varphi_{\underline{V}}^{L,k}(m)} \quad \text{for } m < 0 \ . \tag{9.31c}$$

The Schrödinger equation $u(n+1) + u(n-1) + V(n)u(n) = Eu(n)$ then yields

$$V(m) = \begin{cases} \lambda - x_{m+1}^{-1} - x_m, & m > 0 \\ \lambda - x_1^{-1} - x_{-1}^{-1}, & m = 0 \\ \lambda - x_{m-1}^{-1} - x_m, & m < 0 \end{cases} \tag{9.32}$$

with the understanding that $x_{L+1}^{-1} = x_{-L-1}^{-1} = 0$. Equation (9.32) allows us to compute the Jacobian J with respect to the change of variables $\underline{V} \mapsto (\underline{x}, \lambda)$. It is straightforward to see that

$$\det J = 1 + x_1^{-2}\{1 + x_2^{-2}[1 + \ldots x_{L-1}^{-2}(1 + x_L^{-2})\ldots]\}$$
$$+ x_{-1}^{-2}\{1 + x_{-2}^{-2}[1 + \ldots x_{-L+1}^{-2}(1 + x_{-L}^{-2})\ldots]\}$$
$$= \varphi_{\underline{V}}^{L,k}(0)^{-2} \ ,$$

where we used that $\varphi_{\underline{V}}^{L,k}$ is normalized. Moreover,

$$|\varphi_{\underline{V}}^{L,k}(0)|^{-1}|\varphi_{\underline{V}}^{L,k}(n)| = |x_1^{-1}\cdot x_2^{-1}\cdot\ldots\cdot x_n^{-1}| \ .$$

Hence,

$$\int_{\mathbb{R}^{2L+1}} |\varphi_{\underline{V}}^{L,k}(0)||\varphi_{\underline{V}}^{L,k}(n)| \prod_{n=-L}^{L} r(V(n))\, d^{2L+1}\underline{V}$$
$$= \int_{\Sigma_0} d\lambda \int_{\mathbb{R}^{2L}} |x_1^{-1}x_2^{-1}\cdot\ldots\cdot x_n^{-1}|\left[\prod_{m>0} r(\lambda - x_{m+1}^{-1} - x_m)\right]\cdot r(\lambda - x_1^{-1} - x_{-1}^{-1})$$
$$\otimes\left[\prod_{m<0} r(\lambda - x_{m-1}^{-1} - x_m)\right]dx_{-L}\ldots dx_{-1}\,dx_1\ldots dx_L \ ,$$

where $\Sigma_0 = [-2 - \|V\|_\infty, 2 + \|V\|_\infty]$ and $\|V\|_\infty = \sup\{|\lambda|; \lambda \in \operatorname{supp} r\}$. The possible eigenvalues that occur always lie in this range, so we can restrict the λ integration to this region.

Now we fix λ for awhile and consider

$$\rho_L(0,n;\lambda) = \int \prod_{i=1}^{n} |x_i^{-1}|\left[\prod_{m>0} r(\lambda - x_{m+1}^{-1} - x_m)\right]$$
$$\times\left[\prod_{m<0} r(\lambda - x_{m-1}^{-1} - x_m)\right] r(\lambda - x_1^{-1} - x_{-1}^{-1})\, d^{2L}\underline{x} \ . \tag{9.33}$$

We introduce the integral operators T_0, T_1 by

$$T_0 f(x) := \int r(\lambda - x - y^{-1}) f(y)\, dy \ , \tag{9.34a}$$

$$T_1 f(x) := \int r(\lambda - x - y^{-1}) |y|^{-1} f(y)\, dy \ . \tag{9.34b}$$

Set $\varphi(x) := r(\lambda - x)$ and $U_0 f(x) := |x|^{-1} f(x^{-1})$. Note that U_0 is a unitary operator on $L^2(\mathbb{R})$. From these definitions and (9.33), we see that

$$\rho_L(0,n,\lambda) = \int_{\mathbb{R}} T_0^L \varphi(x_1^{-1})|x_1|^{-1} T_1^{n-1} T_0^{L-n} \varphi(x_1)\, dx_1$$
$$= \langle T_1^{n-1} T_0^{L-n}\varphi, U T_0^L \varphi\rangle \ . \tag{9.35}$$

Observe that both T_0 and T_1 depend on the parameter λ. We now investigate $T_0 = T_0(\lambda)$ and $T_1 = T_1(\lambda)$.

Proposition 9.25:

(a) $\|T_0 f\|_1 \le \|f\|_1$
(b) $\|T_0 f\|_2 = \|T_0(\lambda)f\|_2 \le C\|f\|_1$ uniformly in λ
(c) $\|T_1 f\|_2 \le \|f\|_2$
(d) T_1^2 is a compact operator
(e) $\|T_1^2 f\|_2 \le q\|f\|_2$ with a $q < 1$ uniformly in λ .

Before demonstrating Proposition 9.25, we show how this proposition yields the exponential estimate of $\rho(0, n, \lambda)$.

Proof of Theorem 9.24 (continued). We note that $\|\varphi\|_1 = 1$. From (9.35) and the unitarity of U, we see

$$\begin{aligned}\rho_L(0,n,\lambda) &\le \|T_0^L\varphi\|_2 \|T_1^{n-1}T_0^{L-n}\varphi\|_2 \\ &\le \|T_0\|_{L^1,L^2}\cdot\|T_0^{L-1}\|_{L^1,L^1}\cdot\|T_1^{n-1}\|_{L^2,L^2}\|T_0\|_{L^1,L^2}\cdot\|T_0^{L-n+1}\|_{L^1,L^1} \\ &\le C^2\|T_1^{n-1}\|_{L^2,L^2} \le C^2 q^{(n-2)/2} = \tilde{c}\exp\left(-\frac{1}{2}n|\ln q|\right) .\end{aligned}$$

Hence, $\rho_L(0, n, \lambda)$ decays exponentially in n. Moreover, Proposition 9.25 gives uniformity in λ on compact sets. This finishes the proof of Theorem 9.24, since Σ_0 is bounded. □

Proof (*Proposition 9.25*):

(a) $\|T_0 f\|_1 \le \int\int r(\lambda - x - y^{-1})|f(y)|\,dy\,dx = \int r(x)\,dx \int |f(y)|\,dy = \|f\|_1$
(b) $\|T_0 f\|_2^2 \le \int\int r(\lambda - x - y^{-1})|f(y)|\,dy \cdot \int r(\lambda - x - z^{-1})\cdot|f(z)|\,dz\,dx$
$\le \|r\|_\infty \cdot \|r\|_1 \cdot \|f\|_1^2$
(c) Defining $Kf(x) := \int r(\lambda - x + y)f(y)\,dy = r_\lambda * f(x)$ where $r_\lambda(x) = r(\lambda - x)$ and $\tilde{U}f(x) := |x|^{-1}f(-x^{-1})$, we can write T_1 as $T_1 = K\tilde{U}$, so

$$\|T_1 f\|_2 = \|K\tilde{U}f\|_2 = \|r * \tilde{U}f\|_2 \le \|r\|_1 \cdot \|\tilde{U}f\|_2 = \|r\|_1 \cdot \|f\|_2 .$$

(d) Let F denote the Fourier transformation. We set $\hat{K} := \mathrm{F}K\mathrm{F}^{-1}$ and $\hat{U} := \mathrm{F}\tilde{U}\mathrm{F}^{-1}$. Since $\tilde{U}$—and hence $\hat{U}$—is unitary, we have to show that $\hat{K}\hat{U}\hat{K}$ is compact. Since K is a convolution operator, we have $\hat{K}\hat{f}(p) = \hat{r}_\lambda(p)\hat{f}(p)$ where $\hat{r}_\lambda(p) = \int \exp(-\mathrm{i}xp)r_\lambda(x)\,dx$. Formally we have

$$\hat{U}\hat{f}(k) = \int a(k,p)\hat{f}(p)\,dp \quad \text{with}$$

$$a(k,p) = \frac{1}{2\pi}\int \exp[-\mathrm{i}(kx + px^{-1})]\frac{dx}{|x|} .$$

However, the integral is not absolutely convergent, so $a(k, p)$ requires a careful interpretation: Let g_1 be a C_0^∞-function which is 1 near 0, $g_2 := 1 - g_1$. Set

$U_i\varphi(x) := g_i(x)\tilde{U}\varphi(x)$. Then $\tilde{U}\varphi = U_1\varphi + U_2\varphi$. Define furthermore

$$a_i(k,p) = \frac{1}{2\pi}\int \exp[-\mathrm{i}(kx + px^{-1})]\frac{g_i(x)}{|x|}dx \ .$$

For fixed p, $a_2(k,p)$ is the Fourier transform of $f_p := (2\pi)^{-1}\exp(-\mathrm{i}px^{-1})|x|^{-1}g_2(x)$, which is an L^2-function whose L^2-norm is independent of p. Hence,

$$\sup_p \int |a_2(k,p)|^2\,dk < \infty \ . \tag{9.36a}$$

Moreover, for fixed k

$$\int\limits_{|x|>(1/n)} \exp[-\mathrm{i}(kx + px^{-1})]\frac{g_1(x)}{|x|}dx$$

$$= \int\limits_{|x|<n} \exp[-\mathrm{i}(kx^{-1} + px)]\frac{g_1(x^{-1})}{|x|}dx \ ,$$

which is convergent in the L^2-sense to the Fourier transform of the L^2-function

$$\exp(-\mathrm{i}kx^{-1})\frac{g_1(x^{-1})}{|x|} \ .$$

Hence,

$$\sup_k \int |a_1(k,p)|^2\,dp < \infty \ . \tag{9.36b}$$

Define A_i by $(A_i\varphi)(k) = \int a_i(k,p)\varphi(p)\,dp$. Now we show $\hat{U}_i = A_i$. We will handle freely integrals—such as $\int \exp(-\mathrm{i}kx)f(x)\,dx$ for $f \in L^2$—that exist only in an L^2-sense. The reader can easily verify those manipulations. We have

$$(U_1\varphi)\hat{}\,(k) = \int \exp(-\mathrm{i}kx)\frac{g_1(x)}{|x|}\varphi(-x^{-1})\,dx$$

$$= \int \exp(-\mathrm{i}kx^{-1})\frac{g_1(x^{-1})}{|x|}\varphi(-x)\,dx$$

$$= \frac{1}{2\pi}\int \exp(-\mathrm{i}kx^{-1})\frac{g_1(x^{-1})}{|x|}\int \exp(-\mathrm{i}xp)\hat{\varphi}(p)\,dp\,dx$$

$$= \frac{1}{2\pi}\iint \exp[-\mathrm{i}(kx^{-1} + xp)]\frac{g_1(x^{-1})}{|x|}dx\,\hat{\varphi}(p)\,dp$$

$$= \frac{1}{2\pi} \iint \exp[-i(kx + px^{-1})] \frac{g_1(x)}{|x|} dx\, \hat{\varphi}(p)\, dp$$

$$= \int a_1(k, p) \hat{\varphi}(p)\, dp \ .$$

Thus, $\hat{U}_1 = A_1$. The proof of $\hat{U}_2 = A_2$ is similar (and even simplier). Therefore, $(K\tilde{U}K)\,\hat{}$ has an integral kernel $b(k, p) = \hat{r}_\lambda(k) a_1(k, p) \hat{r}_\lambda(p) + \hat{r}_\lambda(k) a_2(k, p) \hat{r}_\lambda(p)$ since $r \in L^1 \cap L^\infty$ and hence $r \in L^2$, we have $\hat{r} \in L^2 \cap L^\infty$, so

$$\|b(k, p)\|_{L^2(dk) \times L^2(dp)}$$

$$\leq \|\hat{r}\|_2 \sup_k \|a_1(k, p)\|_{L^2(dp)} \|\hat{r}\|_\infty$$

$$+ \|\hat{r}\|_\infty \sup_p \|a_2(k, p)\|_{L^2(dk)} \|\hat{r}\|_2 \ .$$

Thus, $(K\tilde{U}K)\,\hat{}$ is Hilbert Schmidt and consequently T_1^2 is compact.

(e) Since $|T_1^2|$ is positive and compact, $\|T_1^2\|$ is an eigenvalue of $|T_1^2|$. Since $|\hat{r}_\lambda(k)| < 1$ for $k \neq 0$, we have $\|T_1 f\| < \|f\|$ for any $f \in L^2$. Therefore, 1 is not an eigenvalue for $|T_1^2|$. So $\|T_1^2\| < 1$. Moreover, since $\hat{r}_\lambda(k) = \exp(-i\lambda k)\hat{r}_0(k)$, the norm $\|T_1^2\|$ is independent of λ. □

By a refinement of the methods of the above proof, one can prove the following results:

Theorem 9.26 (Delyon, Kunz and Souillard [81]). Suppose that $V_\omega(n)$ satisfies the assumptions of Theorem 9.24. Let $V_0(n)$ be a bounded function on $\mathbb{Z}$. Then

$$H_\omega := H_0 + V_0 + V_\omega$$

has P-a.s. dense point spectrum with exponentially decaying eigenfunctions, and possibly in addition, isolated eigenvalues.

Theorem 9.27 (Simon [333]). Suppose $V_\omega(n)$ satisfies the assumptions of Theorem 9.24. Let a_n be a sequence with $|a_n| \geq C|n|^{-1/2+\delta}$ and set $W_\omega(n) := a_n V_\omega(n)$. Then

$$H_\omega := H_0 + W_\omega(n)$$

has only dense point spectrum.

Remark. (1) Observe that the potentials $V_0 + V_\omega$ and W_ω are not stationary. So, the corresponding H_ω will have a random spectrum in general. The proofs of these theorems can be found in [81] and [333] respectively.

(2) More recently, *Delyon*, *Simon* and *Souillard* [84] and *Delyon* [80] have studied the operators of Theorem 9.27, but with different a_n. If $|a_n| \leq C|n|^{-1/2-\delta}$, then H_ω has no point spectrum [84], and if $a_n \sim \lambda n^{-1/2}$ with λ small, the operator has some singular continuous spectrum [80]!

In the above proof, the assumption that the distribution P_0 of $V(0)$ has a density $r(\lambda)$ was necessary. For example, if $V(0) = 0$ with probability p, and $V(0) = 1$ with probability $1 - p$, the above proof does not apply but a recent paper of R. Carmona, A. Klein and F. Martinelli shows there is also only point spectrum in this case.

We now survey briefly some further results on random potentials.

Brossard [56] proves pure point spectrum (in the continuous case) for certain potentials of the form $V_\omega(x) = V_0(x) + W_\omega(x)$, $x \in \mathbb{R}^1$; where V_0 is a periodic potential and W_ω is a certain random one.

Carmona [60] considers random, but not stationary, potentials (continuous case). For example, suppose $V_\omega^{(1)}(x)$, $x \in \mathbb{R}^1$ is a random potential such that $-(d^2/dx^2) + V_\omega^{(1)}$ has pure point spectrum, and suppose $V^{(2)}(x)$ is periodic. Consider

$$V_\omega(x) = \begin{cases} V_\omega^{(1)}(x) & \text{for } x \le 0 \\ V^{(2)}(x) & \text{for } x > 0 \ . \end{cases}$$

Carmona [60] proves that

$$\sigma_{\text{a.c.}}\left(-\frac{d^2}{dx^2} + V_\omega\right) = \sigma_{\text{a.c.}}\left(-\frac{d^2}{dx^2} + V^{(2)}\right) = \sigma\left(-\frac{d^2}{dx^2} + V^{(2)}\right),$$

$$\sigma_{\text{s.c.}}\left(-\frac{d^2}{dx^2} + V_\omega\right) = \phi \quad \text{and}$$

$$\sigma_{\text{p.p.}}\left(-\frac{d^2}{dx^2} + V_\omega\right) = \overline{\sigma_{\text{p.p.}}\left(-\frac{d^2}{dx^2} + V_\omega^{(1)}\right) \Big\backslash \sigma\left(-\frac{d^2}{dx^2} + V^{(2)}\right)} .$$

There has been large interest in operators with constant electric field and stochastic potential. Suppose q_n are i.i.d. random variables, f a C^2-function with support in $(-\frac{1}{2}, \frac{1}{2})$ and $f \le 0$, $(f \not\equiv 0)$.

$$H_\omega = -\frac{d^2}{dx^2} + Fx + \sum_{n \in \mathbb{Z}} q_n(\omega) f(x - n) .$$

We have seen, using Mourre-estimates, that the spectrum of H_ω is absolutely continuous if $F \neq 0$ (see Chap. 4 and [45]). *Bentosela* et al. prove that for $F = 0$, the operator H_ω has a.s. pure point spectrum (with exponentially decaying eigenfunctions) provided the distribution P_0 of q_0 has continuous density with compact support.

If f is the δ-function, then H_ω has a.s. pure point spectrum even for $F \neq 0$, but $|F|$ small. However, in this case the eigenfunctions are only polynomially localized (but they are exponentially localized for $F = 0$). For $|F|$ large, the spectrum of H_ω is continuous (*Delyon*, *Simon* and *Souillard* [84]).

For the case $\nu > 1$, much less is known than for $\nu = 1$. The physicists' belief is that the nature of the spectrum depends on the magnitude of disorder. For small disorder, one expects that the spectrum is pure point at the boundaries of the spectrum, while it should be continuous (absolutely continuous?) away from the boundary, at least if $v \geq 3$. Those values where the nature of the spectrum changes are called *mobility edges*. If the disorder is increased, the continuous spectrum is supposed to shrink in favor of the pure point one. Finally, at a certain degree of disorder, the spectrum should become a pure point one.

Recently, *Fröhlich* and *Spencer* [119] proved that for the multidimensional Anderson model (with absolutely continuous distribution p_0), the kernel $G(E + i\varepsilon; 0, n)$ of the resolvent $[H_\omega - (E + i\varepsilon)]^{-1}$ decays P-a.s. exponentially in n uniformly as $\varepsilon \to 0$, provided that either $E \to \pm\infty$ (this corresponds to the boundary of the spectrum) or the disorder is large enough.

Martinelli and *Scoppola* [239] observed that the estimates of *Fröhlich* and *Spencer* [119] actually suffice to prove the absence of absolutely continuous spectrum for $|E|$ large or for large disorder. Corresponding results for a continuous model are contained in *Martinelli* and *Holden* [167]. *Fröhlich, Martinelli, Scoppola* and *Spencer* [118] have proven that in the same regime, H_ω has only pure point spectrum, and *Goldsheid* [137] has announced a similar result.

New insight on localized has come from work of *Kotani* [217, 218], *Delyon, Levy* and *Souillard* [82, 83], *Simon* and *Wolf* [345] and *Simon* [343] which has its roots in the work of *Carmona* [60]. The key remark is that the spectral measure averaged over variations of the potential in a bounded region is absolutely continuous with respect to Lebesgue measure, so the sets of measure zero where the Osceledec theorem fails are with probability 1 irrelevant. In any event, the reader should be aware that the state of our understanding of localization was changing rapidly as this book was being completed.

Kunz and Souillard [222] have studied the case of random potentials on the Bethe lattice.

For additional information, see [347].

10. Almost Periodic Jacobi Matrices

This chapter deals with almost periodic Hamiltonians. Those operators have much in common with random Hamiltonians; consequently, Chaps. 9 and 10 are intimately connected. Almost periodic Jacobi matrices, as well as their continuous counterparts, have been the subject of intensive research in the last years. They show surprising phenomena such as singular continuous spectrum, pure point spectrum and absolutely continuous spectrum that is nowhere dense!

Despite much effort, almost periodic Hamiltonians are not well understood. Virtually all the really interesting results concern a small class of examples.

10.1 Almost Periodic Sequences and Some General Results

We consider the space l^∞ of bounded (real-valued) sequences $\{c(n)\}_{n\in\mathbb{Z}^\nu}$. For $c\in l^\infty$, we define c_m to be the sequence $\{c(n-m)\}_{n\in\mathbb{Z}^\nu}$. A sequence c is called almost periodic if the set $\Omega_0=\{c_m|m\in\mathbb{Z}^\nu\}$ has a compact closure in l^∞. The closure of Ω_0 is called the *hull* of c.

A convenient way to construct examples goes as follows: Take a continuous periodic function $F\colon\mathbb{R}\to\mathbb{R}$ with period 1. We can think of F as a function on the torus $\mathbb{T}=\{\exp(2\pi i x)|x\in[0,1)\}$, i.e. $F(x)=\tilde{F}(\exp(2\pi i x))$. Now choose a real number α and define $F^{(\alpha)}(n):=F(\alpha n)$. $F^{(\alpha)}$ as a function on $\mathbb{Z}$ will not be periodic if $\alpha\notin\mathbb{Q}$. It is, however, an almost periodic sequence. To see this, define $F^{(\alpha),\theta}(n):=F(\alpha n+\theta)$. For α fixed, $S:=\{F^{(\alpha),\theta}\}_{\theta\in[0,2\pi]}$ is a continuous image of the circle, and is thus compact. The translates of $F^{(\alpha)}$ lie in S, so their closure is compact. In fact, if α is irrational, S is precisely the hull of $F^{(\alpha)}$.

Similarly, if F is a continuous periodic function on $\mathbb{R}^d$, then for $\alpha\in\mathbb{R}^d$, $F^{(\alpha)}(n):=F(n\alpha)$ defines an almost periodic sequence on $\mathbb{Z}$.

Let us define $\mathbb{T}^d$—the d-dimensional torus—to be the set $\{(\exp(2\pi i x_1),\dots,\exp(2\pi i x_d))|(x_1,\dots,x_d)\in[0,1]^d\}$, i.e. $\mathbb{T}^d$ is $[0,1]^d$ with opposite surface identified. We say that $(c_1,\dots,c_n)$ are *independent over the rationals* $\mathbb{Q}$, if, for $\gamma_i\in\mathbb{Q}$: $\sum\gamma_i c_i=0$ implies that $\gamma_i=0$ for all i. If $(1,\alpha_1,\alpha_2,\dots,\alpha_d)$ are independent over the rationals, then the set $\{[\exp(2\pi i\alpha_1 n),\exp(2\pi i\alpha_2 n),\dots,\exp(2\pi i\alpha_d n)]|n\in\mathbb{Z}\}$ is dense in $\mathbb{T}^d$. From this it is not difficult to see that the hull $\Omega_{F^{(\alpha)}}$ of $F^{(\alpha)}$ (F continuous periodic) is given by $\{F(\alpha n+\theta)|\theta\in[0,2\pi]^d\}\simeq\mathbb{T}^d$, if $(1,\alpha,\dots,\alpha_d)$ are independent over the rationals.

Now let c be an arbitrary, almost periodic sequence on $\mathbb{Z}^\nu$. On $\Omega_0=\{c_m|m\in\mathbb{Z}^\nu\}$ we define an operation $\circ$ by: $c_m\circ c_{m'}:=c_{m+m'}$. By density of Ω_0 in the hull Ω, this operation can be extended to Ω in a unique way. The operation $\circ$

makes Ω a compact topological group. It is well known that any compact topological group Ω carries a unique Baire measure μ, which satisfies

$$\int f(gg')\, d\mu(g') = \int f(g')\, d\mu(g')$$

and $\mu(\Omega) = 1$. This invariant measure is called the *Haar measure* (see [260] or [292] for details). We may (and will) look upon Ω, P as a probability space. We define, for $f \in \Omega$: $T_n f = f_n$. The invariance property of the Haar measure μ tells us that

$$\mu(T_n A) = \mu(A) \ .$$

Thus, the T_n are measure-preserving transformations. It is not difficult to see that any set A with $T_n A = A$ for all A has Haar measure 0 or 1. Indeed, for such an A, $\tilde{\mu}(B) = \int_B \chi_A \, d\mu$ would define another Haar measure on Ω. But up to a constant, the Haar measure is unique. Hence $\{T_n\}$ are ergodic. We may therefore apply Theorems 9.2 and 9.4 to almost periodic Jacobi matrices, i.e. to operators H of the form $H = H_0 + V$ where H_0 is the discretized Laplacian and V is an almost periodic sequence.

Proposition 10.1. Suppose V is an almost periodic sequence.

(i) For all W in the hull Ω of V, the spectrum $\sigma(H_0 + W)$ is the same. The discrete spectrum is empty.

(ii) There is a subset $\tilde{\Omega}$ of Ω of full Haar measure, such that for all $W \in \tilde{\Omega}$ the pure point spectrum (singular continuous, absolutely continuous spectrum) is the same.

That (i) is true for all W rather than merely for a set of measure 1 comes from an easy approximation argument. This argument is not applicable to (ii) since the absolutely continuous spectrum, etc. may change discontinuously under a perturbation.

The above consideration emphasizes some similarity between stochastic and almost periodic Jacobi matrices. However, to get deeper results, more specific methods are required.

Most of the rest of this chapter deals with examples of the type $F^{(\alpha)}(n) = F(\alpha n)$ for a periodic function F (with period 1). The spectral properties of $H = H_0 + \lambda F^{(\alpha)}$ depend on the coupling constant λ, and on "Diophantine" properties of α (an observation of *Sarnak* [306]). More precisely, (suppose $\nu = 1$) if α is rational, $F^{(\alpha)}$ is periodic and we have only absolutely continuous spectrum. If α is irrational but "extremely well approximated" by rational numbers, then $H = H_0 + \lambda F^{(\alpha)}$ has a tendency to singular continuous spectrum for large λ (see Sect. 10.2), while for "typical" irrational α and λ large, the operator should have pure point spectrum (Sect. 10.3). This picture has not been generally proven, but rather for specific examples. We will see such examples below. It is even less well understood what happens in the range between "extremely well approximated" by rationals and "typical" α, and for small λ. Moreover, the spectrum has a tendency to be a Cantor set, that is, a closed set without isolated points, but with empty interior. We will discuss this phenomenon in Sect. 10.4.

We will be able to discuss only a few aspects of the theory in this chapter. Our main goal is to show the flavor and the richness of the field. For further reading, we recommend the survey [335] which has some more material. The reader will, however, realize that some results discussed here were found after the writing of [335], which shows the rapid development of the subject.

10.2 The Almost Mathieu Equation and the Occurrence of Singular Continuous Spectrum

In what follows, we will examine the following one-dimensional example of an almost periodic potential:

$$V_\theta(n) = \sum_{k=1}^{K} a_k \cos[2\pi k(\alpha n + \theta)] \tag{10.1}$$

with $\theta \in [0, 1] \simeq$ the hull of V_0.

For the case $k = 1$, the corresponding (discretized) Schrödinger equation is called the "almost Mathieu equation." It is actually the almost Mathieu equation that we will investigate in detail.

Our first theorem in this section is due to *Herman* [162], and provides an estimate of the Lyaponov exponent γ of (10.1) from below.

Theorem 10.2 (Herman). If $\alpha \notin \mathbb{Q}$, then the Lyaponov exponent γ corresponding to (10.1) satisfies

$$\gamma(E) \geq \ln\left(\frac{|a_K|}{2}\right) .$$

Remarks (1) By Theorem 9.13, we conclude that $H_\theta := H_0 + V_\theta$ for almost all θ has no absolutely continuous spectrum if α is irrational and $|a_K| > 2$.

(2) Prior to Herman, another proof of the case $K = 1$ was given by *Andre* and *Aubry* [15] (with points of rigor clarified by *Avron* and *Simon* [30]).

Proof. For notational convenience, we suppose $K = 1$, i.e.

$$\begin{aligned} V_\theta(n) &= a \cos 2\pi(\alpha n + \theta) \\ &= \frac{a}{2}(e^{2\pi i \alpha n} e^{2\pi i \theta} + e^{-2\pi i \alpha n} e^{-2\pi i \theta}) \\ &= \frac{a}{2}(e^{2\pi i \alpha n} z + e^{-2\pi i \alpha n} z^{-1}) , \end{aligned}$$

where we set $z := \exp(2\pi i\theta)$. For a fixed value E the transfer matrices (see (9.16) $\Phi_L(z)$ are given by

$$\Phi_L(z) = A_L(z)A_{L-1}(z)\cdot\ldots\cdot A_2(z)A_1(z) \quad \text{with}$$

$$A_n(z) = \begin{bmatrix} E - \frac{a}{2}(\mathrm{e}^{2\pi i\alpha n}z + \mathrm{e}^{-2\pi i\alpha n}z^{-1}) & -1 \\ 1 & 0 \end{bmatrix} .$$

We define

$$F_L(z) := z^L\Phi_L(z) = \prod_{n=1}^{L} zA_n(z) .$$

The matrix-valued function $F_L(z)$ is obviously analytic in the whole complex plane, and furthermore satisfies

$$\|F_L(z)\| = \|\Phi_L(z)\|$$

for all z of the form $\exp(i\theta)$. Since $F_L(z)$ is analytic, the function $\ln\|F(z)\|$ is subharmonic (see e.g. *Katznelson* [199] III.3.2), thus

$$\frac{1}{2\pi}\int_0^{2\pi} \ln\|F_L(\mathrm{e}^{i\theta})\|\, d\theta \geq \ln\|F_L(0)\| = L\ln\left(\frac{|a|}{2}\right) . \tag{10.2}$$

Because of $\alpha \notin \mathbb{Q}$, the flow $\tau_n(\theta) = (\theta + \alpha n) \bmod 1$ is ergodic (see Sect. 10.1); hence the subadditive ergodic theorem [200] tells us that for almost all θ

$$\gamma = \lim_{L\to\infty}\frac{1}{L}\ln\|\Phi_L(\mathrm{e}^{i\theta})\| = \lim_{L\to\infty}\frac{1}{L}\int_0^{2\pi}\ln\|\Phi_L(\mathrm{e}^{i\theta})\|\frac{d\theta}{2\pi}$$

$$= \lim_{L\to\infty}\frac{1}{L}\int_0^{2\pi}\ln\|F_L(\mathrm{e}^{i\theta})\|\frac{d\theta}{2\pi} .$$

Therefore, we obtain the bound

$$\gamma \geq \ln\left(\frac{|a|}{2}\right)$$

because of (10.2). □

The next theorem will enable us to exclude also point spectrum for H_θ for special values of α. For those values, H_θ has neither point spectrum nor absolutely continuous spectrum; consequently, the spectrum of H_θ is purely singular continuous!

The theorem we use to exclude eigenvalues is due to *Gordon* [139]. It holds—with obvious modifications—in the continuous case [i.e. on $L^2(\mathbb{R})$] as well (for this case, see *Simon* [335]).

Theorem 10.3 (Gordon). Let $V(n)$ and $V_m(n)$ for $m \in \mathbb{N}$ be bounded sequences on $\mathbb{Z}$ (i.e. $n \in \mathbb{Z}$). Furthermore, let

(i) V_m be periodic, with period $T_m \to \infty$.
(ii) $\sup_{n,m} |V_m(n)| < \infty$.
(iii) $\sup_{|n| \leq 2T_m} |V_m(n) - V(n)| \leq m^{-T_m}$.

Then any solution $u \neq 0$ of

$$Hu = (H_0 + V)u = Eu$$

satisfies

$$\overline{\lim_{|n| \to \infty}} \frac{u(n+1)^2 + u(n)^2}{u(1)^2 + u(0)^2} \geq \frac{1}{4} .$$

Remark. The assumptions of the theorem roughly require that the potential V is extremely well approximated by periodic potentials. The conclusion, in particular, implies that $H = H_0 + V$ does not have (l^2-) eigenfunction, i.e. the point spectrum of H is empty.

Before we give a proof of Gordon's theorem, we apply the theorem to the almost Mathieu equation. We first need the definition:

Definition. A number $\alpha \in \mathbb{R} \backslash \mathbb{Q}$ is called a *Liouville number* if, for any $k \in \mathbb{N}$, there exist $p_k, q_k \in \mathbb{N}$ such that

$$\left| \alpha - \frac{p_k}{q_k} \right| \leq k^{-q_k}.$$

Thus, a Liouville number is an irrational number that is extremely well approximated by rational ones. The set of Liouville numbers is small from an analyst's point of view: It has Lebesgue measure zero. However, from a topologist's point of view, it is rather big: It is a dense G_δ-set. (Recall that F is a G_δ-set if it is a countable intersection of open sets.)

Theorem 10.4 (Avron and Simon [30]). It α is a Liouville number, $|\lambda| > 2$ and

$$V_\theta(n) := \lambda \cos[2\pi(\alpha n + \theta)] ,$$

then $H_\theta = H_0 + V_\theta$ has purely singular continuous spectrum for almost all θ.

Proof. By Theorem 10.2, we know that H_θ does not have absolutely continuous spectrum for a.e. θ. Assume that α is well approximated by p_k/q_k in the sense of the above Definition. By choosing a subsequence $p_{k'}/q_{k'}$ of p_k/q_k, we may assume

$$\left| \alpha - \frac{p_{k'}}{q_{k'}} \right| < q_{k'}^{-1} k^{-q_{k'}} .$$

We set

$$V_k(n) := \lambda \cos\left[2\pi\left(\frac{p_{k'}}{q_{k'}} n + \theta\right)\right].$$

Then $T_k = q_{k'}$ is a period for V_k.

We estimate:

$$\begin{aligned}\sup_{|n|\leq 2q_{k'}} |V_k(n) - V(n)| &= \sup_{|n|\leq 2q_{k'}} \left|\cos\left(2\pi\frac{p_{k'}}{q_{k'}}n\right) - \cos(2\pi\alpha n)\right| \\ &\leq \sup_{|n|\leq 2q_{k'}} 2\pi|n|\left|\frac{p_{k'}}{q_{k'}} - \alpha\right| \\ &\leq 4\pi k^{-T_k}.\end{aligned}$$

Thus, V satisfies the assumptions of Gordon's theorem. □

Now we turn to the proof of Gordon's theorem. We start with an elementary lemma:

Lemma. Let A be an invertible 2×2 matrix, and x a vector of norm 1. Then

$$\max(\|Ax\|, \|A^2x\|, \|A^{-1}x\|, \|A^{-2}x\|) \geq \tfrac{1}{2}.$$

Proof. The matrix A obeys its characteristic equation

$$a_1A^2 + a_2A + a_3 = 0. \tag{10.3}$$

We may assume that $a_i = 1$ for some $i \in \{1, 2, 3\}$ and $|a_j| \leq 1$ for all $j \neq i$.

Let us suppose $a_2 = 1$ and that $|a_1|, |a_3| \leq 1$, the other cases are similar. Then (10.3) gives

$$x = -a_1Ax - a_3A^{-1}x$$

Since x has norm one and $|a_1|, |a_3| \leq 1$, it follows that $\|Ax\| \geq \frac{1}{2}$ or $\|A^{-1}x\| \geq \frac{1}{2}$. □

Proof of Theorem 10.3. Let u be the solution of $(H_0 + V)u = Eu$ with a particular initial condition. Let u_m be the solution of $(H_0 + V_m)u = Eu$ with the same initial condition. Define

$$\phi(n) := \begin{pmatrix} u(n+1) \\ u(n) \end{pmatrix}, \quad \phi_m(n) := \begin{pmatrix} u_m(n+1) \\ u_m(n) \end{pmatrix}$$

and

$$A_n = \begin{pmatrix} E - V(n) & -1 \\ +1 & 0 \end{pmatrix}, \quad A_n^{(m)} = \begin{pmatrix} E - V_m(n) & -1 \\ +1 & 0 \end{pmatrix}.$$

Then

$$\sup_{|n| \le 2T_m} \|\phi_m(n) - \phi(n)\|$$

$$\le \sup_{|n| \le 2T_m} \|A_n A_{n-1} \cdots A_1 - A_n^{(m)} A_{n-1}^{(m)} \cdots A_1^{(m)}\| \left\| \begin{pmatrix} u(1) \\ u(0) \end{pmatrix} \right\|$$

$$\le \sup_{|n| \le 2T_m} |n| e^{C|n|} m^{-T_m} = 2T_m e^{2CT_m} m^{-T_m} .$$

Thus,

$$\max_{a=\pm 1, \pm 2} \|\phi(aT_m) - \phi_m(aT_m)\| \to 0 \quad \text{as } m \to \infty .$$

By the above lemma, we have

$$\max_{a=\pm 1, \pm 2} \|\phi_m(aT_m)\| \ge \tfrac{1}{2} \|\phi_m(0)\| = \tfrac{1}{2}(|u(0)|^2 + |u(1)|^2)^{1/2}$$

Thus

$$\overline{\lim} \frac{|u(n)|^2 + |u(n+1)|^2}{|u(0)|^2 + |u(1)|^2} \ge \overline{\lim} \frac{\max_{a=\pm 1, \pm 2} \|\phi(aT_m)\|^2}{\|\phi(0)\|^2} \ge \frac{1}{4} . \quad \square$$

10.3 Pure Point Spectrum and the Maryland Model

We now turn to an almost periodic (discretized) Schrödinger operator that, to a certain extent, admits an explicit solution. We call this operator the Maryland model, after the place of its creation by Grempel, Prange and Fishman at the University of Maryland. The potential in this model is given by

$$V(n) = V_{\alpha, \theta, \lambda}(n) := \lambda \tan[\pi(\alpha \cdot n) + \theta] \tag{10.4}$$

for $n \in \mathbb{Z}^\nu$. Here $\alpha = (\alpha_1, \ldots, \alpha_\nu) \in \mathbb{Z}^\nu$, $\alpha \cdot n$ denotes the scalar product, and $\theta \in [0, 2\pi]$. To have $V(n)$ finite for all $n \in \mathbb{Z}^\nu$, we require $\theta \ne \pi(\alpha \cdot n) + \pi/2 \bmod \pi$. Then, $V(n)$ will be unbounded (unless all components of α are rational). Therefore V is not an almost periodic function in the sense of Sect. 10.1. We will think of V as a "singular almost periodic function." Since H_0 is a bounded operator, there is no difficulty to define $H = H_0 + V$ properly.

Recently the potential (10.4) was studied extensively by *Fishman, Grempel* and *Prange* [111, 112], *Grempel, Fishman* and *Prange* [142], *Prange, Grempel* and *Fishman* [288], *Figotin* and *Pastur* [110, 272] and *Simon* [338, 340]. We note that Figotin and Pastur even obtain an explicit formula for the Green's function.

There is an explicit expression for the density of states $k_\lambda(E)$ of H. It is not difficult to compute $k_0(E)$, i.e. the density of states for H_0. In momentum space, H_0 is nothing but multiplication by $\phi(k) = 2\sum_{i=1}^{\nu} \cos k_i$; hence, its spectral resolution $P_{(-\infty,E)}$ is multiplication by $\chi_{(-\infty,E)}(\phi(k))$. Using this and the definition of the density of states, one learns

$$k_0(E) = \frac{1}{(2\pi)^\nu} |\{k \in [0, 2\pi]^\nu | \phi(k) < E\}| \ , \tag{10.5}$$

where $|\cdot|$ denotes the Lebesgue measure.

Let us now give the explicit expression for k_λ:

Theorem 10.5. Suppose that $\{1, \alpha_1, \ldots, \alpha_\nu\}$ are independent over the rationals. Then

$$k_\lambda(E) = \frac{1}{\pi} \int \frac{\lambda}{(E - E')^2 + \lambda^2} k_0(E')\, dE' \ . \tag{10.6}$$

Corollary. Suppose $\nu = 1$. Then the Lyaponov exponent $\gamma_\lambda(E)$ of $H_0 + \lambda \tan(\pi\alpha n + \theta)$, $\alpha \notin \mathbb{Q}$ is given by

$$\gamma_\lambda(E) = \frac{1}{\pi} \int \frac{\lambda}{(E - E')^2 + \lambda^2} \gamma_0(E')\, dE' \ , \tag{10.6$'$}$$

where $\gamma_0(E)$ is the Lyaponov exponent of H_0.

Remarks. (1) As long as $(1, \alpha_1, \ldots, \alpha_\nu)$ are independent over the rationals, $k_\lambda(E)$ [and for $\nu = 1$: $\gamma_\lambda(E)$] is independent of α, θ.

(2) $p_\lambda(x) := 1/\pi(\lambda/(x^2 + \lambda^2))$ is the density of a probability measure known as the Cauchy-distribution or the Lorentz-distribution among probabilists and theoretical physicists respectively. Equation (10.6) tells us that $k_\lambda(E)$ is just the convolution $p_\lambda * k_0(E)$. From this, we see that k_λ is a strictly monotone function in E from $(-\infty, +\infty)$ onto $(0, 1)$. Thus, $\sigma(H) = \operatorname{supp} k_\lambda(dE) = (-\infty, \infty)(\lambda \neq 0)$.

(3) For $\nu = 1$, $\gamma_\lambda(E)$ is strictly positive since $\gamma_0(E')$ is positive outside the spectrum of H_0. Thus, H has no absolutely continuous spectrum (for $\nu = 1$).

Proof of the Corollary. The Thouless formula (see Chap. 9) tells us that $\gamma_\lambda(E)$ is the convolution of $f(E) = \ln(E)$ with $dk_\lambda(E)/dE$. Thus,

$$\gamma_\lambda = f * \frac{dk_\lambda}{dE} = f * p_\lambda * \frac{dk_0}{dE} = p_\lambda * \gamma_0. \quad \square$$

To prove the theorem, we will make use of the following lemma:

Lemma. Fix arbitrary reals $\alpha_1, \ldots, \alpha_k$ and positive numbers $\psi_1, \ldots, \psi_k$ with

$\sum_{j=1}^{k} \psi_j = 1$.

Let $\varphi(\theta) := \sum_{j=1}^{k} \psi_j \tan(\alpha_j + \theta)$. Then $\frac{1}{2\pi} \int_0^{2\pi} \mathrm{e}^{\mathrm{i}t\varphi(\theta)}\, d\theta = \mathrm{e}^{-|t|}$.

The proof of the lemma is left to the reader as an exercise in complex integration (for a proof see *Simon* [338]).

Proof of Theorem 10.5. We prove that the Fourier transform $\hat{k}_\lambda(t)$ of k_λ is given by $\exp(-\lambda|t|)\hat{k}_0(t)$ which implies the theorem.

The operators $H_0 + V$ restricted to a finite box are just finite matrices. Thus, we have (for the restricted matrices)

$$\exp[\mathrm{i}t(H_0 + V) = \exp(\mathrm{i}tV) + \mathrm{i} \int_0^t \exp[\mathrm{i}s(H_0 + V)] H_0 \exp[\mathrm{i}(t - s)V]\, ds \ .$$

Iterating this formula, we obtain a series

$$\begin{aligned}
&\exp[\mathrm{i}t(H_0 + V) \\
&\quad = \exp(\mathrm{i}tV) + i \int_0^t \exp(\mathrm{i}s_1 V) H_0 \exp[\mathrm{i}(t - s_1)V]\, ds \\
&\qquad + \mathrm{i}^2 \int_0^t \int_0^{s_1} \exp(\mathrm{i}s_2 V) H_0 \exp[\mathrm{i}(s_2 - s_1)V] H_0 \exp[\mathrm{i}(t - s_1)V]\, ds_2\, ds_1, \\
&\qquad + \cdots
\end{aligned}$$

which is easily seen to be convergent.

Taking expectation of the matrix element $\exp[\mathrm{i}t(H_0 + V)](n, m)$, we see that $\exp[\mathrm{i}t(H_0 + V)]$ is a series of integrals of the type evaluated in the lemma. Therefore,

$$\mathbb{E}(\mathrm{e}^{\mathrm{i}t(H_0+V)}(n, m)) = \mathrm{e}^{-\lambda|t|}\mathrm{e}^{\mathrm{i}tH_0}(n, m) \ . \quad \square$$

Remark. The argument shows that $k_\lambda(e)$ is the density of states for a large variety of models; for example, in the Anderson model with a potential distribution p_λ. This model is known as the Lloyd model, after work of *Lloyd* [235], who computed k_λ in this model. *Grempel, Fishman* and *Prange* [142] obtained Theorem 10.5 for their model by rather different means. Our proof follows *Simon* [340], who investigated the question of why the two models had the same $k(e)$.

We suppose from now on that $(1, \alpha_1, \ldots, \alpha_\nu)$ are independent over $\mathbb{Q}$. While the density of states $k_\lambda(E)$ of $H_0 + \lambda V_{\alpha,\theta}$ does not depend on α, the spectral properties do, at least in dimension $\nu = 1$. We saw already that no absolutely continuous spectrum occurs. If α is a Liouville number, one can apply Gordon's theorem to prove that no (l^2-) eigenfunctions of $H_{\alpha,\theta,\lambda} = H_0 + \lambda V_{\alpha,\theta}$ occur, thus showing that the spectrum is singular continuous in this case. We show now (even in higher dimension) that pure point spectrum occurs for certain other choices of α.

To prove that $H_{\alpha,\theta,\lambda}$ has pure point spectrum for certain values of α, we will transform the eigenvalue equation in a number of steps. Finally, we will arrive at an equation that will make the dependence of the solution on α rather explicit, or more precisely, a sequence ψ_n that determines the solution of our eigenvalue equation will be given by

$$\psi_n = (e^{i\pi(\alpha\cdot n)} - 1)^{-1}\zeta_n ,$$

where ζ_n is a known sequence exponentially decaying in $|n|$. To prevent ψ_n from blowing up, the denominator must approach zero more slowly than ζ_n. This is a typical small divisor problem. Indeed, methods to overcome those problems (KAM-methods) dominate many proofs concerning almost periodic operators. In our case, it is natural to demand the following condition on α.

Definition. We say that α has *typical Diophantine properties* if there exist constants $C, k > 0$ such that

$$\left|\sum_{i=1}^{\nu} m_i\alpha_i - n\right| \geq C\left(\sum_{i=1}^{\nu} m_i^2\right)^{-k/2} \tag{10.7}$$

holds for all $n, m_1, \ldots, m_\nu \in \mathbb{Z}$.

As the name suggests, $\{\alpha | \alpha$ has typical Diophantine properties$\}$ has a complement of Lebesgue measure zero in $\mathbb{R}^\nu$. We will show

Theorem 10.6. If α has typical Diophantine properties, then $H_{\alpha,\theta,\lambda}$ has pure point spectrum for all $\lambda > 0$ and all θ. Moreover, the eigenvalues are precisely the solutions of

$$k_\lambda(E) = \left(\alpha\cdot m + \frac{1}{2} - \frac{\theta}{\pi}\right)_f ,$$

where $(x)_f$ means the fractional part of x, and m runs through $\mathbb{Z}^\nu$. All eigenfunctions decay exponentially.

Theorem 2.9 and 2.10 in Chap. 2 tell us to seek polynomially bounded solutions u of

$$\lambda^{-1}(E - H_0)u(n) = \tan[\pi(\alpha\cdot n) + \theta]u(n) . \tag{10.8}$$

Let us introduce the shorthand notations $A := \lambda^{-1}(E - H_0)$ and $B := \tan[\pi(\alpha\cdot n) + \theta]$. Then formally, (10.8) implies, for $c = (1 + iB)u$

$$\frac{(1 - \mathrm{i}A)}{(1 + \mathrm{i}A)}c = \frac{(1 - \mathrm{i}B)}{(1 + \mathrm{i}B)}c \ . \tag{10.9}$$

The advantage of (10.9) lies in the following simple expression of its right-hand side:

$$\frac{(1 - \mathrm{i}B)}{(1 + \mathrm{i}B)} = \exp(-2\pi\mathrm{i}\alpha n - 2\mathrm{i}\theta) \ .$$

Before we continue, we convince ourselves that the above formal calculation can be justified. Denote by $\mathscr{P}$ the space of all polynomially bounded sequences, i.e. $\mathscr{P} = \{\{u(n)\}_{n\in\mathbb{Z}^\nu} | |u(n)| \leq A(1 + |n|)^k$ for some $A, k\}$.

Proposition 10.7. If $Au = Bu$ has a solution $u \in \mathscr{P}$, then $c = (1 + \mathrm{i}B)u \in \mathscr{P}$ and

$$\frac{1 - \mathrm{i}A}{1 + \mathrm{i}A}c = \frac{1 - \mathrm{i}B}{1 + \mathrm{i}B}c \ .$$

Conversely, if

$$\frac{1 - \mathrm{i}A}{1 + \mathrm{i}A}c = \frac{1 - \mathrm{i}B}{1 + \mathrm{i}B}c \ .$$

has a solution $c \in \mathscr{P}$, then c is of the form $c = (1 + \mathrm{i}B)u$ for a $u \in \mathscr{P}$, and u solves $Au = Bu$.

Remarks. (1) The above equations are a priori to be read pointwise as relations between numbers $u(n)$ rather than as equations in a certain space of sequences. The operator $(1 + \mathrm{i}A)^{-1}$ is well defined on $l^2(\mathbb{Z}^\nu)$. It has a kernel $K(n - m)$ there with K decaying faster than any polynomial, as can be seen by Fourier transform. So $(1 + \mathrm{i}A)^{-1}$ can be defined on $\mathscr{P}$ as well as via its kernel K.

(2) If we consider B as a self-adjoint operator on $l^2(\mathbb{Z}^\nu)$ with domain $D(B)$, and A as an everywhere defined bounded operator, then above $u \in D(B)$ if $c \in l^2(\mathbb{Z}^\nu)$.

Proof. Observe first that both $(1 + \mathrm{i}A)$ and $(1 + \mathrm{i}A)^{-1}$ map $\mathscr{P}$ into $\mathscr{P}$. Thus, $u \in \mathscr{P}$ and $Au = Bu$ implies $c := (1 + B)u = (1 + A)u \in \mathscr{P}$ and

$$\frac{1 - \mathrm{i}A}{1 + \mathrm{i}A}c = \frac{1 - \mathrm{i}B}{1 + \mathrm{i}B}c \ .$$

Suppose now that

$$\frac{1 - \mathrm{i}A}{1 + \mathrm{i}A}c = \frac{1 - \mathrm{i}B}{1 + \mathrm{i}B}c \text{ for a } c \in \mathscr{P} \ .$$

Here $u = (1 + \mathrm{i}B)^{-1}c$ makes perfectly good sense as a (pointwise) equation between sequences (although we do not know $u \in \mathscr{P}$ a priori). Therefore, we obtain

$$\frac{1 - iA}{1 + iA}(1 + iB)u = (1 - iB)u ,$$

thus $Au = Bu$. This, in turn, implies that

$$u = (1 + iB)^{-1}c = (1 + iA)^{-1}c \in \mathscr{P}. \quad \square$$

Now that we know that (10.8) is equivalent to (10.9), we apply a Fourier transform to (10.9). Let us define

$$\hat{f}(k) := \sum_{n \in \mathbb{Z}} f(n) e^{-ink} ,$$

where $\hat{f}$ is well defined if $\sum |f(n)| < \infty$. Moreover, for a continuous function φ on $\mathbb{T}^\nu = [0, 2\pi]^\nu$, we define

$$\check{\varphi}(n) := \frac{1}{2\pi} \int_{\mathbb{T}^\nu} \varphi(k) e^{ink} d^\nu k .$$

If the sequence f is merely in $\mathscr{P}$, we define $\hat{f}$ to be the distribution

$$\langle \hat{f}, \varphi \rangle := \sum f(n) \check{\varphi}(n)$$

for $\varphi \in C^\infty(\mathbb{T}^\nu)$. Here $\mathbb{T}^\nu$ is the ν-dimensional torus. Applying the Fourier transform to (10.9), we obtain in the distribution sense

$$q(k)\hat{c}(k) = e^{-2i\theta}\hat{c}(k + 2\pi\alpha) \quad \text{with} \tag{10.10a}$$

$$q(k) := -\frac{2\sum_{i=1}^{\nu} \cos k_i - E - i\lambda}{2\sum_{i=1}^{\nu} \cos k_i - E + i\lambda} . \tag{10.10b}$$

Here $q(k)$ is an analytic function of $z_i = \exp(ik_i)$ near $|z_i| = 1$, $|q(k)| = 1$ and q does not take the value -1. Thus, $q(k) = \exp[-i\zeta(k)]$ for a function $\zeta(k)$ analytic in $z_i = \exp(ik_i)$ near $|z_i| = 1$ satisfying $-\pi < \zeta(k) < \pi$.

Summarizing, we have shown that if the equation $(H_0 + V)u = Eu$ has a polynomially bounded solution, then

$$e^{-i\zeta(k)}\hat{c}(k) = e^{-2i\theta}\hat{c}(k + 2\pi\alpha) \tag{10.11}$$

has a distributional solution $\hat{c}$.

We will now concentrate on continuous solutions of (10.11) for a while. Since $2\alpha n (\text{mod } 2\pi)$ is dense in $\mathbb{T}^n$, we read off from (10.11) that $|\hat{c}|$ is constant. We may suppose that $|\hat{c}(k)| = 1$. Thus, $\hat{c}(k)$ has the form $\hat{c}(k) = \exp[-im \cdot k - i\psi(k)]$ for a continuous periodic function $\psi(k)$. So (10.11) implies

$$\psi(k + 2\pi\alpha) - \psi(k) = \zeta(k) - 2\theta - 2\pi(m \cdot \alpha) + 2\pi m_0 \tag{10.12}$$

for suitably chosen $m_0 \in \mathbb{Z}$.

Applying the Fourier transform to equation (10.12), we get

$$(e^{-2\pi i \alpha \cdot n} - 1)\check{\psi}_n = \check{\zeta}_n \quad \text{for } n \neq 0 \tag{10.13a}$$

$$\check{\zeta}_0 = 2\theta + 2\pi(m \cdot \alpha) - 2\pi m_0 \ . \tag{10.13b}$$

To solve (10.13b), we observe

Proposition 10.8. $\check{\zeta}_0 = 2\pi k_\lambda(E) - \pi$, $k_\lambda(E)$ being the integrated density of states.

Proof:

$$\frac{\partial}{\partial E}\check{\zeta}_0 = \frac{1}{(2\pi)^\nu}\int_{\mathbb{T}^\nu}\frac{\partial}{\partial E}\zeta(k)d^\nu k = \frac{1}{(2\pi)^\nu}\int_{\mathbb{T}^\nu}\frac{2\lambda}{(2\sum\cos k_i - E)^2 + \lambda^2}d^\nu k$$

[the formula for $\partial\zeta/\partial E$ can be obtained by differentiating (10.10b)]. On the other hand, from (10.6) we know

$$\frac{\partial k_\lambda}{\partial E} = \frac{1}{\pi}\int\frac{\lambda}{(y - E)^2 + \lambda^2}\frac{\partial k_0}{\partial E}(y)\,dy \ .$$

From (10.5) we can read off that $(\partial k_0/\partial E)(y)$ is $1/(2\pi)^\nu$ times the surface measure of the surface $\{k | 2\sum\cos k_i = y\}$. Thus,

$$\frac{\partial k_\lambda}{\partial E} = \frac{1}{\pi}\left(\frac{1}{(2\pi)^\nu}\int\frac{\lambda}{(2\sum\cos k_i - E) + \lambda^2}dk\right)$$

$$= \frac{1}{2\pi}\frac{\partial\zeta}{\partial E} \ .$$

Therefore the assertion follows from $\zeta \to -\pi$ as $E \to -\infty$ while $k_\lambda \to 0$. □

Proposition 10.8 tells us that (10.13b) is equivalent to

$$k_\lambda(E) = \left(\alpha \cdot m + \frac{\theta}{\pi} + \frac{1}{2}\right)_f$$

[$(x)_f$ is the fractional part of x].

Equation (10.13a) is solved, of course, by

$$\check{\psi}_n = (e^{-2\pi i \alpha n} - 1)^{-1}\check{\zeta}_n \ . \tag{10.14}$$

Since ζ is an analytic function, $\check{\zeta}_n$ decays exponentially. Moreover, since α has typical Diophantine properties, we have

$$|e^{-2\pi i\alpha n} - 1| = 2|\sin \pi\alpha \cdot n| \geq \tilde{C}\left(\sum_{i=1}^{\nu} m_i^2\right)^{-k/2} .$$

This follows from the estimate

$$|\sin x| \geq \tilde{\tilde{C}}|x|$$

for $|x| \leq \pi/2$. Therefore, $\check{\psi}_n$ decays exponentially. This, in turn, implies that, for any solution $\check{\psi}_n$ of (10.13a, b), the function $\psi(k) = \sum \check{\psi}_n \exp(-ink)$ is analytic and solves (10.12). Thus, we have shown

Proposition 10.9. The equation

$$e^{-i\zeta(k)}\hat{c}(k) = e^{-2i\theta}\hat{c}(k + 2\pi\alpha) \tag{10.15}$$

has a continuous solution $\hat{c}$ if and only if

$$k_\lambda(E) = \left(\alpha \cdot m + \frac{\theta}{\pi} + \frac{1}{2}\right)_f .$$

Any continuous solution $\hat{c}$ of (10.15) is analytic and of the form $\hat{c}(k) = \exp[-ik \cdot m - i\psi(k)]$, and the Fourier coefficients $\check{\psi}_n$ of ψ decay exponentially.

Now we show that the above solutions are the only ones of interest:

Proposition 10.10. Suppose the sequence c is polynomially bounded, and $\hat{c}$ fulfills (10.15); then $\hat{c}$ is analytic, and c decays exponentially.

Proof. We choose θ_0 such that

$$k_\lambda(E) = \left(\frac{1}{2} + \frac{\theta_0}{\pi}\right)_f .$$

From our considerations above, it follows that there is an analytic function $\hat{d}(k)$ such that $|\hat{d}(k)| = 1$ and

$$\hat{d}(k + 2\pi\alpha) = \exp[-i\zeta(k) + 2i\theta]_0 \hat{d}(k) .$$

Suppose now $\hat{c}$ is a distributional solution of (10.15). Then $l = \hat{c}/\hat{d}$ is also a distribution and satisfies

$$l(k + 2\pi\alpha) = e^{2i(\theta-\theta_0)}l(k) \ ; \quad \text{hence,}$$

$$e^{-2\pi i\alpha n}\check{l}_n = e^{2i(\theta-\theta_0)}\check{l}_n \ ;$$

thus, $\check{l}_n = 0$ for all but one n. Therefore, $l(k) = \exp(-in_0 k)$ for some n_0, i.e.

$$\hat{c}(k) = e^{in_0k}\hat{d}(k) ;$$

thus, $\hat{c}$ is analytic and $c(n) = d(n + n_0)$ decays exponentially. □

We now complete the proof of Theorem 10.6:

Proof (*Theorem 10.6*). Suppose u is a polynomially bounded solution of

$$(H_0 + \lambda V_{\alpha,\theta})u = Eu .$$

Thus $c = (1 + iB)u$ is a polynomially bounded solution of

$$\frac{1 - iA}{1 + iA} c = \frac{1 - iB}{1 + iB} c . \tag{10.16}$$

with $A = \lambda^{-1}(E - H_0)$ and $B = \tan[2\pi(\alpha \cdot n) + \theta]$.

We have shown that any polynomially bounded solution of (10.16) is exponentially decaying. Thus,

$$u = (1 + iB)^{-1} c = (1 - iA)^{-1} c$$

is exponentially decaying.

From Theorem 2.10 in Chap. 2 we know that the spectral measures are supported by $S = \{E | Hu = Eu \text{ has a polynomially bounded solution}\}$. Since any polynomially bounded solution of $Hu = Eu$ is exponentially decaying, S is a countable set; thus, H has pure point spectrum. □

Besides various cleverly chosen transformations of the problem, the very heart of the proof of Theorem 10.6 is the solution of (10.13a), i.e. to control the behavior of

$$\check{\psi}_n = (e^{-2\pi i\alpha \cdot n} - 1)^{-1} \check{\zeta}_n .$$

This is a typical problem of small divisors. Above we ensured that $\check{\psi}_n$ decays exponentially by requiring α to have typical Diophantine properties. Virtually all proofs for pure point spectrum of almost periodic Hamiltonians rely upon handling such small divisor problems. We can only mention some of those works: *Sarnak* [306], *Craig* [68], *Bellissard, Lima* and *Scoppola* [40], *Pöschel* [286]. Those authors construct examples of almost periodic Hamiltonians with dense point spectrum. They use Kolmogoroff-Arnold-Moser (KAM)-type methods to overcome the small divisor problem. Among their examples are, for any $\lambda \in [0, 1]$, almost periodic V's so that $H_0 + V$ has only dense point spectrum and $\sigma(H_0 + V)$ has Hausdorff dimension λ.

The first use of KAM-methods in the present context was made by *Dinaburg* and *Sinai* [87]. They proved that absolutely continuous spectrum occurs for certain almost periodic Schrödinger operators, and moreover, that certain solutions of their Schrödinger equation have Floquet-type structure. Their work was extended considerably by *Russmann* [304] and *Moser* and *Pöschel* [253].

Bellissard, Lima and *Testard* [41] applied KAM-ideas to the almost Mathieu equation (see Sect. 10.2). They proved, for typical Diophantine α and small coupling constant there is some absolutely continuous spectrum. Moreover, for typical Diophantine α and large coupling they found point spectrum of positive Lebesgue measure for almost all values of θ. In neither case could they exclude additional spectrum of other types.

10.4 Cantor Sets and Recurrent Absolutely Continuous Spectrum

General wisdom used to say that Schrödinger operators should have absolutely continuous spectrum plus some discrete point spectrum, while singular continuous spectrum is a pathology that should not occur in examples with V bounded. This general picture was proven to be wrong by *Pearson* [275, 276], who constructed a potential V such that $H = H_0 + V$ has singular continuous spectrum. His potential V consists of bumps further and further apart with the height of the bumps possibly decreasing. Furthermore, we have seen the occurrence of singular continuous spectrum in the innocent-looking almost Mathieu equation (Sect. 10.2).

Another correction to the "general picture" is that point spectrum may be dense in some region of the spectrum rather than being a discrete set. We have seen this phenomenon in Chap. 9 as well as in Sect. 10.3. Thus, so far we have four types of spectra: "thick" point spectrum and singular continuous spectrum, which are the types one would put in the waste basket if they did not occur in natural examples, and "thin" point spectrum and absolutely continuous spectrum, the two types that are expected according to the above picture.

It was *Avron* and *Simon* [28] who proposed a further splitting of the absolutely continuous spectrum into two parts: The transient a.c. spectrum, which is the "expected" one, and the recurrent a.c. spectrum, which is the "surprising" one usually coming along with Cantor sets.

To motivate their analysis, we construct some examples, at the same time fixing notations.

A subset C of the real line is called a *Cantor set* if it is closed, has no isolated points (i.e. is a perfect set), and furthermore, is nowhere dense (i.e. $\bar{C} = C$ has an empty interior). "The" Cantor set is an example for this: Remove from [0, 1] the middle third. From what remains, remove the middle third in any piece, and so on. What finally remains is a perfect set with empty interior. This set is well known to have Lebesgue measure zero.

The construction of "removing the middle third" can be generalized easily. Choose a sequence n_j of real numbers, $n_j > 1$. From $S_0 = [0, 1]$ remove the open interval of size $1/n_1$ about the point $\frac{1}{2}$. The new set is called

$$S_1\colon S_1 = [0,1]\backslash\left(\frac{1}{2} - \frac{1}{2n_1}, \frac{1}{2} + \frac{1}{2n_1}\right) = \left[0, \frac{1}{2}\left(1 - \frac{1}{n_1}\right)\right] \cup \left[\frac{3}{2}\left(1 - \frac{1}{n_1}\right), 1\right].$$

Having constructed S_j, a disjoint union of 2^j closed intervals of size α_j, remove from each of these intervals the open interval of size $\alpha_j n_{j+1}^{-1}$ about the center of the interval. The union of the remaining 2^{j+1} intervals is called S_{j+1}. We define

$$S = S(\{n_i\}) = \bigcap_{j=0}^{\infty} S_j .$$

It is not difficult to see that S is a Cantor set (in the above defined sense; see *Avron* and *Simon* [28]). Moreover, since

$$\alpha_j = 2^{-j} \prod_{k=1}^{j} [1 - (1/n_k)] ,$$

the Lebesgue measure of the set S_j is given by

$$|S_j| = \prod_{k=1}^{j} [1 - (1/n_k)] .$$

Hence

$$|S| = \prod_{k=1}^{\infty} [1 - (1/n_k)] .$$

The infinite product is zero if and only if $\sum_{k=1}^{\infty} (1/n_k) = \infty$. Thus, the above procedure allows us to construct Cantor sets of arbitrary Lebesgue measure (<1). The "middle third" Cantor set, our starting example, has $n_k = 3$ for all k, and thus zero Lebesgue measure. It can be used to construct a singular continuous measure carried by it (see e.g. *Reed* and *Simon* I [292]).

Suppose now S is a Cantor set with $0 < |S| < \infty$. Let χ_S be its characteristic function [$\chi_S(x) = 1$ if $x \in S$, and zero otherwise]. Then $\mu_S := \chi_S(x)\, dx$ defines an absolutely continuous measure (with respect to Lebesgue measure). So μ_S is an absolutely continuous measure with nowhere dense support!

The idea of Avron and Simon was to single out measures like μ_S by looking at their Fourier transform.

It is well known that the Fourier transform $F_\mu(t) = \int \exp(itx)\, d\mu(x)$ goes to zero as $|t| \to \infty$ if μ is an absolutely continuous measure. $F_\mu(t)$ goes to zero at least in the averaged sense that $1/2T \int_{-T}^{T} F_\mu(t)\, dt \to 0$ as $T \to \infty$ if μ is a continuous (a.c. or s.c.) measure. We will now distinguish two types of a.c. measures by the fall-off of their Fourier transform. We call two measures, μ and ν, *equivalent* if they are mutually absolutely continuous, that is to say, there exists functions $f \in L^1(\mu)$ and $g \in L^1(\nu)$ such that $d\nu = f\, d\mu$ and $\mu = g\, d\nu$.

Proposition 10.11. (1) Suppose μ is an absolutely continuous measure supported by a Cantor set S, then $F_\mu(t)$ is not in L^1.

(2) Consider the measure $\nu = \chi_A\, dx$ where ∂A has Lebesgue measure zero. Then there exists a measure $\tilde{\nu}$ equivalent to ν such that $F_{\tilde{\nu}}(t) = O(t^{-N})$ for all $N \in \mathbb{N}$.

Proof. (1) $\mu = f(x)\,dx$ for a function f supported by S. Since S is a Cantor set, f cannot be continuous. But if $F_\mu(t) = \int \exp(itx) f(x)\,dx$ were in L^1, then f would be continuous.

(2) There exists a function $f \in \mathrm{S}(\mathbb{R})$, the Schwartz functions, with $\operatorname{supp} f = A$ and $f > 0$ on the interior A^{int} of A, such that $\tilde{\nu} := f(x)\,dx$ is equivalent to ν. Then $F_{\tilde{\nu}}(t) = \int f(x) \exp(itx)\,dx = \mathrm{O}(t^{-N})$, which can be seen by integration by parts. □

The above considerations motivate the following definition:

Definition. Let H be a self-adjoint operator on a separable Hilbert space H. The quantity $\varphi \in \mathrm{H}$ is called a *transient vector* for H if

$$\langle \varphi, \mathrm{e}^{-tH} \varphi \rangle = \mathrm{O}(t^{-N}) \quad \text{for all } N \in \mathbb{N} \;.$$

The closure of the set of transient vectors is called H_{tac} (*transient absolutely continuous subspace*). Thus, φ is a transient vector if the spectral μ_φ measure associated with φ has rapidly decaying Fourier transform. Proposition 10.11 would equally well suggest to define φ as a transient vector if the Fourier transform of its spectral measure is L^1. Fortunately, this leads to the same set H_{tac}.

Proposition 10.12:

(i) H_{tac} is a subspace of H,
(ii) $\mathrm{H}_{\text{tac}} \subset \mathrm{H}_{\text{ac}}$
(iii) $\mathrm{H}_{\text{tac}} = \overline{\{\varphi \mid F_{\mu_\varphi} \in L^1\}}$.

For a proof, see *Avron* and *Simon* [28].

Definition. We define $\mathrm{H}_{\text{rac}} = \mathrm{H}_{\text{tac}}^{\perp} \cap \mathrm{H}_{\text{ac}}$. H_{rac} is called the *recurrent absolutely continuous subspace.* Both H_{tac} and H_{rac} are invariant subspaces under H. We can therefore define $\sigma_{\text{tac}}(H) = \sigma(H|_{\mathrm{H}_{\text{tac}}})$ and $\sigma_{\text{rac}}(H) = \sigma(H|_{\mathrm{H}_{\text{rac}}})$.

As the reader might expect, the occurrence of σ_{rac} and Cantor sets are intimately related:

Proposition 10.13. Suppose that H has nowhere dense spectrum. Then $\sigma_{\text{tac}}(H) = \phi$.

This is actually a corollary to Proposition 10.11. It is, of course, easy to construct operators with $\sigma_{\text{rac}} \neq \phi$. Take $\mathrm{H} = L^2(\mathbb{R})$ and consider the operator $T_A = x\chi_A(x)$ where $A \in B(\mathbb{R})$. We have $\sigma(T_A) = \bar{A}$. If A is an interval $[a, b]$, $(a < b)$, then the spectrum is purely transient absolutely continuous. There are, however, vectors φ with bad behavior of $\int \exp(itx)\,d\mu_\varphi(x)$. For example, take $\varphi = \chi_S$, S a Cantor set of positive Lebesgue measure in $[a, b]$. This shows clearly that not all $\varphi \in \mathrm{H}_{\text{tac}}$ show fast decay of $\int \exp(itx)\,d\mu_\varphi(x)$, but rather a dense subset of φ's does. If A is a Cantor set of positive Lebesgue measure, then the spectrum is recurrent absolutely continuous.

One might think that σ_{rac} is always a nowhere dense set. This is wrong!

Avron and *Simon* [28] constructed a set A such that $\sigma_{rac}(T_A) = (-\infty, +\infty)$. This means, in particular, that $(-\infty, \infty)$ is the support of an absolutely continuous measure $d\mu = f(x)\,dx$, $\operatorname{supp} f = \mathbb{R}$, but dx is not absolutely continuous with respect to $d\mu$.

So far, we worked in a quite abstract setting, and one might think that Cantor sets and recurrent absolutely continuous spectrum do not occur for Schrödinger operators. However, there is some evidence that Cantor sets as spectra of one-dimensional almost periodic operators are very common, although recurrent absolutely continuous spectrum might be less generic.

Chulaevsky [64], *Moser* [252] and *Avron* and *Simon* [29] have constructed examples of limit periodic potentials whose spectra are Cantor sets. A sequence $\{c_n\}_{n \in \mathbb{Z}}$ is called *limit periodic* if it is a uniform limit (i.e. a limit in l^∞) of periodic sequences. For example,

$$V(n) = \sum_{j=-\infty}^{+\infty} a_j \cos\left(\frac{2\pi n}{2^j}\right), \tag{10.17}$$

for $\sum |a_j| < \infty$ is such a sequence.

We denote by L the space of all limit periodic sequences, and by L_0 the space of all sequences as in (10.17). Limit periodic sequences are particular examples of almost periodic ones as one easily verifies. The definition can be carried over to higher dimensions, but we consider only sequences indexed by $\mathbb{Z}^1$ here.

L and L_0 are closed subspaces of the Banach space l^∞, so that topological notions like dense, closed and G_δ (countable intersection of open sets) make sense.

Theorem 10.14 (SCAM).

(i) For a dense G_δ in L, the spectrum $\sigma(H_0 + V)$ is a Cantor set.
(ii) The same is true for a dense G_δ in L_0.

Remark. The name SCAM-theorem is a (linguistic) permutation of initials: *Avron* and *Simon* [29], *Chulaevsky* [64] and *Moser* [252]. Those authors actually worked in the continuous case, i.e. with Schrödinger operators on $L^2(\mathbb{R})$.

Avron and *Simon* [29] and *Chulaevsky* [64] proved—for a perhaps smaller set—the occurrence of recurrent absolutely continuous spectrum:

Theorem 10.15 (Avron and Simon, Chulaevsky). For a dense subset of L, the spectrum $\sigma(H_0 + V)$ is both a Cantor set and absolutely continuous. The same is true for L_0.

Notice that the above theorem does not claim that the dense set in question is a G_δ. We do not even believe that this is true.

We learn from Theorem 10.15 and Proposition 10.13 that the spectrum of $H_0 + V$ is recurrent absolutely continuous.

There is another result by *Bellissard-Simon* [42] establishing Cantor spectrum, this time for the almost Mathieu equation:

Theorem 10.16 (Bellissard-Simon). The set of pair (λ, α) for which $\sigma(H_0 + \lambda\cos(2\pi\alpha n + \theta))$ is a Cantor set is a dense G_δ in $\mathbb{R}^2$.

The physical significance of the distinction between recurrent and transient absolutely continuous spectrum comes from the intuitive connection of the long-time behavior of $\exp(-itH)$ and transport phenomena in the almost periodic structure. Fast decay of $\langle\varphi, \exp(-itH)\varphi\rangle$ means that the wave packet φ will spread out rapidly, while slow decay means that it will have anomalous long-time behavior. Hence, fast decay of $F(t) = \int \exp(-itx)\,d\mu_\varphi$ for the spectral measure μ_φ indicates good transport properties of the medium (think of electric transport via electrons moving in an imperfect crystal); slow or no decay of $F(t)$ indicates bad transport.

In this respect, recurrent absolutely continuous spectrum behaves much more like singular continuous spectrum than like a transient absolutely continuous one.

11. Witten's Proof of the Morse Inequalities

Thus far, we have described the study of Schrödinger operators for their own sake. In this chapter and the next, we will discuss some rather striking applications of the Schrödinger operators to analysis on manifolds. In a remarkable paper, *Witten* [370] showed that one can obtain the strong Morse inequalities from the semiclassical analysis of the eigenvalues of some appropriately chosen Schrödinger operators on a compact manifold M. The semiclassical eigenvalues theorems are discussed in Sect. 11.1, and Witten's choice of operators in Sect. 11.4. The Morse inequalities are stated in Sect. 11.2 and proven in Sect. 11.5. Some background from Hodge theory is described in Sect. 11.3.

Supersymmetric ideas play a role in the proof of the Morse index theorem, and played an even more significant role in Witten's motivation.

11.1 The Quasiclassical Eigenvalue Limit

We begin by discussing the quasiclassical eigenvalue limit for Schrödinger operators acting in $L^2(\mathbb{R}^\nu)$. We will consider self-adjoint operators of the form

$$H(\lambda) = -\Delta + \lambda^2 h + \lambda g$$

defined as the closure of the differential operator acting on $C_0^\infty(\mathbb{R}^\nu)$. Here h, $g \in C^\infty(\mathbb{R}^\nu)$, g is bounded, $h \geq 0$ and $h > \text{const} > 0$ outside a compact set. Furthermore, we assume that h vanishes at only finitely many points $\{x^{(a)}\}_{a=1}^k$, and that the Hessian

$$[A_{ij}^{(a)}] = \frac{1}{2}\left[\frac{\partial^2 h}{\partial x_i \partial x_j}(x^{(a)})\right]$$

is strictly positive definite for every a. The goal is to estimate the eigenvalues of $H(\lambda)$ for large λ. The idea is that for large λ the potential $\lambda^2 h + \lambda g$ should look like finitely many harmonic oscillator wells centered at the zeros of h and separated by large barriers. Thus, one expects that for large λ the spectrum of $H(\lambda)$ should look like the spectrum of a direct sum of operators of the form

$$H^{(a)}(\lambda) = -\Delta + \lambda^2 \sum_{ij} A_{ij}^{(a)}(x - x^{(a)})_i (x - x^{(a)})_j + \lambda g(x^{(a)}) \ .$$

Let $T(b)$ for $b \in \mathbb{R}^\nu$ and $D(\lambda)$ for $\lambda > 0$ be the unitary translation and dilation operators on $L^2(\mathbb{R}^\nu)$ given by

$$(T(b)f)(x) = f(x - b)$$
$$(D(\lambda)f)(x) = \lambda^{n/2}f(\lambda x) \ .$$

Also, let

$$K^{(a)} = -\Delta + \sum_{ij} A_{ij}^{(a)} x_i x_j + g(x^{(a)}) \ .$$

Then it is easily seen that

$$D(\lambda^{1/2})T(\lambda^{1/2}x^{(a)})\lambda K^{(a)}T(-\lambda^{1/2}x^{(a)})D(\lambda^{-1/2}) = H^{(a)}(\lambda) \ , \tag{11.2}$$

so that $H^{(a)}(\lambda)$ is unitarily equivalent to $\lambda K^{(a)}$.

Now the eigenvalues of $K^{(a)}$ are easy to compute, since up to the constant $g(x^{(a)})$, $K^{(a)}$ is a harmonic oscillator. First we find the eigenvalues $\{(\omega_i^{(a)})^2\}_{i=1}^{\nu}$ with $\omega_i^{(a)} > 0$ of the matrix $[A_{ij}^{(a)}]$. Then the eigenvalues of $K^{(a)}$ are sums of the eigenvalues of the one-dimensional oscillators $-(d^2/dx^2) + (\omega_i^{(a)})^2 x^2$ shifted by the constant $g(x^{(a)})$, i.e.

$$\sigma(K^{(a)}) = \left\{ \left[\sum_{i=1}^{\nu} \omega_i^{(a)}(2n_i + 1) \right] + g(x^{(a)}) | n_1, \ldots, n_\nu \in \{0, 1, 2, \ldots\} \right\} .$$

By collecting all the eigenvalues of $K^{(a)}$ for $a = 1, \ldots, k$ we obtain the spectrum of $\bigoplus_a K^{(a)}$ acting in $\bigoplus_a L^2(\mathbb{R}^\nu)$, namely

$$\sigma\left(\bigoplus_a K^{(a)}\right) = \bigcup_a \sigma(K^{(a)}) \ .$$

For future reference, we note that eigenfunctions of $K^{(a)}$ are functions of the form

$$p(x)\exp\left(-\frac{1}{2}\sum_{i=1}^{\nu} \omega_i^{(a)} \langle x, v_i^{(a)} \rangle^2\right) ,$$

where p is a polynomial and $v_i^{(a)} \in \mathbb{R}^\nu$, $i = 1, \ldots, \nu$ are the orthonormal eigenvectors of $[A_{ij}^{(a)}]$. Thus, the eigenfunctions of $H^{(a)}(\lambda)$ look like

$$p(\lambda^{1/2}(x - x^{(a)}))\exp\left(-\frac{1}{2}\sum_{i=1}^{\nu} \omega_i^{(a)} \langle \lambda^{1/2}(x - x^{(a)}), v_i^{(a)} \rangle^2\right) \tag{11.3}$$

by virtue of (11.2). We now present the theorem on the quasiclassical limit of eigenvalues. This theorem can be found in *Simon* [337], and was proven previously for the one-dimensional case by *Combes*, *Duclos* and *Seiler* [66]; see also *Combes* [65]. It can also be proven by a method of *Davies* [76].

Theorem 11.1. Let $H(\lambda)$ and $\bigoplus_a K^{(a)}$ be as above. Let $E_n(\lambda)$ denote the nth eigenvalue of $H(\lambda)$ counting multiplicity. Denote by e_n the nth eigenvalue of $\bigoplus_a K^{(a)}$ counting multiplicity. Then for fixed n and large λ, $H(\lambda)$ has at least n eigenvalues and

$$\lim_{\lambda\to\infty} E_n(\lambda)/\lambda = e_n \ .$$

Proof. We begin by proving

$$\overline{\lim_{\lambda\to\infty}}\, E_n(\lambda)/\lambda \leq e_n \ . \tag{11.4}$$

Let J be a $C_0^\infty(\mathbb{R}^\nu)$ function with $0 \leq J \leq 1$, $J(x) = 1$ for $|x| \leq 1$ and $J(x) = 0$ for $|x| \geq 2$. Define

$$J_a(\lambda) = J(\lambda^{2/5}(x - x^{(a)})) \quad \text{for } a = 1, \ldots, k \ ,$$

$$J_0(\lambda) = \sqrt{1 - \sum_{a=1}^{k} [J_a(\lambda)]^2} \ .$$

Clearly, for large enough λ, we have

$$\sum_{a=0}^{k} J_a(\lambda)^2 = 1 \ ,$$

so that $\{J_a\}_{a=0}^k$ is a partition of unity in the sense of Definition 3.1.

We claim that for any $a \in \{1, \ldots, k\}$

$$\|J_a(\lambda)[H(\lambda) - H^{(a)}(\lambda)]J_a(\lambda)\| = O(\lambda^{4/5}) \ , \tag{11.5}$$

where $H^{(a)}(\lambda)$ is given by (11.1). To see this, we note that $J_a(\lambda)[H(\lambda) - H^{(a)}(\lambda)]J_a(\lambda)$ is given by two terms which we examine in turn. First,

$$\begin{aligned}
&|\lambda^2 J_a(\lambda)[h - \textstyle\sum A_{ij}^{(a)}(x - x^{(a)})_i(x - x^{(a)})_j]J_a(\lambda)| \\
&= \lambda^2 J_a^2(\lambda)|x - x^{(a)}|^3 \frac{|h - \sum A_{ij}^{(a)}(x - x^{(a)})_i(x - x^{(a)})_j|}{|x - x^{(a)}|^3} \\
&\leq \lambda^2 \cdot \lambda^{-6/5} \cdot O(1) \\
&= O(\lambda^{4/5}) \ .
\end{aligned} \tag{11.6}$$

Here we used that $J_a(\lambda)$ has support where $\lambda^{2/5}|x - x^{(a)}| < 2$ and that $\sum A_{ij}^{(a)}(x - x^{(a)})_i(x - x^{(a)})_j$ is the second-order Taylor expansion for h about $x^{(a)}$. Similarly,

$$\lambda J_a(\lambda)[g(x) - g(x^{(a)})]J_a(\lambda) = O(\lambda^{3/5}) \ , \tag{11.7}$$

and thus by (11.6) and (11.7), we see that (11.5) holds.

Now fix a nonnegative integer n. Then there is an $a(n)$ and ψ_n, so that

$$H^{(a(n))}(\lambda)\psi_n = \lambda e_n \psi_n \ . \tag{11.8}$$

Define

$$\varphi_n = J_{a(n)}(\lambda)\psi_n \ . \tag{11.9}$$

Then for nonnegative integers m and n

$$\langle \varphi_n, \varphi_m \rangle = \delta_{nm} + \mathrm{O}(\exp(-c\lambda^{1/5})) \tag{11.10}$$

for some positive constant c. This is clear for the case when $a(n) \neq a(m)$, since in this case $J_{a(n)}(\lambda)$ and $J_{a(m)}(\lambda)$ have disjoint supports for large enough λ so that $\langle \varphi_n, \varphi_m \rangle = 0$. If $a(n) = a(m)$, then $\langle \psi_n, \psi_m \rangle = \delta_{nm}$. Thus,

$$\begin{aligned} |\langle \varphi_n, \varphi_m \rangle - \delta_{nm}| &= \int [1 - J^2_{a(n)}(\lambda)] \psi_n \psi_m \, d^\nu x \\ &\leq \int_{|x - x^{(a)}| \geq \lambda^{-2/5}} \left| p_n p_m \exp\left(-\sum_{i=1}^{\nu} \omega_i^{(a)} \langle \lambda^{1/2}(x - x^{(a)}), v_i^{(a)} \rangle^2 \right) \right| d^\nu x \\ &\leq \text{const.} \int_{\lambda^{-2/5}}^{\infty} \exp(-c\lambda u^2)\, du \\ &\leq \text{const.} \int_{\lambda^{-2/5}}^{\infty} \exp(-c\lambda^{3/5} u)\, du \\ &= \mathrm{O}(\exp(-c\lambda^{1/5})) \ . \end{aligned}$$

Here we used that ψ_n and ψ_m are of the form (11.3); p_n and p_m are polynomials in the components of $\lambda^{1/2}(x - x^{(a)})$. We now use the localization formula (see Theorem 3.2)

$$J_a H^{(a)} J_a = \tfrac{1}{2}(J_a^2 H^{(a)} + H^{(a)} J_a^2) + (\nabla J_a)^2$$

and the estimate

$$\sup_x |\nabla J_a(\lambda)|^2 = \mathrm{O}(\lambda^{4/5}) \tag{11.11}$$

to estimate $\langle \varphi_n, H(\lambda)\varphi_m \rangle$. Again, if $a(n) \neq a(m)$, then $\langle \varphi_n, H(\lambda)\varphi_m \rangle = 0$ for large λ. On the other hand, if $a(n) = a(m) = a$,

$$\begin{aligned} \langle \varphi_n, H(\lambda)\varphi_m \rangle &= \langle \psi_n, J_a(\lambda) H(\lambda) J_a(\lambda) \psi_m \rangle \\ &= \langle \psi_n, J_a(\lambda) H^{(a)}(\lambda) J_a(\lambda) \psi_m \rangle + \mathrm{O}(\lambda^{4/5}) \\ &= \lambda \frac{e_n + e_m}{2} \langle \varphi_n, \varphi_m \rangle + \mathrm{O}(\lambda^{4/5}) \ . \end{aligned}$$

Here we used (11.5) and (11.8). Now using (11.10), we obtain

$$\langle \varphi_n, H(\lambda)\varphi_m \rangle = \lambda e_n \delta_{nm} + O(\lambda^{4/5}) \ . \tag{11.12}$$

Let $\mu_n(\lambda)$ for $n = 1, 2, \ldots$ be the number given by the minimax formula

$$\mu_n(\lambda) = \sup_{\zeta_1, \ldots, \zeta_{n-1}} Q(\zeta_1, \ldots, \zeta_{n-1}; \lambda) \ ,$$

where

$$Q(\zeta_1, \ldots, \zeta_{n-1}; \lambda) \\ = \inf\{\langle \psi, H(\lambda)\psi \rangle | \psi \in D(H(\lambda)), \|\psi\| = 1, \psi \in [\zeta_1, \ldots, \zeta_{n-1}]^\perp\} \ .$$

Then either $H(\lambda)$ has n eigenvalues and $\mu_n(\lambda) = E_n(\lambda)$ or $\mu_n(\lambda) = \inf \sigma_{\text{ess}}(H(\lambda))$ [295]. Fix $\varepsilon > 0$, and for each λ, choose $\zeta_1^\lambda, \ldots, \zeta_{n-1}^\lambda$ so that

$$\mu_n(\lambda) \le Q(\zeta_1^\lambda, \ldots, \zeta_{n-1}^\lambda; \lambda) + \varepsilon \ . \tag{11.13}$$

From (11.10), it follows that for large λ, $\varphi_1, \ldots, \varphi_n$ span an n-dimensional space. Thus, we can find, for each sufficiently large λ, a linear combination φ of $\varphi_1, \ldots, \varphi_n$ such that $\varphi \in [\zeta_1^\lambda, \ldots, \zeta_{n-1}^\lambda]^\perp$. Thus, using (11.12), we see that

$$Q(\zeta_1^\lambda, \ldots, \zeta_{n-1}^\lambda; \lambda) \le \langle \varphi, H(\lambda)\varphi \rangle \le \lambda e_n + O(\lambda^{4/5}) \ ,$$

so that by (11.13) and the fact that ε was arbitrary, we have

$$\mu_n(\lambda) \le \lambda e_n + O(\lambda^{4/5}) \ . \tag{11.14}$$

Since $h > \text{const} > 0$ outside a compact set, and g is bounded, it follows from Persson's formula (Theorem 3.12)

$$\inf \sigma_{\text{ess}}(\mathrm{H}(\lambda)) \ge c\lambda^2$$

for some positive constant c. Hence, in view of (11.14), we have $\mu_n(\lambda) \ne \inf \sigma_{\text{ess}}(H(\lambda))$ for λ sufficiently large. Thus, $\mu_n(\lambda) = E_n(\lambda)$ and (11.14) implies (11.4).

We now will prove the opposite inequality

$$\underline{\lim}_{\lambda \to \infty} E_n(\lambda)/\lambda \ge e_n \ . \tag{11.15}$$

To prove this inequality, it suffices to show that for any e not in the spectrum of $\bigoplus_a K^{(a)}$, say $e \in (e_m, e_{m+1})$,

$$H(\lambda) \ge \lambda e + R + o(\lambda) \ , \tag{11.16}$$

where R is a rank m symmetric operator. To see this, suppose (11.16) holds for each m with $e_{m+1} > e_m$ and $e \in (e_m, e_{m+1})$. Pick a vector ψ in the span of the first $m + 1$ eigenfunctions of $H(\lambda)$ with $\|\psi\| = 1$ such that $\psi \in \mathrm{Ker}(R)$. Then

$$E_{m+1} \geq \langle \psi, H(\lambda)\psi \rangle \geq \lambda e + o(\lambda) ,$$

which implies (11.15) for $n = m + 1$ with $e_{m+1} > e_m$. The degenerate case $e_{m+1} = e_m$ follows easily from this.

So let e satisfy $e \in (e_m, e_{m+1})$. By the IMS localization formula (Theorem 3.2), we have

$$\begin{aligned} H(\lambda) &= \sum_{a=0}^{k} J_a H(\lambda) J_a - \sum_{a=0}^{k} (\nabla J_a)^2 \\ &= J_0 H(\lambda) J_0 + \sum_{a=1}^{k} J_a H^{(a)}(\lambda) J_a + O(\lambda^{4/5}) . \end{aligned} \tag{11.17}$$

Here we used (11.5) and (11.11). Now $J_0 = J_0(\lambda)$ has support away from balls of radius $\lambda^{-2/5}$ about the zeros of h. Since h vanishes quadratically at each of its zeros, $h(x) \geq c(\lambda^{-2/5})^2 = c\lambda^{-4/5}$ on the support of J_0. Thus, $J_0 H(\lambda) J_0$ grows like $\lambda^2 \cdot \lambda^{-4/5} = \lambda^{6/5}$ and for large λ

$$J_0 H(\lambda) J_0 \geq \lambda e J_0^2 . \tag{11.18}$$

For $a \neq 0$, we have

$$J_a(H^{(a)}(\lambda))J_a \geq J_a R^{(a)} J_a + \lambda e J_a^2 , \tag{11.19}$$

where $R^{(a)}$ is the restriction $H^{(a)}$ to the span of all eigenvalues of $H^{(a)}$ lying below λe. We have $\mathrm{Rank}(J_a R^{(a)} J_a) \leq \mathrm{Rank}(R^{(a)}) = \#\{$eigenvalues of $K^{(a)}$ below $e\}$. Since $\mathrm{Rank}(A + B) \leq \mathrm{Rank}(A) + \mathrm{Rank}(B)$, this implies

$$\begin{aligned} \mathrm{Rank}\left[\sum_{a=1}^{k} J_a R^{(a)}_{(\lambda)} J_a\right] &\leq \sum_{a=1}^{k} \mathrm{Rank}[J_a R^{(a)}(\lambda) J_a] \\ &= \#\{\text{eigenvalues of } \textstyle\bigoplus_a K^{(a)} \text{ below } e\} \\ &= n . \end{aligned}$$

Hence, by (11.17) through (11.19), we have

$$\begin{aligned} H(\lambda) &= \lambda e J_0^2 + \lambda e \sum_{a=1}^{k} J_a^2 + \sum_{a=1}^{k} J_a R^{(a)}(\lambda) J_a + 0(\lambda^{4/5}) \\ &= \lambda e + R + O(\lambda^{4/5}) , \end{aligned}$$

where R has rank at most n. This implies (11.16), and completes the proof. □

Theorem 11.1 gives the leading order behavior of the eigenvalues $E_n(\lambda)$. These ideas are extended in [337] to give an asymptotic expansion for $E_n(\lambda)$ in powers of λ^{-m}, $m \geq -1$. The following is a sketch of some of the ideas involved. Let $a = a(n)$ as in (11.8), and define

$$\tilde{H}(\lambda) = \lambda^{-1}(U^{(a)})^{-1}H(\lambda)U^{(a)} ,$$

where $U^{(a)} = T(-\lambda^{1/2}x^{(a)})D(\lambda^{-1/2})$, T and D being given by (11.3) and (11.4). Then

$$\tilde{H}(\lambda) = K^{(a)} + V(\lambda) ,$$

where the asymptotic behavior of $V(\lambda)$ can be controlled, given some additional restriction on g and h. In the case that e_n is a simple eigenvalue, we can choose ε small enough to ensure that

$$P(\lambda) = (2\pi \mathrm{i})^{-1} \oint_{|z-e_n|=\varepsilon} [z - \tilde{H}(\lambda)]^{-1}\, dz$$

is one-dimensional for large λ. Then one shows that $\langle \varphi_n, P(\lambda)\varphi_n \rangle \to 1$ and that $P(\lambda)\varphi_n$ has an L^2-asymptotic expansion where φ_n is given by (11.9). The asymptotic expansion for $E_n(\lambda)$ then follows from the formula

$$\lambda^{-1}E_n(\lambda) = \langle \tilde{H}(\lambda)\varphi_n, P(\lambda)\varphi_n \rangle / \langle \varphi_n, P(\lambda)\varphi_n \rangle .$$

This is done in [327] where, in addition, asymptotic expansions for $E_n(\lambda)$ in the degenerate case, and for the eigenvectors in the simple, and certain degenerate cases are obtained.

Of additional interest is the fact that often several eigenvalues are separated by an amount which is $O(\lambda^{-N})$ for an N, and that the exact exponential splitting is determined by tunneling. See *Harrell* [149], *Simon* [339] and *Helffer* and *Sjöstrand* [152].

11.2 The Morse Inequalities

The Morse inequalities describe a relationship between the Betti numbers of a smooth compact manifold M and the critical point behavior of any one of a large class of smooth functions on M called Morse functions. To illustrate informally the nature of these inequalities, consider the class of compact two-dimensional surfaces which look like deformed n-hole doughnuts. The Betti numbers of such a surface M are given by

$$b_0(M) = b_2(M) = 1 ,$$

$$b_1(M) = 2 \times (\text{number of holes in } M) .$$

Now imagine these surfaces imbedded in three-dimensional space with some choice of coordinate axes. Then every point $m \in M$ has the representation $m = (m_1, m_2, m_3)$ and we can define the height function $f\colon M \to \mathbb{R}$ given by $f(m) = m_3$. Suppose that matters are arranged in such a way that there are

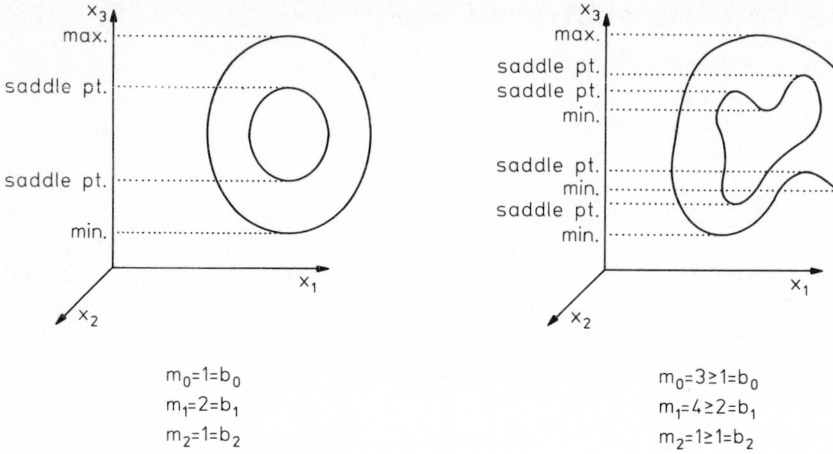

Fig. 11.1. Examples of critical points

only finitely many points at which $df = 0$, and that each of these critical points is either a local maximum, a local minimum, or a saddle point. In this context, the Morse inequalities are

$$m_0 := \text{number of local minima of } f \geq b_0(M) \ ,$$

$$m_1 := \text{number of saddle points of } f \geq b_1(M) \ ,$$

$$m_2 := \text{number of local maxima of } f \geq b_2(M) \ .$$

The first and third of these inequalities are clearly true since $b_0(M) = b_2(M) = 1$ and f must have a global maximum and a global minimum. It is also intuitively clear that each hole in M should produce at least two saddle points for f, at the top and bottom of the hole. This implies the second inequality. Figure 11.1 illustrates two possibilities in the case that M has one hole.

To give a statement of the Morse inequalities for a general manifold M we must define the Betti numbers $b_p(M)$ and the numbers $m_p(f)$ for any Morse function $f: M \to \mathbb{R}$. We will assume knowledge of the calculus of differential forms (see, for example, [348]).

Let M be a v-dimensional smooth compact orientable manifold without boundary. Let $\bigwedge^p(M)$ for $0 \leq p \leq v$ denote the space of smooth p forms on M, and let $d: \bigwedge^p \to \bigwedge^{p+1}$ denote exterior differentiation. Since $d \circ d = 0$, it follows that $\mathrm{Ran}(d: \bigwedge^{p-1} \to \bigwedge^p) \subseteq \mathrm{Ker}(d: \bigwedge^p \to \bigwedge^{p+1})$. Thus, we can define the quotient space for $0 \leq p \leq v$

$$H^p = \mathrm{Ker}(d: \textstyle\bigwedge^p \to \bigwedge^{p+1})/\mathrm{Ran}(d: \bigwedge^{p-1} \to \bigwedge^p)$$

(in this definition, $\bigwedge^{-1} = \bigwedge^{v+1} = \{0\}$). In the next section, we will show that

H^p, called the pth *de Rham cohomology group* of M, is a finite dimensional vector space.

Definition 11.2. The *Betti numbers* of M are given by

$$b_p(M) := \dim(H^p) \tag{11.20}$$

for $0 \leq p \leq v$.

Remark. This is an analyst's definition of b_p. There is a more usual topologist's definition [360] in which case (11.20) is de Rham's theorem [Parenthetically, we note three advantages of the topological definition: (i) It shows that b_p is independent of the choice of differentiable structure in cases where many such structure exist. (ii) Maps between topological cohomology classes are induced by arbitrary continuous maps rather than just by smooth maps. (iii) It is often easier to compute b_p from the combinatorial-topological definition]. By using the analytic definition, we will require no topology. In some sense, we will only prove Morse inequalities "modulo de Rham's theorem."

Suppose $f: M \to \mathbb{R}$ is a smooth function. Then a point $m \in M$ is called a critical point for f if $df(m) = 0$. This means in any local coordinate system $x_1, \ldots, x_v$ about m, $(\partial f/\partial x_1)(m) = \cdots = (\partial f/\partial x_v)(m) = 0$. A critical point is called nondegenerate if the Hessian matrix $[\partial^2 f/\partial x_i \partial x_j(m)]/2$ is nondegenerate. The index of a critical point is defined to be the number of negative eigenvalues of the Hessian matrix. Thus, local nondegenerate maxima and minima have index v and 0, respectively, while saddle points have some intermediate index. It is easy to see that the nondegeneracy and index of a critical point do not depend on the coordinate system used to define the Hessian matrix.

Definition 11.3. A smooth function $f: M \to \mathbb{R}$ is called a *Morse function* if it has finitely many critical points and each critical point is nondegenerate. For a Morse function f and $0 \leq p \leq v$, $m_p(f)$ is defined to be the number of critical points of f with index p.

The Morse inequalities are the statement that, for any Morse function f

$$m_p(f) \geq b_p \qquad 0 \leq p \leq v \; . \tag{11.21}$$

Actually, the Morse inequalities hold in the following sharper form (cf. Fig. 11.1).

Theorem 11.4. Let M be a compact orientable manifold without boundary, and suppose f is a Morse function on M. Then,

$$\sum_{p=0}^{k} (-1)^{k-p} m_p(f) \geq \sum_{p=0}^{k} (-1)^{k-p} b_p, \quad 0 \leq k < v \; , \tag{11.22}$$

$$\sum_{p=0}^{v} (-1)^p m_p(f) = \sum_{p=0}^{v} (-1)^p b_p \; . \tag{11.23}$$

Remarks (1) These are sometimes called the *strong Morse inequalities.* For an account of the standard proof of these inequalities, see [245]. Witten's proof proceeds via Hodge theory, which we examine in the next section.

(2) Equation (11.22) implies (11.21) by summing the inequalities (11.22) for $k = p$ and $k = p - 1$.

(3) Equation (11.23) is sometimes called the *Morse Index Theorem.* The right side of (11.23) is the Euler-Poincaré characteristic of M, usually denoted $\chi(M)$.

11.3 Hodge Theory

In this section, we suppose that we have chosen an orientation for our compact orientable manifold M, i.e. a nowhere vanishing smooth ν-form o. (To say that M is orientable means that such an o exists.) In addition, we assume that M has been given a Riemannian metric, i.e. for each $m \in M$, we are given an inner product $g_m(\cdot, \cdot)$ on the tangent space $T_m(M)$, which varies smoothly in the sense that for smooth vector fields $X(m)$ and $Y(m)$, $g_m(X(m), Y(m))$ is a smooth function. The results of the next two sections hold for any metric g_m. It will be convenient in Sect. 11.5 to pick a particular metric for computations.

Using the metric, we will define an L^2 inner product on $\bigwedge^p(M)$ for each p and the Laplace Beltrami operator L. The fundamental result of Hodge theory is that the space H^p is isomorphic to the space of p forms annihilated by L. Thus,

$$b_p \equiv \dim(H^p) = \dim[\mathrm{Ker}(L \restriction \textstyle\bigwedge^p)] .$$

Recall that b_p is defined independently of the Riemannian structure. Thus, $\dim[\mathrm{Ker}(L \restriction \bigwedge^p)]$ does not depend on which of the many possible Riemannian structures on M is used to define L.

In the remainder of this section, we give a brief exposition of Hodge theory.

The inner product on $T_m(M)$ induces in a natural way an isomorphism i from $T_m(M)$ onto $T_m^*(M)$ by $[i(X)](Y) = g_m(X, Y)$. We use i to transfer the inner product from T_m to T_m^*. Thus, the inner product in T_m^* which we denote by $\langle \cdot, \cdot \rangle_m^1$ is given by $\langle \omega_1, \omega_2 \rangle_m^1 = g_m(i^{-1}(\omega_1), i^{-1}(\omega_2))$. This inner product on $T_m^*(M)$ can be used in turn to define an inner product on $\bigwedge_m^p(M)$, the pth exterior power of $T_m^*(M)$. Given an orthonormal basis $\omega_1^m, \ldots, \omega_\nu^m$ for $T_m^*(M)$ the elements

$$\omega_{i_1}^m \wedge \cdots \wedge \omega_{i_p}^m \qquad 1 \le i_1 < i_2 < \cdots < i_p \le \nu$$

form a basis for $\bigwedge_m^p(M)$. By declaring this basis to be orthonormal, we obtain an inner product on $\bigwedge_m^p(M)$. To see that this inner product, which will be denoted $\langle \cdot, \cdot \rangle_m^p$, is independent of the orthonormal basis of $T_m^*(M)$ used in its definition, note that by multilinearity and antisymmetry one finds, that for any $2p$, vectors $\lambda_1, \dots, \lambda_p, \rho_1, \dots, \rho_p$ in $T_m^*(M)$ we have (by expanding in the basis ω_i^m)

$$\langle \lambda_1 \wedge \cdots \wedge \lambda_p, \rho_1 \wedge \cdots \wedge \rho_p \rangle_m^p = \det(\langle \lambda_i, \rho_j \rangle_m^1) ,$$

which is a basis independent formula.

When dealing with the Riemannian structure, it is convenient to work with an oriented orthonormal frame. This is a sequence $\omega_1, \dots, \omega_\nu$ of 1-forms with the properties that for each m, $\omega_1(m), \dots, \omega_\nu(m)$ form an orthonormal basis for $T_m^*(M)$ and $\omega_1(m) \wedge \cdots \wedge \omega_\nu(m)$ is a positive multiple of $o(m)$, the form defining the orientation. Such a frame can be obtained locally by applying the Gram-Schmidt procedure to $dx^1, \dots, dx^\nu$ where $x^1, \dots, x^\nu$ are local coordinate functions, and then re-ordering the resulting 1-forms if necessary.

The volume form is the ν-form given by

$$\omega = \omega_1 \wedge \cdots \wedge \omega_\nu$$

where $\omega_1, \dots, \omega_\nu$ is an oriented orthonormal frame. Since any other oriented orthonormal frame can be written $\mathrm{O}\omega_1, \dots, \mathrm{O}\omega_\nu$ for some special orthogonal transformation O(depending on m) and

$$\begin{aligned} \mathrm{O}\omega_1 \wedge \cdots \wedge \mathrm{O}\omega_\nu &= \det(\mathrm{O})\omega_1 \wedge \cdots \wedge \omega_\nu \\ &= \omega_1 \wedge \cdots \wedge \omega_\nu \end{aligned}$$

we see that ω is independent of choice of frame and hence globally defined on M. In terms of coordinate forms $dx^1, \dots, dx^\nu$ and the $\nu \times \nu$ matrix-valued function, $g_{ij}(m) = g_m(\partial/\partial x^i, \partial/\partial x^j)$, it is easy to see that

$$\omega = (\det g)^{1/2}\, dx^1 \wedge \cdots \wedge dx^\nu .$$

The integral of a ν-form α is defined locally by means of an orientation preserving coordinate map $\varphi\colon M \to U$, where U is an open subset of $\mathbb{R}^\nu$ and $\varphi(m) = (x^1(m), \dots, x^\nu(m))$. Orientation preserving means that $dx^1(m) \wedge \cdots \wedge dx^\nu(m)$ is a positive multiple of o(m), the orientation form, for each $m \in \varphi^{-1}(U)$. Writing $\alpha = f dx^1 \wedge \cdots \wedge dx^\nu$ for some real-valued function f on M, we define

$$\int_{\varphi^{-1}(U)} \alpha := \int_U f \circ \varphi^{-1} d^\nu x .$$

Using a partition of unity, it is now straightforward to define the integral of α over M. Sometimes we will abuse notation and write f for $f \circ \varphi^{-1}$, calling this function "f written in the coordinates $x^1, \dots, x^\nu$".

Now we can define the L^2 inner product of two p-forms, α and β. It is given by

$$\langle \alpha, \beta \rangle = \int_M \langle \alpha(m), \beta(m) \rangle_m^p \omega .$$

The completion of each $\bigwedge^p(M)$ under this L^2 inner product is a Hilbert space which we will denote by $\overline{\bigwedge}^p(M)$. Exterior differentiation is defined as a closed operator on $\overline{\bigwedge}^p$ by taking the operator closure of d defined in $\bigwedge^p$. Thus,

$$d\colon \mathrm{D}(d) \subset \overline{\bigwedge}^p \to \overline{\bigwedge}^{p+1} .$$

The adjoint of d is denoted d^*, i.e.

$$d^*\colon \mathrm{D}(d^*) \subset \overline{\bigwedge}^{p+1} \to \overline{\bigwedge}^p \quad \text{with}$$

$$\langle \alpha, d\beta \rangle = \langle d^*\alpha, \beta \rangle$$

for α and β in the respective domains. It is easy to see that $\bigwedge^{p+1} \subset D(d^*)$. Also, it follows easily from $d^2 = 0$ that $d^{*2} = 0$.

Definition 11.5 The *Laplace-Beltrami operator* L in $\overline{\bigwedge}^p(M)$ is defined as the operator closure of $d^*d + dd^*$ acting on smooth p-forms.

It is known (see *Chernoff* [63], *Strichartz* [351] or Sect. 12.6) that L is a self-adjoint operator. Also, L is positive since

$$\langle \alpha, L\alpha \rangle = \langle d\alpha, d\alpha \rangle + \langle d^*\alpha, d^*\alpha \rangle \geq 0 . \tag{11.24}$$

We now define three subspaces of $\overline{\bigwedge}^p(M)$.

Definition 11.6. A form α is called *harmonic* if $L\alpha = 0$ (equivalently, by (11.24), if $d\alpha = 0$ and $d^*\alpha = 0$). Define

$$\overline{\bigwedge}{}^p_H(M) := \{\text{harmonic } p\text{-forms}\}$$

$$\overline{\bigwedge}{}^p_d(M) := \mathrm{Ran}[d\colon \mathrm{D}(d) \subset \overline{\bigwedge}^{p-1} \to \overline{\bigwedge}^p]$$

$$\overline{\bigwedge}{}^p_{d^*}(M) := \mathrm{Ran}[d^*\colon \mathrm{D}(d^*) \subset \overline{\bigwedge}^{p+1} \to \overline{\bigwedge}^p] .$$

These subspaces can be used to decompose $\overline{\bigwedge}^p(M)$ as the following proposition shows.

Proposition 11.7. Let $\overline{\bigwedge}^p(M)$, $\overline{\bigwedge}{}^p_H(M)$, $\overline{\bigwedge}{}^p_d(M)$ and $\overline{\bigwedge}{}^p_{d^*}(M)$ be as defined above and let $\bigwedge^p_H(M)$, etc. denote the subspaces of the corresponding spaces with a bar consisting of all smooth forms in the subspaces. Then

(i) $\overline{\bigwedge}^p(M) = \overline{\bigwedge}{}^p_H(M) \oplus \overline{\bigwedge}{}^p_d(M) \oplus \overline{\bigwedge}{}^p_{d^*}(M)$,
(i)′ $\bigwedge^p(M) = \bigwedge^p_H(M) \oplus \bigwedge^p_d(M) \oplus \bigwedge^p_{d^*}(M)$,
(ii) $\mathrm{Ker}(d\colon \bigwedge^p \to \bigwedge^{p+1}) = \bigwedge^p_H(M) \oplus \bigwedge^p_d(M)$,
(iii) $\bigwedge^p_H(M) = \overline{\bigwedge}{}^p_H(M)$ is finite dimensional.

Before proving this proposition, we note that it implies the basic result of Hodge theory mentioned earlier:

Theorem 11.8. $b_p(M) = \dim(\mathrm{Ker}(L{\restriction}\bigwedge^p))$.

Proof. $\bigwedge_H^p(M)$ is, by definition, $\mathrm{Ker}(L \restriction \bigwedge^p)$, and by (ii) of Proposition 11.7

$$\begin{aligned} H^p &:= \mathrm{Ker}(d\colon \textstyle\bigwedge^p \to \bigwedge^{p+1})/\mathrm{Ran}(d\colon \bigwedge^{p-1} \to \bigwedge^p) \\ &= \textstyle\bigwedge_H^p(M) \oplus \bigwedge_d^p(M)/\bigwedge_d^p(M) \\ &\cong \textstyle\bigwedge_H^p(M) \ . \end{aligned}$$

Thus, by (iii) of Proposition 11.7, H^p is finite-dimensional with dimension equal to that of the kernel of $L \restriction \bigwedge^p$, thought of either as a self-adjoint operator acting in the Hilbert space $\overline{\bigwedge}^p$, or as a differential operator acting on smooth p-forms. □

We will prove Proposition 11.7 modulo two well-known analytic facts proven in Sect. 12.6: (i) Elliptic regularity, which asserts that

$$\bigcap_{k=1}^{\infty} \mathrm{D}((L \restriction \overline{\textstyle\bigwedge}^p)^k) = \textstyle\bigwedge^p \ .$$

This can be proven by using the pseudo-differential calculus (e.g. *Taylor* [353]), or by localizing, using local coordinates and the simple analysis of Sect. IX.6 of *Reed* and *Simon* [293]; (ii) $(L+1)^{-1}$ is compact. This can be proven using the localization formula (11.37) below and the compactness of resolvents of Dirichlet Laplacians on bounded regions of $\mathbb{R}^\nu$.

Proof of Proposition 11.7. It follows easily from $d^2 = 0$ and $d^{*2} = 0$ and the definition of harmonic forms that the three subspaces on the right of (i) are pairwise orthogonal. So to prove (i), we must decompose an arbitrary p-form into a sum of three pieces, one from each of these subspaces. To begin, we show

$$\overline{\textstyle\bigwedge}^p(M) = \mathrm{Ker}(L) \oplus \mathrm{Ran}(L) \ . \tag{11.25}$$

This follows from the compactness of $(L+1)^{-1}$, for this implies $(L+1)^{-1} = \sum \lambda_i P_i$ where the P_i are projections onto finite dimensional subspaces of $\overline{\bigwedge}^p$ and where $\{\lambda_i\}$ have 0 as their only possible accumulation point. Hence, $L = \sum (\lambda_i^{-1} - 1) P_i$ which implies that L has a finite dimensional kernel and closed range. Then $\overline{\bigwedge}_H$ is finite dimensional and (11.25) holds. Now, given a form α, we can write

$$\alpha = \alpha_H + L\gamma = \alpha_H + d^*(d\gamma) + d(d^*\gamma) \ ,$$

where α_H is harmonic. Thus (i) holds. To show that (i)$'$ holds, we must show that smooth forms have a smooth decomposition. By elliptic regularity, $\mathrm{Ker}(L)$ is spanned by smooth functions. Since $\mathrm{Ker}(L)$ is finite dimensional, α is smooth implies that α_H is smooth, which implies that $L\gamma$ is smooth which, again by elliptic regularity, implies that γ is smooth. Thus, smooth forms have a smooth decomposition and $\overline{\bigwedge}_H = \bigwedge_H$. Finally, to prove (ii), note that $\bigwedge_H \oplus \bigwedge_d \subseteq \mathrm{Ker}(d)$ and that $\bigwedge_{d^*} \cap \mathrm{Ker}(d) = \{0\}$ which, in view of (i)$'$, proves (ii). □

While we do not need it in our proofs of the Morse inequalities, having given the Hodge machinery, we want to note a quick proof of the Poincaré duality theorem that $b_{\nu-p} = b_p$ for a compact orientable manifold of dimension ν. The rest of the section, except for the final paragraph, may be omitted without loss. First, we need to introduce the Hodge $*$ operator:

Proposition 11.9. Let ω be the volume form on an oriented Riemannian manifold. Given any $\beta \in \bigwedge^p(T_m^*(M))$, there is a unique element $*\beta \in \bigwedge^{\nu-p}(T_m^*(M))$ with

$$\langle \alpha, *\beta \rangle_m^{\nu-p} = \langle \beta \wedge \alpha, \omega \rangle_m^\nu \tag{11.26}$$

for every $\alpha \in \bigwedge^{\nu-p}(T_m^*(M))$. Moreover, $*$ has the following properties:

(a) $*(\alpha + \beta) = *\alpha + *\beta$; $*(f\alpha) = f(*\alpha)$,
(b) $\langle *\alpha, *\beta \rangle_m^{\nu-p} = \langle \alpha, \beta \rangle_m^p$,
(c) $\alpha \in \bigwedge^p$, $\beta \in \bigwedge^{\nu-p}$ implies $\langle *\alpha, \beta \rangle = (-1)^{p(\nu-p)} \langle \alpha, *\beta \rangle$,
(d) If $\alpha \in \bigwedge^p$, then $*(*\alpha) = (-1)^{p(\nu-p)}\alpha$,
(e) $\alpha \wedge (*\beta) = \langle \alpha, \beta \rangle_m^p \omega$ for all $\alpha, \beta \in \bigwedge^p$.

Proof. The right side of (11.26) clearly defines a linear functional on $\bigwedge^{\nu-p}$ and so, by duality, there is a unique element of $\bigwedge^{\nu-p}$, $*\beta$, obeying (11.26). Linearity of $*$ is obvious, and (c) is just the assertion that if $\alpha \in \bigwedge^p$ and $\beta \in \bigwedge^{\nu-p}$, then $\beta \wedge \alpha = (-1)^{p(\nu-p)} \alpha \wedge \beta$. To prove (b), let $\omega_1, \ldots, \omega_n$ be an orthonormal basis of $T_m^*(M)$ with $\omega = \omega_1 \wedge \cdots \wedge \omega_n$. It is easy to see that $*(\omega_{i_1} \wedge \cdots \wedge \omega_{i_p}) = (-1)^\pi \omega_{j_1} \wedge \cdots \wedge \omega_{j_{\nu-p}}$ where $i_1 < \cdots < i_p$ and $j_1 < \cdots < j_{\nu-p}$ is the complementary set of $\{i_1, \ldots, i_p\}$ and where π is the sign of the permutation taking $1, \ldots, p$ into $i_1, \ldots, i_p$ and $p+1, \ldots, \nu$ into $j_1, \ldots, j_{\nu-p}$. Thus, $*$ takes an orthonormal basis of $\bigwedge^p$ into an orthonormal basis for $\bigwedge^{\nu-p}$, so (b) holds. Then (d) follows from (b) and (c). Finally, $\alpha \wedge *\beta = c\omega$ for some c and so $c = \langle \alpha \wedge *\beta, \omega \rangle = \langle *\beta, *\alpha \rangle = \langle \beta, \alpha \rangle = \langle \alpha, \beta \rangle$ proving (e). $\square$

While $*$ is defined on $\bigwedge^p(T_m^*(M))$ as a pointwise map, we use the same symbol for the map from $\bigwedge^p(M)$ to $\bigwedge^{\nu-p}(M)$ given by using the point maps. The key is

Theorem 11.10. (a) $d^*\alpha = (-1)^{\nu+\nu p+1} *[d(*\alpha)]$ if $\alpha \in \bigwedge^p$.
(b) $L(*\alpha) = *(L\alpha)$.

Proof. (a) Let α be a p-form and β an arbitrary $p-1$ form. Then, by Stokes' theorem, the $\nu - 1$ form $\beta \wedge *\alpha$ obeys $\int d(\beta \wedge *\alpha) = 0$, i.e.

$$\int d\beta \wedge (*\alpha) + (-1)^{p-1} \int \beta \wedge d(*\alpha) = 0 \ . \tag{11.27}$$

But

$$\int d\beta \wedge *\alpha = \int \langle d\beta, \alpha \rangle_m \omega = \langle d\beta, \alpha \rangle = \langle \beta, d^*\alpha \rangle \ , \tag{11.28}$$

while

$$\begin{aligned}\int \beta \wedge d(*\alpha) &= (-1)^{(p-1)(v-p+1)} \int \beta \wedge *(*d*\alpha) \\ &= (-1)^{(p-1)(v-p+1)} \int \langle \beta, *d*\alpha \rangle_m \omega \\ &= (-1)^{v+vp+p-1} \langle \beta, *d*\alpha \rangle \ .\end{aligned} \tag{11.29}$$

Since $(p-1)(v-p+1) - v - vp - p + 1 = -2v - p(p-1)$ is even, (11.27, 28 and 29) imply part (a).

(b) If v is even, $L = -(d*d* + *d*d)$ and $L_{v-p}* = -[*d*d* + (-1)^{p(v-p)} d*d] = *L_p$. Here L_p denotes $L \restriction \bigwedge^p$. If v is odd, then $L_p = (-1)^p [d*d* - *d*d]$ and $L_{v-p}* = (-1)^{v-p}[(-1)^{p(v-p)} d*d - *d*d*] = -(-1)^p[(-1)^{p(v-p)} d*d - *d*d*] = *L_p$. □

Corollary 11.11 (Poincaré Duality Theorem). Any compact orientable manifold M of dimension v obeys $b_{v-p} = b_p$ for all p.

Proof. Pick a Riemann metric and note that by (b) of the last theorem, $*$ is a unitary map from $\mathrm{Ker}(L_p)$ to $\mathrm{Ker}(L_{v-p})$. □

Remark. If $v = 4k$, then $*\colon \bigwedge^{2k} \to \bigwedge^{2k}$ obeys $(*)^2 = 1$ and by (b), one can simultaneously diagonalize L and $*$. $\sigma := \dim\{u \mid Lu = 0, *u = u\} - \dim\{u \mid Lu = 0, *u = -u\}$ is the celebrated Hirzebruch signature of M.

To conclude this section, we note that if $Q := d + d^*$ acting in $\mathrm{H} := \bigoplus_{p=0}^{v} \overline{\bigwedge}^p$ and P is defined on H by $P \restriction \overline{\bigwedge}^p = (-1)^p$, then $Q^2 = L$, $P^2 = 1$ and $\{Q, P\} = 0$ so that the theory of supersymmetry applies (see Sect. 6.3). In particular, if $E \neq 0$,

$$\sum_{p\,\mathrm{odd}} \dim\{\mathrm{Ker}[(L-E) \restriction \bigwedge^p]\} = \sum_{p\,\mathrm{even}} \dim\{\mathrm{Ker}[(L-E) \restriction \bigwedge^p]\} \ . \tag{11.30}$$

11.4 Witten's Deformed Laplacian

The considerations of the previous section show that the Betti numbers of M are given by the number of zero eigenvalues of L acting on p-forms. In practice, however, it is difficult to compute the spectrum of L. It was Witten's observation that the results of the previous section hold when d, d^* and L are replaced by

$$\begin{aligned} d_t &:= \mathrm{e}^{-tf} d \mathrm{e}^{tf} \\ d_t^* &:= \mathrm{e}^{tf} d^* \mathrm{e}^{-tf} \\ L_t &:= d_t d_t^* + d_t^* d_t \ , \end{aligned}$$

where $t \in \mathbb{R}$ and f is a Morse function. Indeed, if $\bigwedge_{d_t}^p$, $\bigwedge_{d_t^*}^p$ and $\bigwedge_{H_t}^p$ and their L^2

closures $\bar{\bigwedge}^p_{d_t}$, $\bar{\bigwedge}^p_{d_t^*}$ and $\bar{\bigwedge}^p_{H_t}$ are defined analogously to the corresponding spaces in the previous section, we find

$$\bar{\bigwedge}^p_{H_t} \cong \mathrm{Ker}(d_t\colon \bigwedge^p \to \bigwedge^{p+1})/\mathrm{Ran}(d_t\colon \bigwedge^{p-1} \to \bigwedge^p) \ .$$

However, it is easy to see that

$$\mathrm{Ker}(d_t) = \mathrm{e}^{-tf}\mathrm{Ker}(d)$$

$$\mathrm{Ran}(d_t) = \mathrm{e}^{-tf}\mathrm{Ran}(d), \quad \text{so that}$$

$$\bar{\bigwedge}^p_{H_t} \cong \mathrm{Ker}(d)/\mathrm{Ran}(d) = H^p \ .$$

We have thus proven the following result.

Theorem 11.12. $b_p = \dim[\mathrm{Ker}(L_t \ \bigwedge^p)]$.

Note that the supersymmetry framework also applies to this deformed situation; if $Q_t := d_t + d_t^*$ acting in $\mathrm{H} = \bigoplus_{p=0}^{\nu} \bar{\bigwedge}^p$ and P is defined by $P \upharpoonright \bar{\bigwedge}^p = (-1)^p$, then $Q_t^2 = L_t$ and the results of Sect. 6.3 imply that (11.30) holds with L_t in place of L.

From Theorem 11.12 we see that to estimate b_p it suffices to estimate the dimensions of $\bar{\bigwedge}^p_{H_t} = \mathrm{Ker}(L_t \upharpoonright \bar{\bigwedge}^p)$ for any t. The crux of Witten's method is the observation that the $t \to \infty$ limit is the quasiclassical limit for the operator L_t. Thus, the asymptotic estimates for the eigenvalues of L_t can be used to estimate the dimension of $\mathrm{Ker}(L_t \upharpoonright \bar{\bigwedge}^p)$ for large t. The consequent estimates on the Betti numbers reproduce the Morse inequalities. To be explicit, let us describe how (11.21) is proven. By realizing $L_t \upharpoonright \bar{\bigwedge}^p$ as a quasiclassical limit, we will be able to compute that the number of eigenvalues $E_k(t)$ obeying $\lim_{t\to\infty} E_k(t)/t = 0$ is precisely m_p. Since $E_k(t) = 0$ for $k = 1, \ldots, b_p$ this implies $m_p \geq b_p$.

To proceed further, we need to find an expression for L_t which exposes the role of the critical point behavior of f. Let $x^1, \ldots, x^\nu$ be a coordinate system in some neighborhood in M. Define locally the operator $(a^i)^*$ acting on p-forms by

$$(a^i)^*\alpha = dx^i \wedge \alpha \ . \tag{11.31}$$

Let a^i be the adjoint of $(a^i)^*$. By a direct calculation

$$a^i\, dx^{j_1} \wedge \cdots \wedge dx^{j_p} = \sum_{k=1}^{p} (-1)^k g^{ij_k}\, dx^{j_1} \wedge \cdots dx^{j_{k-1}} \wedge dx^{i_{k+1}} \wedge \cdots dx^{j_p} \ , \tag{11.32}$$

where g^{ij} is the metric on T^*, i.e. $g^{ij} = \langle dx^i, dx^j \rangle$. (Note that g^{ij} and g_{ij} are inverse matrices.) The operators $(a^i)^*$ and a^i are zeroth order operators. We will always assume that they act on forms with support in the region where the coordinate system x is defined. Using (11.31) and (11.32), one easily sees that

$$\{a^i, (a^j)^*\} := a^i(a^j)^* + (a^j)^*a^i = g^{ij} \ .$$

Notice the formal similarity to Fermion creation and annihilation operators in physics. The remainder of this section is devoted to proving the following proposition:

Proposition 11.13. L_t has the form

$$L_t = L + t^2 \|df\|^2 + tA \ ,$$

where A is a zeroth-order operator. If the metric is flat in some neighborhood, then in a local orthonormal co-ordinate system $x^1, \ldots, x^\nu$

$$A = \sum_{ij} (\partial^2 f/\partial x^i \partial x^j)[(a^i)^*, a^j] \ . \tag{11.33}$$

Proof. The proof of this proposition is a long calculation. First note that

$$d_t\alpha = e^{-tf} d e^{tf} \alpha = d\alpha + t\, df \wedge \alpha$$

and

$$df = \sum_i \frac{\partial f}{\partial x^i} dx^i \ .$$

Thus, letting $f_i = \partial f/\partial x^i$

$$d_t = d + t \sum_{i=1}^{\nu} f_i (a^i)^* \ ,$$

and therefore

$$d_t^* = d^* + t \sum_{i=1}^{\nu} f_i a^i \ ,$$

so that (with $\{A, B\} = AB + BA$)

$$L_t = \{d_t, d_t^*\} = L + t^2 \sum_{ij} f_i f_j \{a^i, (a^j)^*\} + tA \ ,$$

where

$$A = A_1 + A_1^*$$

$$A_1 = \left\{ d, \sum_j f_j a^j \right\} \ .$$

To show that A is zeroth-order at a general point, we introduce the local coordinate derivative operator ∂_i, defined by

$$\partial_i \left(\sum u_{j_1 \cdots j_p} dx^{j_1} \wedge \cdots \wedge dx^{j_p} \right) = \sum \frac{\partial u_{j_1 \cdots j_p}}{\partial x^i} dx^{j_1} \wedge \cdots \wedge dx^{j_p} \ ,$$

so that $d = \sum_i (a^i)^* \partial_i$. Thus,

$$A_1 = \sum_{i,j} \{(a^i)^*, f_j a^j\} \partial_i + (a^i)^* [\partial_i, f_j a^j]$$
$$= \sum_{i,j} g^{ij} f_j \partial_i + A_2 \ . \tag{11.34}$$

Now A_2 is a zeroth-order operator involving second derivatives of f and derivatives of g^{ij} [since (11.32) shows a^i has derivatives in it], so

$$A = A_1 + A_1^* = -\sum_{i,j} [\partial_i, g^{ij} f_j] + A_2 + A_2^*$$

is a zeroth-order operator.

To compute the formula (11.33) at points m where the metric is flat, i.e. $g^{ij} = \delta^{ij}$ in a neighborhood of m in a suitable coordinate system, we return to (11.34). In that case, $[\partial_i, a^j] = 0$, so

$$A_1 = \sum_i f_i \partial_i + \sum_{i,j} (a^i)^* a^j f_{ij} \ ,$$

where $f_{ij} = \partial^2 f / \partial x^i \partial x^j$. Thus A_2 is Hermitian since f_{ij} is symmetric, and so

$$A = \sum_{i,j} f_{i,j} [2(a^i)^* a^j - \delta^{ij}] \ .$$

This yields (11.33) if we note that $\{(a^i)^*, a^j\} = \delta^{ij}$ in this flat case. □

The above calculation of A at general points is not wholly satisfactory, although it does suffice for our purposes here since we only need to know that A is zeroth order, and its explicit form at special points. The formalism of covariant derivatives introduced in the next chapter and the relations between the a's and covariant derivatives (11.14, 15, 17, 28) make it fairly easy to compute A in terms of the second covariant derivative (Definition 12.24); explicitly:

$$A = \sum (\nabla^2_{(i,j)} f) [(a^i)^*, a^j] \ . \tag{11.35}$$

11.5 Proof of Theorem 11.4

Given an orientable compact manifold M without boundary, and a Morse function f on M, we wish to prove the inequalities (11.22, 23). To do this we will choose a Riemannian metric on M and apply the considerations of Sect. 11.1 to the operator

$$L_t = L + t^2 \|df\|^2 + tA$$

acting on p-forms to estimate the multiplicity of the eigenvalue 0, i.e. the dimension of the kernel, for large t. The function $\|df\|^2$ vanishes at only finitely many points since f is a Morse function, and A is a smooth zeroth-order operator. Thus, some variant of Theorem 11.1 should apply. We are going to choose the Riemannian metric in a special way depending on f.

Suppose $\{m^{(a)}\}_{a=1}^{k}$ are the critical points of f with the index of $m^{(a)}$ equal to $\mathrm{ind}(a)$. Then there exists, in a neighborhood of each $m^{(a)}$, coordinate system x_1, $\ldots$, x_ν such that

$$f = x_1^2 + \cdots + x_{\nu-\mathrm{ind}(a)}^2 - x_{\nu-\mathrm{ind}(a)+1}^2 - \cdots - x_\nu^2$$

when written in these coordinates. These are called Morse coordinates [245]. By stipulating that dx_1, $\ldots$, dx_ν be orthonormal, we obtain a metric in some neighborhood of $m^{(a)}$. Since the critical points of f are isolated, we can patch together such metrics in a neighborhood of each critical point with an arbitrary metric defined in other regions, using partition of unity. Thereby we obtain a metric on all of M.

It is easy to write down the expression for the operator L_t given by this metric, in Morse coordinates in a neighborhood of the critical point $m^{(a)}$. By Proposition 11.10,

$$L_t = -\Delta + 4t^2x^2 + 2t \sum_{i=1}^{\nu-\mathrm{ind}(a)} [a_i^*, a_i] - 2t \sum_{i=\nu-\mathrm{ind}(a)+1}^{\nu} [a_i^*, a_i] \ .$$

Here $x^2 = \sum_{i=1}^{\nu} x_i^2$ and Δ acts on p-forms as follows

$$\Delta(f dx_{i_1} \wedge \cdots \wedge dx_{i_p}) = \sum_{i=1}^{\nu} \frac{\partial^2 f}{\partial x_i^2} dx_{i_1} \wedge \cdots \wedge dx_{i_p} \ .$$

Define $K^{(a)}$ acting in $\bar{\bigwedge}^p(\mathbb{R}^n)$ by

$$K^{(a)} = -\Delta + 4x^2 + A^{(a)}$$

$$A^{(a)} = 2 \sum_{i=1}^{\nu-\mathrm{ind}(a)} [a_i^*, a_i] - 2 \sum_{i=\nu-\mathrm{ind}(a)+1}^{\nu} [a_i^*, a_1] \ .$$

The asymptotic values of the eigenvalues of L_t will be given in terms of the spectrum of $\bigoplus_a K^{(a)}$ which we now compute.

On p-forms, $-\Delta + 4x^2$ acts as a scalar operator, i.e. it acts the same way on all $dx^{j_1} \wedge \cdots \wedge dx^{j_p}$. Thus, its eigenvalues are the harmonic oscillator eigenvalues $\{\sum_{i=1}^{\nu} 2(1 + 2n_i): n_1, \ldots, n_\nu \in \{0, 1, 2, \ldots\}\}$. For each of these eigenvalues, there are $\nu!/(\nu - p)!p!$ independent eigenforms given by

$$\psi\, dx_{i_1} \wedge \cdots \wedge dx_{i_p} \quad 1 \le i_1 < \cdots < i_p \le \nu \ ,$$

where ψ is the corresponding harmonic oscillator eigenfunction. Since

$$[a_i^*, a_i] f\, dx_{i_1} \wedge \cdots \wedge dx_{ip} = \begin{cases} f\, dx_{i_1} \wedge \cdots \wedge dx_{ip} & \text{if } i \in \{i_1, \ldots, i_p\} \\ -f\, dx_{i_1} \wedge \cdots \wedge dx_{ip} & \text{if } i \notin \{i_1, \ldots, i_p\} \ , \end{cases}$$

we see that $A^{(a)}$ acts diagonally on these eigenforms

$$A^{(a)}\psi\, dx_{i_1} \wedge \cdots \wedge dx_{i_p} = \lambda_a(i_1, \ldots, i_p)\psi\, dx_{i_1} \wedge \cdots \wedge dx_{i_p}$$

where, with $I = \{i_1, \ldots, i_p\}$, $J = \{1, \ldots, v\}\backslash I$, $K = \{1, \ldots, v - \text{ind}(a)\}$ and $L = \{v - \text{ind}(a) + 1, \ldots, v\}$

$$\lambda_a(i_1, \ldots, i_p) = \#(I \cap K) - \#(J \cap K) - \#(I \cap L) + \#(J \cap L) \; .$$

Thus,

$$\sigma(K^{(a)}) = \left\{ \sum_{i=1}^{v} 2(1 + 2n_i) + 2\lambda_a(i_1, \ldots, {}_{ip})\colon n_1, \ldots, n_v \in \{0, 1, 2, \ldots\} \right.$$
$$\left. \text{and} \quad i \leq i_1 < \cdots i_p \leq v \right\}$$

$$\sigma\left(\bigoplus_a K^{(a)}\right) = \bigcup_a \sigma(K^{(a)}) \; .$$

We are interested in the multiplicity of the eigenvalue zero. A little thought shows that $\lambda_a(i_1, \ldots, i_p) \geq -v$ and $\lambda_a(i_1, \ldots, i_p) = -v$ only when $\text{ind}(a) = p$ and $i_1, \ldots, i_p = v - p + 1, \ldots, v$. Thus, $\text{Ker}(K^{(a)} \restriction \bar{\bigwedge}^p) = 0$ unless $\text{ind}(a) = p$, in which case $\dim(K^{(a)} \restriction \bar{\bigwedge}^p) = 1$. So the dimension of $\text{Ker}(\bigoplus_a K^{(a)} \restriction \bar{\bigwedge}^p)$ is the number of critical points of f of index p, i.e.

$$\dim(\text{Ker}(K^{(a)} \restriction \bar{\bigwedge}^p)) = m_p \; .$$

Now Theorem 11.1 as stated does not apply to L_t acting on p-forms. However, the reader can check that only minor changes are needed in the proof to show that if $E_n^p(t)$ are the eigenvalues of $L_t \restriction \bar{\bigwedge}^p$, counting multiplicity, and e_n^p are the eigenvalues of $\bigoplus_a K^{(a)} \restriction \bar{\bigwedge}^p$, counting multiplicity, then

$$\lim_{t \to \infty} E_n^p(t)/t = e_n^p \; . \tag{11.36}$$

The manifold version of the IMS localization formula which one needs reads

$$L = \sum_a J_a L J_a - \sum_a \|dJ_a\|^2 \; . \tag{11.37}$$

For large t, it is clear from (11.36) that there cannot be more $E_n^p(t)$ equal to zero than there are e_n^p equal to zero. Thus,

$$\begin{aligned} b_p &= \dim[\text{Ker}(L_t \restriction \bar{\bigwedge}^p)] \\ &\leq \dim\left[\text{Ker}\left(\bigoplus_a K^{(a)} \restriction \bar{\bigwedge}^p\right)\right] \\ &= m_p \; . \end{aligned}$$

This proves (11.21), the weak Morse inequalities.

We can now prove the Morse index theorem (11.23) using supersymmetry ideas. We know $e^p_{m_p+1}$ is the first of the $\{e^p_n\}_{n=1}^{\infty}$ not equal to zero. Hence for large t, $E^p_n(t)$ grows like t for $n \geq m_p + 1$. Of the remaining eigenvalues $\{E^p_1(t), \ldots, E^p_{m_p}(t)\}$ we know that the first b_p are zero. We will call eigenvalues $\{E^p_{b_{p+1}}(t), \ldots, E^p_{m_p}(t)\}$ the low-lying eigenvalues. They are non-zero, but are $o(t)$ as $t \to \infty$. The way that supersymmetry enters is that from Equation (11.26), for L_t it follows that the low-lying eigenvalues occur in even-odd pairs, i.e. for each low-lying E^p_n with p odd there is a low-lying $E^{p'}_{n'}$ with p' even. This implies that

$$\sum_{p\,\text{odd}} (m_p - b_p) = \sum_{p\,\text{even}} (m_p - b_p) \;, \quad \text{so that}$$

$$\sum_{p=0}^{v} (-1)^p m_p = \sum_{p=0}^{v} (-1)^p b_p + \sum_{p\,\text{even}} (m_p - b_p) - \sum_{p\,\text{odd}} (m_p - b_p)$$

$$= \sum_{p=0}^{v} (-1)^p b_p \;,$$

which is (11.23).

To prove (11.22), we need to analyze the supersymmetric cancellation more closely. Let Ξ^p_t denote the $m_p - b_p$-dimensional space of low-lying eigenvalues. As above, let $Q_t = d_t + d_t^*$ acting on $\bigoplus_{p=0}^{v} \overline{\wedge}^p$. Since $Q_t^2 = L_t$, Q_t preserves the eigenspaces of L_t. Thus,

$$Q_t\colon \Xi^p_t \to \Xi^{p-1}_t \oplus \Xi^{p+1}_t \;.$$

Furthermore, since the kernels of Q_t and L_t coincide, Q_t is one to one on $\bigoplus_{p=0}^{v} \Xi^p_t$. Thus, Q_t is a one to one map

$$Q_t\colon \bigoplus_{\substack{l\,\text{odd}\\ l=1}}^{2j-1} \Xi^l_t \to \bigoplus_{\substack{l\,\text{even}\\ l=0}}^{2j} \Xi^l_t$$

$$Q_t\colon \bigoplus_{\substack{l\,\text{even}\\ l=0}}^{2j} \Xi^l_t \to \bigoplus_{\substack{l\,\text{odd}\\ l=1}}^{2j+1} \Xi^l_t \;, \tag{11.38}$$

which implies that the dimensions of the spaces on the right of (11.38) are greater than the dimensions of the spaces on the left. Hence,

$$(m_1 - b_1) + \cdots + (m_{2j-1} - b_{2j-1}) \leq (m_0 - b_0) + \cdots + (m_{2j} - b_{2j})$$

$$(m_0 - b_0) + \cdots + (m_{2j} - b_{2j}) \leq (m_1 - b_1) + \cdots + (m_{2j+1} - b_{2j+1}) \;,$$

which are precisely the strong Morse inequalities (11.22). Note that the Morse index theorem also follows from (11.38), since in the case $2j + 1 = v$ or $2j = v$, Q_t is bijective in the appropriate equation in (11.38), which implies the dimension of the space on the right equals that of the space on the left. □

12. Patodi's Proof of the Gauss-Bonnet-Chern Theorem and Superproofs of Index Theorems

Witten's remarkable paper [370] and a companion paper [369] contained not only the ideas of the last chapter, but also extensive remarks on the form of the Atiyah-Singer index theorem in supersymmetric quantum theory. This suggested that supersymmetry might provide the framework for a simple proof of the classical Atiyah-Singer theorem. This hope was realized by *Alvarez-Gaumez* [12] and subsequently by *Friedan-Windy* [113] (see also Windy [368]). These theoretical physicists relied on formal manipulations inside path integrals (including "fermion" path integrals) so their proofs were certainly not rigorous. *Getzler* [129] found a rigorous version of their arguments which, since it relied on some pseudodifferential operator machinery and the theory of supermanifolds, was still rather involved. More recently, *Getzler* [130] found a proof whose gometric and algebraic parts are especially elementary and transparent. Independent of this work, *Bismut* [49] found a related proof. The analytic part of these two proofs [49, 130] is somewhat sophisticated, relying on Brownian motion estimates on manifolds.

One of our goals here is to describe some elementary Schrödinger operator theory on manifolds to provide the analytic steps. But more basically, we want to expose these ideas in pedagogic detail. For this reason, we concentrate on an especially simple, special case of the general index theorem: the Gauss-Bonnet-Chern (GBC) theorem. Surprisingly, when one specializes the Getzler proof to this case, one obtains a proof originally found in 1971 (!) by *Patodi* [273], who did not know that he was speaking supersymmetry. Except for a rather distinct analytic machine in Sect. 12.5–7, our proof in Sect. 12.1–8 follows the general ideas of strategy in Patodi's magnificent paper (we also provide considerable background in differential geometry for the reader's convenience). We return to some remarks on the general case of the index theorem in Sect. 12.9 and 10.

12.1 A Very Rapid Course in Riemannian Geometry

In order to discuss the GBC theorem, we need some elementary facts and definitions in Riemannian geometry. In fact, to be brief, we will not discuss the geometrical content of these notions, so perhaps we should say "Riemannian analysis" rather than "Riemannian geometry." We will describe Riemannian connections and then curvature.

Recall that vector fields, i.e. smooth functions X from a manifold M to the tangent bundle $T(M)$ with $X(p) \in T_p(M)$, act on functions by writing $X(p) =$

$\sum a^i(p)\partial/\partial x^i$ in local coordinates and defining

$$(Xf)(p) = \sum a^i(p)\frac{\partial f}{\partial x^i}(p) \ .$$

It is not hard to show that, given any two vector fields X, Y, $[X, Y] \equiv XY - YX$ is the operator associated to a unique vector field also denoted by $[X, Y]$.

In general, we will denote the set of vector fields as $\Gamma(T(M))$ and $\Gamma(\bigwedge^p T^*(M))$ will be the set of p-forms which we also denoted as $\bigwedge^p(M)$ in the last chapter. Given $X \in \Gamma(T(M))$, one defines a *Lie derivative*, L_X on $\Gamma(T(M))$, etc. by

$$\begin{aligned} &L_X f = Xf \qquad f \in C^\infty(M) \\ &L_X Y = [X, Y] \quad Y \in \Gamma(T(M)) \\ &(L_X\omega)(Y) = L_X((\omega, Y)) - \omega(L_X Y), \ \omega \in \Gamma(T^*(M)), \ X \in T(\textstyle\bigwedge) \end{aligned} \tag{12.1}$$

and L_X is defined on p-forms so that

$$L_X(\alpha \wedge \beta) = L_X\alpha \wedge \beta + \alpha \wedge L_X\beta \ ;$$

[of course, to check that (12.1) defines a map from 1-forms to 1-forms, we must check that the value of $(L_X\omega)(Y)$ at p only depends on $Y(p)$, i.e. that $(L_X\omega)(fY) = f(L_X\omega)(Y)$, which is easy to see]. Lie derivatives have at least two major defects: (i) L_X is not tensorial in X, i.e. L_X depends not only on $X(p)$ but on X near p, equivalently $L_{fX} \neq fL_X$ for $L_{fX}Y \equiv fL_XY - (Yf)X$. (ii) If there is a Riemann metric, it is *not* true that $L_X\langle Y, Z\rangle = \langle L_X Y, Z\rangle + \langle Y, L_X Z\rangle$.

From an analytic point of view, covariant derivatives try to remedy (i).

Definition 12.1. A *covariant derivative* or *connection* is a map $\nabla_X\colon \Gamma(T(M)) \to \Gamma(T(M))$ for each $X \in T(M)$ obeying:

(i) $\nabla_X Y$ is linear in X and Y over the real numbers,
(ii) $\nabla_{fX} Y = f\nabla_X Y$ if $f \in C^\infty(M)$,
(iii) $\nabla_X(fY) = f\nabla_X Y + (Xf)Y$.

Definition 12.2. The connection is called *Torsion-free* if and only if

$$\nabla_X Y - \nabla_Y X = [X, Y] \ . \tag{12.2}$$

Now we consider Riemannian manifolds, i.e. smooth manifolds with a Riemann metric.

Theorem 12.3 (The Fundamental Theorem of Riemannian Geometry). There exists a unique connection on any Riemannian manifold which is torsion free and obeys

$$X\langle Y, Z\rangle = \langle \nabla_X Y, Z\rangle + \langle Y, \nabla_X Z\rangle \tag{12.3}$$

for all vector fields X, Y, Z.

Sketch of Proof. Suppose we have a connection obeying (12.2, 3). Then

$$\begin{aligned}
\langle \nabla_X Y, Z\rangle &= X\langle Y,Z\rangle - \langle Y, \nabla_X Z\rangle && \text{[by (12.3)]}\\
&= X\langle Y,Z\rangle - \langle Y,[X,Z]\rangle - \langle Y, \nabla_Z X\rangle && \text{[by (12.2)]}\\
&\equiv t_1 - \langle Y, \nabla_Z X\rangle \\
&= t_1 - Z\langle Y,X\rangle + \langle \nabla_Z Y, X\rangle && \text{[by (12.3)]}\\
&= t_1 - Z\langle Y,X\rangle + \langle [Z,Y],X\rangle + \langle \nabla_Y Z, X\rangle && \text{[by (12.2)]}\\
&\equiv t_1 + t_2 + \langle \nabla_Y Z, X\rangle \\
&= t_1 + t_2 + Y\langle Z,X\rangle - \langle Z, \nabla_Y X\rangle && \text{[by (12.3)]}\\
&= t_1 + t_2 + Y\langle Z,X\rangle - \langle Z,[Y,X]\rangle - \langle Z, \nabla_X Y\rangle\ . && \text{[by (12.2)]}
\end{aligned}$$

Thus, any connection obeying (12.2, 3) must obey the formula

$$\begin{aligned}
\langle \nabla_X Y, Z\rangle = \tfrac{1}{2}\{&X\langle Y,Z\rangle + Y\langle X,Z\rangle - Z\langle X,Y\rangle - \langle Y,[X,Z]\rangle \\
&- \langle X,[Y,Z]\rangle + \langle [X,Y],Z\rangle\}\ . \qquad (12.4)
\end{aligned}$$

This establishes uniqueness.

It is straightforward to verify that the right side of (12.4) is tensorial in Z (i.e. $\langle \nabla_X Y, fZ\rangle = f\langle \nabla_X Y, Z\rangle$), so that it only depends on $Z(p)$. Since it is linear in Z, we can use (12.4) to define $\nabla_X Y$. Further straightforward manipulations from (12.4) verify that $\nabla_X Y$ obeys the axioms of a connection and obeys (12.2, 3). □

If we pick a local coordinate system, let $X_i = \partial/\partial x^i$, $g_{ij} = \langle X_i, X_j\rangle$ and g^{ij} the inverse matrix to g_{ij}. Then since $[X_i, X_j] = 0$, (4) says that ($\nabla_{X_i} \equiv \nabla_i$)

$$\langle \nabla_i X_j, X_k\rangle = \frac{1}{2}\left\{\frac{\partial}{\partial x^i} g_{jk} + \frac{\partial}{\partial x^j} g_{ik} - \frac{\partial}{\partial x^k} g_{ij}\right\}\ , \qquad (12.5a)$$

so

$$\nabla_i X_j = \sum \Gamma_{ij}^k X_k \qquad (12.5b)$$

$$\Gamma_{ij}^k = \sum_l g^{kl}\langle \nabla_i X_j, X_l\rangle\ . \qquad (12.5c)$$

Here Γ is called the *Christoffel symbol* of the connection.

The fact that the connection is torsion free and the X_i Lie commute says that $\nabla_i X_j = \nabla_j X_i$, so Γ_{ij}^k is symmetric under $i \mapsto j$.

In local coordinates, if $B = \sum b^i(\partial/\partial x^i)$ and $C = \sum c^i(\partial/\partial x^i)$, then $\nabla_B = \sum b^i \nabla_i$ and $\nabla_i C = \sum_j e^j(\partial/\partial x^j)$, where

$$e^j = \frac{\partial c^j}{\partial x^i} + \sum_l \Gamma_{il}^j c^l\ . \qquad (12.6)$$

We warn the reader that the above formulae are in terms of a set of *coordinate* vector fields X_i. In some references, covariant derivatives are described in terms of an *orthonormal* frame of vector fields as mentioned in Sect. 11.3. Thus, the Γ used in those places is *not* symmetric in i, j.

The *Christoffel symbol* does not transform as a tensor, since $\langle \nabla_X Y, Z\rangle$ is only tensorial in X and Z. However, the combination

$$R(X, Y) = \nabla_X \nabla_Y - \nabla_Y \nabla_X - \nabla_{[X,Y]}$$

called the *curvature* is tensorial, i.e.

$$K(X, Y, Z, W) = \langle R(X, Y)Z, W\rangle$$

only depends on the values of X, Y, Z, W at p. $K \in \Gamma(\otimes^4 T^*(M))$ is called the *Riemann curvature tensor.*

Theorem 12.4. (a) (First Bianchi Identity) $R(X, Y)Z + R(Y, Z)X + R(Z, X)Y = 0$.

(b) K is antisymmetric under interchanging X and Y or under interchanging W and Z.

(c) $K(X, Y, Z, W) = K(Z, W, X, Y)$.

Proof (a) Since R is tensorial, we can always extend X, Y, Z defined at p to Lie commute, i.e. without loss, we can suppose that $[X, Y] = [X, Z] = [X, Y] = 0$. Then,

$$\begin{aligned} \text{LHS of (a)} &= \nabla_X \nabla_Y Z - \nabla_Y \nabla_X Z + 4 \text{ others} \\ &= \nabla_X(\nabla_Y Z - \nabla_Z Y) + 2 \text{ others} \\ &= \nabla_X([Y, Z]) + 2 \text{ others} \quad &&[\text{by (2)}] \\ &= 0 &&(\text{since } [Y, Z] = \cdots = 0)\ . \end{aligned}$$

(b) Antisymmetry in X and Y is obvious. To prove antisymmetry in Z, W one needs only transfer the derivatives from Z to W using $\langle \nabla_A B, C\rangle = A\langle B, C\rangle - \langle B, \nabla_A C\rangle$ repeatedly.

(c) By tensoriality, we can pick X_i obeying $[X_i, X_j] = 0$. Let

$$R_{ijkl} = K(X_k, X_l, X_i, X_j)\ .$$

Then

$$\begin{aligned} R_{ijkl} &= -R_{jkil} - R_{kijl} &&[\text{by (a)}] \\ &= R_{jkli} + R_{kilj} &&[\text{by (b)}] \\ &= -R_{klji} - R_{ljki} - R_{ilkj} - R_{lkij} &&[\text{by (a)}] \\ &= R_{klij} + R_{ljik} + R_{iljk} + R_{klij} &&[\text{by (b)}] \\ &= 2R_{klij} - R_{jilk} &&[\text{by (a)}] \\ &= 2R_{klij} - R_{ijkl} &&[\text{by (b)}] \end{aligned}$$

proving the required symmetry. □

A straightforward calculation in local coordinates shows that

$$R(X_i, X_j)X_k \equiv \sum_l R_k{}^l{}_{ij} X_l$$

with

$$R_k{}^l{}_{ij} = \frac{\partial}{\partial x_i}\Gamma^l_{jk} - \frac{\partial}{\partial x_j}\Gamma^l_{ik} + \sum_r (\Gamma^r_{jk}\Gamma^l_{ir} - \Gamma^r_{ik}\Gamma^l_{jr}) \ . \qquad (12.7)$$

It will be convenient later to know the following:

Theorem 12.5. Let m be a point on a Riemannian manifold M. There exists á coordinate system x^i in a neighborhood of m so that

(a) $x^i(m) = 0$,
(b) $g_{ij}(m) = \delta_{ij}$,
(c) $\Gamma^k_{ij}(m) = 0$,
(d) $\frac{\partial}{\partial x^i} g_{jk}(m) = 0$,
(e) $\det g_{ij} \equiv 1$ near m.

One gets such a *normal coordinate system* (*centered at* m) by picking an orthonormal basis e_i in T_m and letting p have coordinates x^i, if the geodesic from m to p has length $|x|$ and tangent $\sum x^i e_i$ at m. For details see *Boothby* [52], pp. 331–335. Boothby does not note (d), but by (12.5)

$$\frac{\partial}{\partial x^i} g_{jk} = \sum_l (g_{kl}\Gamma^l_{ij} + g_{jl}\Gamma^l_{ik}) \ .$$

Condition (e) may not hold in the geodesic coordinates just mentioned, and indeed most authors do not require it for "normal coordinates." We want to explain how, given a coordinate system obeying (a)–(d), we can modify it to obtain (e). Given such a coordinate system x, let

$$y^1 = \int_0^{x^1} \sqrt{\det g(s, x^2, \ldots, x^n)}\, ds, \ y^i = x^i \text{ if } i \geq 2 \ .$$

In the y coordinate system, $d^n y = \sqrt{\det g}\, d^n x$ so the metric $\tilde{g}$ of the new coordinate system obeys $\sqrt{\det \tilde{g}} \equiv 1$. A straightforward calculation shows that $\tilde{g}_{ij} = \delta_{ij} + 0(|x|^2)$ so the y coordinate system obeys (a)–(d).

Normal coordinates are useful for the following reason. Tensors can be computed in any coordinate system, and since Γ's often enter in a complicated way, it is useful to be able to do the calculations in a convenient coordinate system. In particular, we will often show that two covariant objects are equal by verifying their equality at m in a normal coordinate system centered at m. Given (12.5) and (12.7), one obtains an elementary formula for R_{ijkl} at the center of a normal coordinate system:

$$R_{ijkl}(m) = \tfrac{1}{2}(G_{ik;jl} + G_{jl;ik} - G_{il;jk} - G_{jk;il}) \tag{12.8a}$$

where

$$G_{ij;kl} = \partial^2 g_{ij}/\partial x^k \partial x^l(m) \ . \tag{12.8b}$$

We emphasize that his formula is *only* valid in normal coordinates. As a bonus, we note that symmetry under $(ij) \leftrightarrow (kl)$ is evident given the symmetry of $G_{ij;kl}$ in $i \leftrightarrow j$ and $k \leftrightarrow l$. Condition (e) has an important consequence: $\det g_{ij} = 1 + Tr(g_{ij} - \delta_{ij}) + O(|g_{ij} - \delta_{ij}|^2)$ and thus

$$\frac{\partial^2 \det g}{\partial x^k \partial x^l} = \sum_i G_{ii;kl} \ .$$

Thus,

$$\det g \equiv 1 \Rightarrow \sum_i G_{ii;kl}(m) = 0 \ . \tag{12.9a}$$

This means that the total contraction of R_{ijkl}, called the scalar curvature, has an especially simple expression

$$R := \sum_{i,j} R_{ij}{}^{ji} = \sum_{i,j} G_{ij;ij} \textit{ in normal coordinates} \ , \tag{12.9b}$$

where we emphasize once again that normal includes $\det g \equiv 1$.

We need to transfer covariant derivatives to forms. Given a one-form ω, and $X \in \Gamma(T(M))$, we define a one-form $\nabla_X \omega$ so that

$$(\nabla_X \omega)(Y) = X[\omega(Y)] - \omega(\nabla_X Y) \ .$$

One checks that the right side of this expression is tensorial in Y. In local coordinates,

$$\nabla_i(dx^j) = -\sum \Gamma^j_{il} dx^l \ , \tag{12.10a}$$

and one then defines ∇_X on p-forms so that

$$\nabla_X(\alpha \wedge \beta) = (\nabla_X \alpha) \wedge \beta + \alpha \wedge (\nabla_X \beta) \ . \tag{12.10b}$$

If

$$\alpha = \sum_{i_1 < \cdots < i_p} \alpha_{i_1 \ldots i_p} dx^{i_p} \wedge \cdots \wedge dx^{i_p}, \text{ then}$$

$$(\nabla_j \alpha)_{i_1 \ldots i_p} = \frac{\partial \alpha_{i_1 \ldots i_p}}{\partial x_j} - \sum_{l,r} \Gamma^l_{ji_r} \alpha_{i_1 \ldots i_{r-1} l i_{r+1} \ldots i_p}$$

with the convention that α is extended from $i_1 < \cdots < i_r$ to all indices by requiring antisymmetry. Let $(a^i)^*$ be defined from p-forms to $(p+1)$-forms by

$$(a^i)^*\omega = dx_i \wedge \omega \ .$$

Interpret ∇_X on $C^\infty(M)$ by $\nabla_X f = Xf$. Then

Theorem 12.6. $d = \sum (a^i)^* \nabla_i$.

Proof. Let ∂_i be the operator defined by components with differentiation, i.e.

$$\partial_i\left(\sum_{j_1<\cdots<j_p} a_{j_1\ldots j_p} dx_{j_1} \wedge \cdots \wedge dx_{j_p}\right)$$

$$= \sum_{j_1<\cdots<j_p} \frac{\partial a_{j_1\ldots j_p}}{\partial x_i} dx_{j_1} \wedge \cdots \wedge dx_{j_p} \ .$$

Then d is defined by

$$d = \sum (a^i)^* \partial_i \ .$$

Equations (12.10a, b) imply that ∇_i is given on p-forms by

$$\nabla_i = \partial_i - \sum_{j,m,l} \Gamma^j_{il} g_{jm} (a^l)^* a^m \tag{12.10c}$$

[since $(\sum_m g_{jm} a^m)\, dx^k = \delta_{jk}$]. Thus, the theorem is equivalent to

$$\sum_{i,j,l} \Gamma^j_{il} g_{jm} (a^i)^* (a^l)^* a^* = 0 \ .$$

This equation holds if one notes that $(a^i)^*(a^l)^* = -(a^l)^*(a^i)^*$ is antisymmetric in i, l, while Γ^j_{il} is symmetric in i, l. □

Curvature on forms is further discussed in Sect. 12.2.

Since R_{ijkl} is a covariant tensor, we can form various mixed tensors by raising indices via $R_i{}^j{}_{kl} = \sum_a g^{aj} R_{iakl}$, etc. ($R_i{}^j{}_{kl}$ and R_{ijkl} defined above are related in this way). Because of the antisymmetry of R_{ijkl} in $i \leftrightarrow j$ and $k \leftrightarrow l$, the only interesting contractions of R are the *Ricci tensor*

$$S_{ij} = \sum_l R_{lij}{}^l$$

and the *total curvature*

$$R := \sum_{i,l} R_l{}^i{}_i{}^l = \sum_i S^i{}_i = -\sum R^{il}{}_{il} \ .$$

Given any two-dimensional subspace π in $T_m(M)$, let e_1, e_2 be an orthonormal basis in π and extend them to vector fields, X, Y. Then

$$K(\pi) := -R(X, Y, X, Y)$$

is the *sectional curvature* of π. Here is an equivalent definition: Define the *curvature operator* $R^{(2)}$ on $\bigwedge_m^2(M)$ to be the linear map with

$$R^{(2)}(dx_i^i \wedge dx_j^j) = \sum R^{ij}{}_{lk}\, dx^k \wedge dx^l \ . \tag{12.11}$$

If ω, μ are dual to e_1, e_2 and $\alpha_\pi = \omega \wedge \mu$ is the two form determined by ω, μ, then

$$K(\pi) = (\alpha_\pi, R^{(2)}(\alpha_\pi)) \ .$$

Note that $\mathrm{tr}(R^{(2)}) = R$, the total curvature.

While we do not intend to say much about the geometric content of curvature, we should at least indicate that curvature measures whether geodesics tend to spread out or not. Positive curvature produces a focusing of geodesics as is illustrated by the fact that if the Ricci curvature is bounded below by a fixed positive multiple κ of the identity, then M must be compact and there is an a priori κ-dependent bound on the distance between any two points of M. On the other hand, negative curvature implies a spreading of geodesics from a point. For example, if all sectional curvatures are bounded above by a fixed negative number, $-\kappa$, and if M is simply connected, then there is a unique geodesic between any pair of points and M is diffeomorphic to R^ν. These things are discussed, for example, in Chapter III of *Chavel* [62].

Finally, we want to discuss the divergence of a vector field X. Given such a field, we can define a map ∇X from $\Gamma(T(M))$ to itself by $(\nabla X)(Y) = \nabla_Y X$. The tensoriality of $\nabla_Y X$ in Y shows that ∇X actually defines a matrix function, i.e. there is a linear map $(\nabla X)(m)$ from $T_m(M)$ to $T_m(M)$ so that $(\nabla_Y X)(m) = (\nabla X)(m)[Y(m)]$. The trace of this transformation (which is a function) is called the *divergence* of X, written $\mathrm{div}(X)$. This is an invariant definition. In coordinates, if $X = \sum b^i(\partial/\partial x^i)$, then $(X_j = \partial/\partial x^j)$

$$(\nabla X)(X_j) = \sum_i \frac{\partial b^i}{\partial x^j} X_i + \sum_{i,k} b^i \Gamma_{ji}^k X_k \ ,$$

and, thus,

$$\mathrm{div}(X) = \sum_i \frac{\partial b^i}{\partial x^i} + \sum_{i,j} b^i \Gamma_{ji}^j \ . \tag{12.12}$$

Theorem 12.7. (a) For any $f \in C_0^\infty(M)$ and $X \in \Gamma(T(M))$, we have that

$$\int (Xf)\, dx = -\int f(\mathrm{div}\, X)\, dx \ ,$$

where dx is the natural measure on M [which in local coordinates is $\sqrt{g}\, dx_1^1 \ldots dx_n^n$ with $g = \det(g_{ij})$].

(b) Acting on smooth p-forms,

$$\nabla_X^* = -\nabla_X - \operatorname{div} X \tag{12.13}$$

$$\text{(c)}\ \frac{1}{2}\frac{\partial}{\partial x^i}\ln(g) = \sum_j \Gamma_{ji}^j\ .$$

Proof. Given a coordinate neighborhood N with coordinates x_i, define the function

$$(\operatorname{div})\tilde{}_x(X) = (\sqrt{g})^{-1} \sum_i \frac{\partial}{\partial x^i}(b^i\sqrt{g})\ .$$

Then by integrating by parts, we see that if $f \in C_0^\infty(N)$, then

$$\int (Xf)\,dx = -\int f((\operatorname{div})\tilde{}_x X)\,dx\ .$$

Given two coordinate systems x, y in neighborhoods N, L, for $f \in C_0^\infty(N \cap L)$ we have that

$$\int f\{[(\operatorname{div})\tilde{}_x(X) - (\operatorname{div})\tilde{}_y(X)]\}\,dx = 0\ ;$$

so $(\operatorname{div})\tilde{}(X)$ is coordinate independent. Since $\operatorname{div}(X)$ and $(\operatorname{div})\tilde{}(X)$ are invariantly defined, we need only show they are equal at each point m in some coordinate system at m. For normal coordinates centered at m, $\Gamma = 0$ and $(\partial/\partial x^i)(\sqrt{g}) = 0$, so $(\operatorname{div})\tilde{}(X)(m) = \operatorname{div}(X)(m)$. This equality proves (c), and shows that (a) holds for f's supported in a small coordinate neighborhood. Inserting a suitable partition of unity in front of f proves (a) in general.

Finally, let $u, v \in \Gamma(T^*(M))$. Then

$$\begin{aligned}&\int \{\langle \nabla_X u, v\rangle + \langle u, \nabla_X v\rangle + (\operatorname{div} X)\langle u, v\rangle\}\,dx\\ &\quad = \int \{X(\langle u, v\rangle) + \langle u, v\rangle \operatorname{div} X\}\,dx = 0\end{aligned}$$

proving (b). □

For a more direct proof of (c), see *Spain* [346], pp. 27–28.

Equation (12.12) can be rewritten in another suggestive way. Let u_X be the one-form dual to X, i.e. the linear functional given by $Y \mapsto \langle X, Y\rangle_m$. Then $\int (Xf)\,dx = \langle df, u_X\rangle$ so (12) says that

$$\operatorname{div}(X) = -d^* u_X\ .$$

In terms of the Hodge star operator of Sect. 11.3 (see Theorem 11.10), we have that

$$\operatorname{div}(X) = *d(*u_X)\ .$$

We have already seen (Theorem 12.6) that

$$d = \sum_i (a^i)^* \nabla_i\ . \tag{12.14}$$

We claim that also

$$d = -\sum_i (\nabla_i)^*(a^i)^* \ , \quad \text{so that} \tag{12.15a}$$

$$d^* = -\sum_i a^i \nabla_i \ . \tag{12.15b}$$

To prove (12.15), we note that by (12.12) and (12.13)

$$\nabla_i^* = -\nabla_i - \sum_j \Gamma_{ji}^j \ . \tag{12.16}$$

On the other hand, (12.10a, b) imply that

$$[\nabla_i, (a^j)^*] = -\sum_k \Gamma_{ik}^j (a^k)^* \ . \tag{12.17}$$

Thus, by (12.14)

$$d = \sum_i \nabla_i (a_i^i)^* - \sum_i [\nabla_i, (a^i)^*] = \sum_i \nabla_i (a^i)^* + \sum_{i,k} \Gamma_{ik}^i (a^k)^*$$

$$= \sum_i \left(\nabla_i + \sum_j \Gamma_{ji}^j \right) (a^i)^* = -\sum_i (\nabla_i)^* (a^i)^*$$

by (12.17). This proves (12.15a).

12.2 The Berezin-Patodi Formula

The machinery of supersymmetry will enter in our proof of GBC in two ways. One is the pairing of eigenvalues, which we have already discussed. The other is through an elementary piece of linear algebra implicit in *Berezin* [47], called Berezin's formula in the physics literature. At about the same time as Berezin's work, *Patodi*, in the context of his work on GBC [273], proved a formula which is essentially equivalent to Berezin's, so we dub the formula the "Berezin-Patodi" formula. We should also note that related formulae appear in *Atiyah* and *Bott* [17].

Let V be a finite dimensional inner product space, and let $\bigwedge^p(V)$ be the space of antisymmetric p tensors on V. Let

$$\bigwedge^*(V) = \bigoplus_{p=1}^{n} \bigwedge^p(V)$$

where $n = \dim(V)$. Let $(-1)^p$ be the operator which acts on $\bigwedge^*(V)$ by multiplying $\omega \in \bigwedge^p(V)$ by $(-1)^p$. Given any A in $\mathrm{L}(\bigwedge^*(V))$, a linear operator on $\bigwedge^*(V)$, define its supertrace, $\mathrm{str}(A)$, by

$$\mathrm{str}(A) \equiv \mathrm{tr}[(-1)^p A] \ .$$

Let $e_1, \ldots, e_n$ be an orthonormal basis for V, and let a_i^* be identical with $e_i \wedge \cdot$ and a_i with the adjoint of a_i^* as operators on $\bigwedge^*(V)$. Given $I \subset \{1, \ldots, n\}$, let a_I be $\prod_{i \in I} a_i$ ordered by the natural order on $\{1, \ldots, n\}$ and a_I^* be $(a_I)^*$. Notice that $\mathrm{L}(\bigwedge^*(V))$ has dimension $(2^n)^2$ and this is precisely the number of $a_I^* a_J$. Thus, the following result, which we will prove below, should not be too surprising:

Proposition 12.8. Any $A \in \mathrm{L}(\bigwedge^*(V))$ can be written uniquely in the form

$$A = \sum \alpha_{I,J} a_I^* a_J \ .$$

The key fact we need is

Theorem 12.9 (The Berezin-Patodi Formula). For any $A \in \mathrm{L}(\bigwedge^*(V))$

$$\mathrm{str}(A) = (-1)^n \alpha_{\{1, \ldots, n\}, \{1, \ldots, n\}} \ .$$

Proofs. We pass from fermion creation operators to Dirac matrices (in mathematicians' language, from alternating algebras to Clifford algebras). Define

$$\gamma_{2k-1} = a_k + a_k^*, \quad \gamma_{2k} = i^{-1}(a_k - a_k^*) \ .$$

Since the a's obey $\{a_i, a_j\} = 0$, $\{a_i^*, a_j\} = \delta_{ij}$, the γ's obey $\{\gamma_\mu, \gamma_\nu\} = 2\delta_{\mu\nu}$ and obviously $\gamma_\mu^* = \gamma_\mu$. For $A = \{\mu_1, \ldots, \mu_l\} \subset \{1, \ldots, 2n\}$ with $\mu_1 < \cdots < \mu_l$ define

$$\gamma_A = i^{l(l-1)/2} \gamma_{\mu_1} \cdots \gamma_{\mu_l} \ ,$$

so $\gamma_A^2 = 1$, $\gamma_A = \gamma_A^*$.

We first claim that

$$\mathrm{tr}(\gamma_A) = 0 \qquad A \neq \phi \ , \tag{12.18a}$$

$$= 2^n \qquad A = \phi \ , \tag{12.18b}$$

for if $\#(A)$ is even and nonzero, let $\mu_1 = \min\{i | i \in A\}$ and let $\tilde{A} = A \backslash \{\mu_1\}$. Then,

$$\mathrm{tr}(\gamma_A) = \mathrm{tr}(\gamma_{\mu_1} \gamma_{\tilde{A}}) = -\mathrm{tr}(\gamma_{\tilde{A}} \gamma_{\mu_1}) = -\mathrm{tr}(\gamma_{\mu_1} \gamma_{\tilde{A}}) = -\mathrm{tr}(\gamma_A) \ ,$$

since $(\gamma_{\mu_1}, \gamma_A) = 0$ and trace is cyclic. If $\#(A)$ is odd, find $\mu \notin A$ and write

$$\mathrm{tr}(\gamma_A) = \mathrm{tr}(\gamma_\mu \gamma_\mu \gamma_A) = -\mathrm{tr}(\gamma_\mu \gamma_A \gamma_\mu) = -\mathrm{tr}(\gamma_A) \ ,$$

since $\{\gamma_\mu, \gamma_\mu \gamma_A\} = 0$.

Given (12.18) and $\gamma_\mu^2 = 1$, we immediately have

$$\mathrm{tr}(\gamma_A^* \gamma_B) = 0 \qquad \text{if } A \neq B \ , \tag{12.19a}$$

$$= 2^n \qquad \text{if } A = B \ , \tag{12.19b}$$

and thus $\{\gamma_A\}$ form an orthogonal set in $L(\wedge^*(V))$ with its natural Hilbert-Schmidt inner product. But the number of such A's is equal to $\dim(L(\wedge^*(V))) = 2^{2n}$, so the γ_A are a basis for $L(\wedge^*(V))$. Since any γ_A is a sum of $a_I^* a_J$'s, we see that the $a_I^* a_J$ are also a basis, proving Proposition 12.8.

Next, we claim that

$$(-1)^p = (-1)^n \gamma_{\{1, \ldots, 2n\}} \ . \tag{12.20}$$

Let the right side of (12.20) be denoted by α. Then clearly $\{\gamma_\mu, \alpha\} = 0$, so $\{a_i^*, \alpha\} = 0$, and thus it suffices that

$$\alpha\psi_0 = \psi_0 \ , \tag{12.21}$$

where ψ_0 is the vector in $\wedge^0(V)$. But

$$\gamma_{2k-1}\gamma_{2k} = i^{-1}(a_k^* a_k - a_k a_k^*) \ , \quad \text{so} \tag{12.22}$$

$$\gamma_{\{1, \ldots, 2n\}}\psi_0 = i^n i^{(2n)(2n-1)/2}\psi_0 = i^{2n^2}\psi_0 = (-1)^{n^2}\psi_0 = (-1)^n\psi_0$$

proving (12.21).

Equations (12.19, 20) immediately imply that

$$\operatorname{str}(\gamma_A) = 0 \qquad A \neq \{1, \ldots, 2n\} \tag{12.23a}$$

$$= (-1)^n 2^n \qquad A = \{1, \ldots, 2n\} \ . \tag{12.23b}$$

Finally, by using

$$a_k = \tfrac{1}{2}(\gamma_{2k-1} + i\gamma_{2k}), \quad a_k^* = \tfrac{1}{2}(\gamma_{2k-1} - i\gamma_{2k})$$

any $a_I^* a_J$ can be expanded into γ_A's. If $\#(I) + \#(J) < 2n$, $A = \{1, \ldots, 2n\}$ cannot appear, so (12.23) implies that

$$\operatorname{str}(a_I^* a_J) = 0 \quad \text{if } I \neq \{1, \ldots, n\} \text{ or } J \neq \{1, \ldots, n\} \ .$$

By (12.22)

$$a_k^* a_k = \tfrac{1}{2}(1 + i\gamma_{2k-1}\gamma_{2k}) \ , \quad \text{so}$$

$$\begin{aligned} a_{\{1, \ldots, n\}}^* a_{\{1, \ldots, n\}} &= (a_1^* a_1)(a_2^* a_2)\ldots(a_n^* a_n) \\ &= (\tfrac{1}{2})^n i^n i^{-(2n)(2n-1)/2}\gamma_{\{1, \ldots, 2n\}} + \text{other } \gamma_A\text{'s} \\ &= (\tfrac{1}{2})^n \gamma_{\{1, \ldots, 2n\}} + \text{other } \gamma_A\text{'s} \ , \end{aligned}$$

so that, by (12.19)

$$\operatorname{str}(a_{\{1, \ldots, n\}}^* a_{\{1, \ldots, n\}}) = (-1)^n$$

proving Theorem 12.9. □

We want to give Patodi's version and proof of Theorem 12.9 (modulo Proposition 12.8). While this alternate proof can be skipped, the notation introduced next and Theorem 12.10 and Proposition 12.11 will be needed later. As is standard in the study of exterior algebra, given an operator A on V, there are two operators on $\bigwedge^*(V)$ called $\bigwedge^*(A)$ and $d\bigwedge^*(A)$. They are defined on p-forms by requiring that

$$\textstyle\bigwedge^p(A)(u_1 \wedge \cdots \wedge u_p) = Au_1 \wedge \cdots \wedge Au_p \tag{12.24a}$$

$$(d\textstyle\bigwedge)^p(A)(u_1 \wedge \cdots \wedge u_p) = \sum_{j=1}^{p} u_1 \wedge \cdots \wedge u_{j-1} \wedge Au_j \wedge u_{j+1} \wedge \cdots \wedge u_p \ . \tag{12.24b}$$

The relation between them is that

$$\frac{d}{dt}[\textstyle\bigwedge^*(e^{tA})]|_{t=0} = (d\bigwedge^*)(A) \ , \quad \text{so that}$$

$$\textstyle\bigwedge^*(e^{tA}) = \exp[td\bigwedge^*(A)] \ . \tag{12.25}$$

In terms of the operators a_i^*, a_j, we have:

Theorem 12.10. Let $A: V \to V$ have matrix elements A_{ij} defined by $Ae_j = \sum A_{ij}e_i$. Then

$$d\textstyle\bigwedge^*(A) = \sum_{i,j} A_{ij}a_i^* a_j \ . \tag{12.26}$$

Proof. Let B denote the right side of (12.26). Since it is clearly a linear operator, we need only check that $B\restriction\bigwedge^1(V) = d\bigwedge^*(A)\restriction(V) = A$ and that

$$B(u \wedge v) = (Bu) \wedge v + u \wedge (Bv) \tag{12.27}$$

for all $u, v \in \bigwedge^*(V)$. The first follows by noting that

$$Be_k = \sum_{i,j} A_{ij}a_i^* a_j a_k^* \psi_0 = \sum_i A_{ik}e_i = Ae_k \ .$$

To prove (12.27), we note that if $u \in \bigwedge^p$, then

$$a_i^*(u \wedge v) = (a_i^* u) \wedge v = (-1)^p u \wedge (a_i^* v)$$

and that

$$a_i(u \wedge v) = (a_i u) \wedge v - (-1)^p u \wedge (a_i v) \ .$$

These formulae, which can be checked by taking u, v to be elements of the bases $e_{i_1} \wedge \cdots \wedge e_{i_j}$, imply that $\tilde{B} = a_i^* a_j$ obeys (12.26) and so B does by linearity. □

Remark 12.11. The above discussion assumed the fact from linear algebra that there exist linear transformations obeying (12.24a, b). However, one can prove this by *defining* $d\bigwedge^*(A)$ by (12.26), checking (12.24b) (the above proof) and then defining $\bigwedge^*(B)$ by (12.25) (at least for invertible B and then taking limits).

As a typical example of the use of these notions, let ∇_i denote the covariant derivative associated to a coordinate vector field $\partial/\partial x^i$ acting on p-forms. Then

Proposition 12.12:

$$[\nabla_i, \nabla_j] = -\sum_{k,l} R_{ijkl}(a^k)^* a^l \ . \tag{12.28}$$

Proof. On p-form valued functions, ∇_i is defined via (12.10b), from which one sees (note: $\nabla_i \nabla_j$ means the product of ∇_i and ∇_j as operators; as we will discuss in Sect. 12.4, this symbol is often used for another object) that

$$\nabla_i \nabla_j(\alpha \wedge \beta) = (\nabla_i \nabla_j \alpha) \wedge \beta + \alpha \wedge \nabla_i \nabla_j \beta + \nabla_i \alpha \wedge \nabla_j \beta + \nabla_j \alpha \wedge \nabla_i \beta \ .$$

This implies that the unwanted terms cancel in $[\nabla_i, \nabla_j]$ and

$$[\nabla_i, \nabla_j](\alpha \wedge \beta) = [\nabla_i, \nabla_j]\alpha \wedge \beta + \alpha \wedge [\nabla_i, \nabla_j]\beta \ ,$$

and so $[\nabla_i, \nabla_j]$ is $d\bigwedge^*([\nabla_i, \nabla_j]$ on 1-forms). Thus, we need only prove (12.28) on 1-forms.

Let $X_i = \partial/\partial x^i$, $\omega^l = dx^l$. Then $0 = X_i X_j \langle X_k, \omega^l \rangle$ so

$$0 = \langle \nabla_i \nabla_j X_k, \omega^l \rangle + \langle X_k, \nabla_i \nabla_j \omega^l \rangle + \langle \nabla_i X_k, \nabla_j \omega^l \rangle + \langle \nabla_j X_k, \nabla_i \omega^l \rangle \ .$$

Again the unwanted terms cancel from the commutation, so

$$\langle X_k, [\nabla_i, \nabla_j]\omega^l \rangle = -\langle [\nabla_i, \nabla_j] X_k, \omega^l \rangle = -R_k{}^l{}_{ij} \ .$$

Given the symmetry $R_{klij} = R_{ijkl}$, this implies (12.28) on 1-forms. □

The remaining remarks on Patodi's work play no role in the rest of the chapter. Patodi does not discuss a_i and a_i^*'s and so he does not state or prove Theorem 12.9 in the form we gave it. Rather, he proves that

$$\mathrm{str}(d\textstyle\bigwedge^*(B^{(1)}) \dots d\bigwedge^*(B^{(l)})) = 0 \quad \text{if } l < n \tag{12.29a}$$

$$= (-1)^n \sum_{\pi,\sigma} (-1)^\pi (-1)^\sigma B^{(1)}_{\pi(1)\sigma(1)} B^{(2)}_{\pi(2)\sigma(2)} \dots B^{(n)}_{\pi(n)\sigma(n)} \quad l = n \ , \tag{12.29b}$$

where the sum is over all permutations π, σ on $\{1, \dots, n\}$. We leave it to the reader to check that (given the fact that if $|I| \neq |J|$, the $a_I^* a_J$ takes each $\bigwedge^p$ into $\bigwedge^q$, $q \neq p$ so that $\mathrm{str}(a_I^* a_J)$ is then 0), (12.19) implies $\mathrm{str}(\sum \alpha_{I,J} a_I^* a_J) = (-1)^n \alpha_{\{1,\dots,n\},\{1,\dots,n\}}$.

To prove (12.28), Patodi notes that

$$\det(I + A) = \sum_{p=0}^{n} \mathrm{Tr}_{\wedge^{p}(V)}(\wedge^{p}(A)) \;, \quad \text{so}$$

$$\frac{\partial^l}{\partial x_1 \ldots \partial x_l} det\left(I - \prod_{j=1}^{l} \exp[x_j B^{(j)}]\right)$$

$$= \frac{\partial^l}{\partial x_1 \ldots \partial x_l} \sum_{p=0}^{n} (-1)^p \mathrm{Tr}_{\wedge_p}\left[\prod_{j=1}^{l} \wedge^p(\exp x_j B^{(j)})\right]$$

$$= \mathrm{str}(d\wedge^*(B^{(1)}) \ldots d\wedge^*(B^{(l)})) \tag{12.30}$$

at $x_1 = \cdots = x_l = 0$.

Letting $\|x\| = (\sum_1^l x_j^2)^{1/2}$, we see that

$$I - \prod_{j=1}^{l} \exp(x_j B^{(j)}) = -\prod_{j=1}^{l} x_j B^{(j)} + \mathrm{O}(\|x\|^2) \;,$$

so $\det[I - \prod_{j=1}^{l} \exp(x_j B^{(1)})]$ has leading order $\|x\|^n$ which proves (12.29a) from (12.30). Moreover, the $\|x\|^n$ term is just $(-1)^n \det(x_1 B^{(1)} + \cdots + x_n B^{(n)})$ which, given (12.30), yields (12.29b).

12.3 The Gauss-Bonnet-Chern Theorem: Statement and Strategy of the Proof

In this section, we will state the Gauss-Bonnet-Chern theorem and give an overview of the strategy of the proof we will present. Let M be a manifold of even dimension $n = 2k$. Let R be the Riemann curvature tensor. Then

Proposition 12.13. The quantity [the sum is over all pairs of permutations of $(1, \ldots, n)$]

$$E(x) = \frac{(-1)^k}{(4\pi)^k k! 2^k} \sum_{\pi,\sigma} (-1)^\pi (-1)^\sigma R^{\pi(1)\pi(2)}{}_{\sigma(1)\sigma(2)} \cdots R^{\pi(n-1)\pi(n)}{}_{\sigma(n-1)\sigma(n)}$$

is a scalar (independent of coordinate systems).

We will slightly defer the proof of this proposition when we will also give a coordinate free definition of the Euler n-form, $E(x)\,dx$, with dx the natural measure on M. If $n = 2$, there are four non-zero terms in the sum, all equal by the symmetry of R, so $E(x) = -(2\pi)^{-1} R^{12}{}_{12}$ but the scalar curvature $R = \sum_{k,l} R^{kl}{}_{lk} = -2R^{12}{}_{12}$, so

$$E(x) = (4\pi)^{-1} R(x) \quad (n = 1) \;. \tag{12.31}$$

The GBC theorem says that:

Theorem 12.14 (Gauss-Bonnet-Chern-Theorem). Let M be a compact orientable manifold of even dimension $n = 2k$. Then

$$\chi(M) \equiv \sum_p (-1)^p b_p = \int_M E(x)\,dx \ .$$

In particular, given (12.31) and the fact that for $n = 2$, $b_1 = 2g$ with g the genus of the surface, we obtain the classical Gauss-Bonnet theorem: $2 - 2g = (4\pi)^{-1} \int R(x)\,dx$. An interesting corollary of this theorem is that if M, N are C^∞ compact orientable manifolds and $M \overset{\pi}{\to} N$ is a smooth l-fold covering map, then $\chi(M) = l\chi(N)$ since picking a metric g on N and letting $\tilde{g}$ be its pullback on M, we have that $E_N(\pi(x)) = E_M(x)$. For a history of the GBC theorem, as well as the bundle theoretic (Chern class) proof, see *Spivak* [349]. As a warm-up for the proof of Proposition 12.13, we define Pffafians and explore their properties.

Definition 12.15. Let $n = 2k$. Let A_{ij} be a real antisymmetric $n \times n$ matrix. The Pffafian, Pff(A), is defined by

$$\mathrm{Pff}(A) = (2^k k!)^{-1} \sum_\pi (-1)^\pi A_{\pi(1)\pi(2)} A_{\pi(3)\pi(4)} \cdots A_{\pi(n-1)\pi(n)} \ , \tag{12.32a}$$

the sum being over all permutations on $\{1, \ldots, n\}$.

The $(2^k k!)^{-1}$ factor is not mysterious. The summand is the same for two permutations π, σ where the pairings $\{\pi(1)\pi(2)\}, \ldots, \{\pi(n-1)\pi(n)\}$ are the same. Given any π, there are $2^k k!$ such pairings, so in fact

$$\mathrm{Pff}(A) = \sum_{\substack{i_1 < \cdots < i_k \\ i_1 < j_1, \ldots, i_k < j_k}} (-1)^\pi A_{i_1 j_1} \cdots A_{i_k j_k} \ , \tag{12.32b}$$

π being the obvious permutation.

Proposition 12.16. (a) If B is an arbitrary matrix and

$$\bar{A}_{ij} = \sum_{k,l} B_{ik} B_{jl} A_{kl} \ , \tag{12.33}$$

then $\mathrm{Pff}(\bar{A}) = \det(B)\,\mathrm{Pff}(A)$.
(b) $\mathrm{Pff}(A)^2 = \det(A)$.

Proof. (a) Let $e_1, \ldots, e_n$ be the canonical basis for $\mathbb{C}^n$ and view e_i as elements of $\bigwedge^*(\mathbb{C}^n)$. Let

$$a = \sum_{i,j} A_{ij} e_i \wedge e_j \ .$$

Then (12.32a) says that

$$a \wedge \cdots \wedge a (k \text{ times}) = (2^k k!)\,\mathrm{Pff}(A) e_1 \wedge \cdots \wedge e_n \ . \tag{12.34}$$

Let B be the operator whose matrix is B_{ij}, i.e. $Be_i = \sum B_{ji} e_j$. Then,

$$\bar{a} \equiv \sum_{i,j} \bar{A}_{ij} e_i \wedge e_j = \sum_{i,j} A_{ij} Be_i \wedge Be_j = \wedge^*(B)a \ ,$$

and so

$$\bar{a} \wedge \cdots \wedge \bar{a} = \wedge^n(B)(a \wedge \cdots \wedge a) \ .$$

But $\wedge^n(B)$ on the n dimensional space $\wedge^n(\mathbb{C}^n)$ is just multiplication by $\det(B)$, so (12.34) yields the necessary invariance.

(b) $\mathrm{i}A$ is a self-adjoint matrix whose eigenvalues are pure imaginary. If $f \in \mathbb{C}^n$ is an eigenvector for $\mathrm{i}A$, with $(\mathrm{i}A)f = \lambda f$ (λ real), then $(\mathrm{i}A)\bar{f} = \overline{-\mathrm{i}Af} = -\lambda \bar{f}$ so A has eigenvalue $\pm \mathrm{i}\lambda_1, \ldots, \pm \mathrm{i}\lambda_k$ with orthonormal eigenvectors $f_2 = \bar{f}_1$, $f_4 = \bar{f}_3, \ldots$. Letting $g_{2j-1} = (f_{2j} + f_{2j-1})/\sqrt{2}$, $g_{2j} = (\mathrm{i}f_{2j} - \mathrm{i}f_{2j-1})_2$, we obtain *real* orthonormal vectors g_j with

$$Ag_{2j} = \lambda_j g_{2j-1}; \quad Ag_{2j-1} = -\lambda_j g_{2j} \ ;$$

so there exists an orthogonal matrix B so that

$$BAB^{-1} = \begin{pmatrix} 0 & \lambda_1 & & & \\ -\lambda_1 & 0 & & & \\ & & 0 & \lambda_2 & \\ & & -\lambda_2 & 0 & \\ & & & & \ddots \end{pmatrix} = \bar{A} \ .$$

Since B is orthogonal, $\det(B) = \pm 1$ and $(B^{-1})_{ij} = B_{ji}$, so $\bar{A}$ and A are related by (12.33). Thus, $\mathrm{Pff}(A) = \pm \mathrm{Pff}(\bar{A}) = \pm \lambda_1 \lambda_2 \ldots \lambda_2$ (by (12.32b)). Thus, $\mathrm{Pff}(A)^2 = \prod_j \lambda_j^2 = \prod_j [(\mathrm{i}\lambda_j)(-\mathrm{i}\lambda_j)] = \det(A)$. □

Thus, we have the remarkable fact that for antisymmetric matrices, $\det(A)$ as a polynomial in the matrix elements is a perfect square!

Proof of Proposition 12.13. Let $A_{ij;kl}$ be an array which is $n \times n \times n \times n$ antisymmetric in i, j and in k, l. Given arbitrary B, C, let

$$\bar{A}_{ij;kl} = \sum_{a,b,c,d} B_{ia} B_{jb} C_{kc} C_{ld} A_{ab;cd} \tag{12.35}$$

and let

$$\mathrm{DPf}(A) = \sum_{\pi,\sigma} (-1)^\pi (-1)^\sigma A_{\pi(1)\pi(2);\sigma(1)\sigma(2)} \cdots A_{\pi(n-1)\pi(n);\sigma(n-1)\sigma(n)}$$

be the "double Pffafian." We claim that

$$\mathrm{DPf}(\bar{A}) = \det(B)\det(C)\,\mathrm{DPf}(A) \ . \tag{12.36}$$

The proof is essentially the same as that in Proposition 12.16(c). Take two independent copies of $\mathbb{C}^n$ and form $\bigwedge^*(\mathbb{C}^n) \oplus \bigwedge^*(\mathbb{C}^n)$ with $\mathbb{C}^n$ bases e_i and f_i..Let $a = \sum_{ijkl} A_{ij;kl}(e_i \wedge e_j) \oplus (f_i \wedge f_j)$ so

$$\mathrm{DPf}(A)(e_1 \wedge \cdots \wedge e_n) \oplus (f_1 \wedge \cdots \wedge f_n) = a \wedge \cdots \wedge a \quad (k \text{ times}) .$$

Then $\mathrm{DPf}(\bar{A})(e_1 \wedge \cdots \wedge e_n) \oplus (f_1 \wedge \cdots \wedge f_n) = \bar{a} \wedge \cdots \wedge \bar{a} = \mathrm{DPf}(A) \times (Be_1 \wedge \cdots \wedge Be_n) \oplus Cf_1 \wedge \cdots \wedge \mathbb{C}f_n = \det(B)\det(C)\,\mathrm{DPf}(A)(e_1 \wedge \cdots \wedge e_n) \oplus (f_1 \wedge \cdots \wedge f_n)$, proving (12.36).

But if we shift local coordinates from x to $\bar{x}$, then by tensoriality of R

$$\bar{R}^{ij}{}_{kl} = \sum_{a,b,c,d} B^i{}_c B^j{}_b C_k{}^c C_l{}^d R^{ab}{}_{cd} , \quad \text{where}$$

$$B^i{}_a = \frac{\partial \bar{x}^i}{\partial x^a} \text{ and } C_k{}^c = \frac{\partial x^c}{\partial \bar{x}^k} .$$

Thus, by (12.35)

$$\mathrm{DPf}(\bar{R}) = \det(B)\det(C)\,\mathrm{DPf}(R) .$$

But B and C are inverse matrices (up to transposes) and so $\det(B)\det(C) = 1$. Thus, $E(x)$ is a scalar. □

Of course, as a scalar, $E(x)$ must have an invariant definition. Here it is: The Pffafian defined by (12.32a) can be defined for a matrix of elements in an algebra not just for complex matrices [although (12.32b) will not hold any more if the algebra is non-Abelian], and is an invariant of antisymmetric operators on an inner product space. For each X, Y, $R(X, Y)_m$ defines a map of $T_m(M)$ to $T_m(M)$ which is antisymmetric, and it is bilinear and antisymmetric in X, Y. Thus, $R(\cdot, \cdot)_m$ is a two-form of antisymmetric maps from $T_m(M)$ to $T_m(M)$, and so $\mathrm{Pff}[R(\cdot, \cdot)]$ defines an object which, as a sum of product of k 2-forms, is an n-form. In fact, $(-1)^k(4\pi)^{-k}\,\mathrm{Pff}[R(\cdot, \cdot)] = E(x)\,dx$.

The proof we will give of Theorem, 12.14 will depend on the fact that $\exp(-tL_p)$ is an operator with a smooth integral kernel, i.e. there exists a linear map $\exp(-tL_p)(x, y)\colon \bigwedge^p T_y^* \to \bigwedge^p T_x^*$ depending smoothly on x, y so that, if u is p-form,

$$(\mathrm{e}^{-tL_p}u)(x) = \int \mathrm{e}^{-tL_p}(x, y)u(y)\,dy . \tag{12.37}$$

In (12.37), $u(y) \in \bigwedge^p T_y^*$ and $w(x) = \exp(-tL_p)(x, y)u(y)$ denotes the action of the map on $u(y)$, so $w(x) \in \bigwedge^p T_x^*$; dy is the natural measure on M. We will prove this regularity of $\exp(-tL)$ in Sect. 12.6. Since $\exp(-tL_p)(x, y)$ is smooth, its Hilbert Schmidt norm (as a map from $\bigwedge^p T_y^*$ to $\bigwedge^p T_x^*$) is bounded, and so since M is compact, $\exp(-tL_p)$ as a map on the infinite dimensional space $\bigwedge^p T^*$ is Hilbert-Schmidt. By the semigroup property, $\exp(-tL_p)$ is trace class, and by the continuity of the kernel

$$\mathrm{Tr}(\mathrm{e}^{-tL_p}) = \int \mathrm{tr}(\mathrm{e}^{-tL_p}(x, x))\,dx , \tag{12.38}$$

where Tr is the trace on $\bigwedge^p T^*$ and tr on $\bigwedge^p T_x^*$. Let str be the supertrace on $\bigwedge^*(T_x^*)$ and Str the supertrace on $\bigwedge^*(T^*)$, i.e. $\mathrm{Str}(A) = \sum(-1)^p \mathrm{Tr}(A_p)$. Then (12.38) yields

$$\mathrm{Str}(\mathrm{e}^{-tL}) = \int \mathrm{str}[\mathrm{e}^{-tL}(x,x)]\, dx \ .$$

Given any λ which is an eigenvalue of some L_p, let $n_p(\lambda)$ be the multiplicity of λ for L_p on $\bigwedge^p$. In Chap. 11, we proved (11.30) that, for $\lambda \neq 0$,

$$\sum_p (-1)^p n_p(\lambda) = 0 \ . \tag{12.39}$$

Thus, (the fact that $\exp(-tL_p)$ is trace class lets us interchange sums):

$$\begin{aligned} \mathrm{Str}(\mathrm{e}^{-tL}) &= \sum_p (-1)^p \sum_\lambda n_p(\lambda) \mathrm{e}^{-t\lambda} \\ &= \sum_\lambda \mathrm{e}^{-t\lambda} \sum_p (-1)^p n_p(\lambda) \\ &= \sum_p (-1)^p n_p(0) = \sum_p (-1)^p b_p \equiv \chi(M) \ . \end{aligned}$$

We have thus proven:

Theorem 12.17 (McKean and Singer [240]). For any compact orientable manifold,

$$\chi(M) = \mathrm{Str}(\mathrm{e}^{-tL})$$

independently of t.

We can now describe the strategy of the proof of the GBC theorem in a semi-historical context. In 1948, *Minakshisundaram* and *Pleijel* [246] discussed the small t asymptotics of $\mathrm{Tr}[\exp(-tL_0)]$ for the Laplacian on functions on a compact Riemannian manifold (see Corollary 12.59), thereby generalizing Weyl's theorem. In a famous paper on "Can you hear the shape of a drum?" *Kac* [188] asked what aspects of a manifold are determined by the spectrum of L_0, and in particular looked at $\mathrm{Tr}[\exp(-tL_0)]$ for small t in two dimensions. For certain planar regions with boundary, he showed the number of holes could be read off of the small t behavior of the trace. *McKean* and *Singer* [240] then studied $\exp(-tL_p)$, and, in particular, they proved (12.39) and Theorem 12.17 in the context of proving the GBC theorem. They noted that one expects an asymptotic expansion (recall that $n = 2k$):

$$\mathrm{tr}_{\bigwedge^p}[\mathrm{e}^{-tL_p}(x,x)] \sim t^{-k} \sum_{m=0} c_m^{(p)}(x) t^m \ .$$

They remarked that an interesting proof of the GBC theorem would result from Theorem. 12.17 if there were a remarkable cancellation:

$$\sum_{p=0}^{n}(-1)^p c_m^p(x) = 0, \quad \text{for } m = 0, 1, \ldots, k-1 \tag{12.40}$$

and

$$\sum_{p=0}^{n}(-1)^p c_k^{(p)}(x) = E(x) \ . \tag{12.41}$$

They verified this in case $n = 2$ and raised the question in general. It was *Patodi* [273], in a brilliant paper, who verified the conjecture of McKean and Singer and established (12.40) and (12.41). Patodi proved his version of the Berezin-Patodi formula precisely to control the cancellations in (12.40). Indeed, the key to our proving (12.40) and (12.41) will be to show that, as operators from $\bigwedge^* T_x^*$ to itself

$$(e^{-tL})(x,x) = \text{operators involving fewer than } n\ a\text{'s}$$

$$+\ g(x,t)a^*_{\{1,\ldots,n\},\{1,\ldots,n\}} \tag{12.42a}$$

and

$$g(x,t) = E(x) + O(t) \ . \tag{12.42b}$$

Equations (12.41, 42) will then immediately follow from Theorem 12.9. Our proof closely follows the strategy of Patodi, except that we have an alternative machine which we hope makes the proof of (12.42) more transparent than that in [273].

We will prove (12.42) by showing first that one need only consider the case $M = R^n$ [i.e. $\exp(-tL)(x,x)$ is a quasilocal object, so its asymptotics only depend on how M looks near x; thus, we can cut a neighborhood of x out of M and paste it onto R^n. This cutting and pasting is justified in Sect. 12.7]. On R^n, one can write

$$L = B_0 + R_{(4)}(0) + A_{(2)} + A_{(4)} \ , \tag{12.43}$$

where B_0 is the operator which acts like L_0 on the coordinate functions, $R_{(4)}$ involves the curvature at zero and has $2a$'s and $2a^*$'s $A_{(2)}$ (resp. $A_{(4)}$) has two (resp. four) a's or a^*'s, and in suitable coordinates has some vanishing coefficients vanishing at $x = 0$. Because of this vanishing, we will be able to prove that the $g(x,t)$ for $\exp(-tL)$ and that for $\exp\{-t[B_0 + R_{(4)}(0)]\}$ are equal up to terms of order t. But since B_0 acts only on functions and R_4 only on $\bigwedge^* T_x$,

$$\operatorname{str}[\exp\{-t[B_0 + R_{(4)}(0)]\}(0,0)] = \exp(-tB_0)(0,0)\operatorname{str}\{\exp[-tR_{(4)}(0)]\} \ .$$

We will show that $\exp(-tB_0)(0,0) \sim (4\pi t)^{-n}[1 + O(t)]$, while, since $R_{(4)}$ has 4 a's or a^*'s, we cannot get $2n$ a's and a^*'s until we get $R_{(4)}(0)^{n/2}$, i.e.

$$\operatorname{str}\{\exp[-tR_{(4)}(0)]\} = \frac{(-t)^k}{k!}\operatorname{str}(R_{(4)}^k) + O(t^{k+1}) \ .$$

This will precisely yield (12.42). Notice that by exploiting the Berezin-Patodi formula, one need not check the cancellations (12.40) by hand, and that we are always looking only at leading non-zero asymptotics:

The invariant formula which will yield (12.43) is proven in Sect. 12.4, and a calculus for computing asymptotics of heat kernels can be found in Sect. 12.5, 6. The proof of the GBC theorem putting all the previous material together appears in Sect. 12.8.

We should close with a few remarks on the "competing" heat kernel proof of these theorems due to *Gilkey* [132–134] and *Atiyah*, *Bott* and *Patodi* [18] (see the further remarks in Sect. 12.10). It is noted [although *Gilkey* [133] does prove (12.40)] that (12.40) is irrelevant (!), for since $\mathrm{Str}[\exp(-tL)]$ is t independent, one is guaranteed that $\int dx[\sum_{p=0}^{n}(-1)^p c_m^p(x)] = 0$ for $m \le k-1$ and thus (12.41) suffices. Equation (12.41) is not proven by a direct calculation, but rather in two steps: (1) One shows that the left side of (12.41) must be a multiple of $E(x)$ by showing that (i) the left side has certain invariances under coordinate changes, and is only a function of g_{ij} and its first two derivatives, and (ii) that there is only one such invariant. (2) Since the left side of (12.41) is a local invariant, the constant must be universal, and so it can be computed in any convenient case. We will give the flavor of this proof at the end of Sect. 12.6. In fact, this proof is not much to our taste: The necessary machinery to prove that only second derivatives of g^{ij} enter, properly organized, is essentially enough to directly compute (12.41) without the appeal to a lengthy and indirect invariant theory argument.

12.4 Bochner Laplacian and the Weitzenböck Formula

The Laplace-Beltrami operator on p-forms, $L = (d + d^*)^2$, has the nice features of connecting to Betti numbers via Hodge theory and a doubling of non-zero eigenvalues via sypersymmetry. There is a second natural Laplacian, often called the Bochner or flat Laplacian, which can be defined not only on $\bigwedge^p$, but on any Hermitian bundle over a Riemannian manifold. In this section, we want to describe this Laplacian and prove the beautiful formula of Weitzenböck relating the two Laplacians. We will describe here two applications of the interplay between the Laplacians; the GBC theorem will be a third.

Definition 12.18. The *Bochner Laplacian B* is the map of smooth p-forms $\bigwedge^p$ to itself which, in local coordinates about the point $m \in M$, is given by

$$(Bu)(m) = \sum_{i,j} (\nabla_i^* g^{ij} \nabla_j u)(m) \ . \tag{12.44}$$

If x^i and $\bar{x}^i$ are two sets of local coordinates at m, and $X_i = \partial/\partial x^i$, then $\bar{X}_i = \sum_j A_i^{\ j} X_j$, where $A_i^{\ j} = \partial x^j/\partial \bar{x}^i$. Since $\nabla_{\bullet}$ is tensorial, we have that

$$\bar{\nabla}_i = \sum_j A_i^{\ j} \nabla_j \ .$$

and taking adjoints

$$\bar{V}_i^* = \sum V_j^* A_i^j \ .$$

(Note that A_i^j is not constant, so V and A do not commute.) Since $\sum_{i,j} A_i^k \bar{g}^{ij} A_j^l = g^{kl}$, the Bochner Laplacian is coordinate independent. For those who are fond of coordinate-free definitions of coordinate-free objects, we note the following coordinate-free definition of B: $V_{\bullet}u$ defines a $(p+1)$ rank covariant tensor via $(V_{\bullet}u)(X, Y_1, \ldots, Y_p) = (V_X u)(Y_1, \ldots, Y_p)$. This tensor is antisymmetric in $Y_1, \ldots, Y_p$ but not in X. The metric induces a natural inner product, $\langle \ , \ \rangle_m$, on $\bigotimes^{p+1} T_m^*$ and B obeys

$$\langle u, Bu \rangle = \int \langle V_{\bullet}u, V_{\bullet}u \rangle_m \, dx \ . \tag{12.45}$$

This formula defines B as a quadratic form on $\bigwedge^p$. At the end of this section, we will describe the more usual definition of B in terms of second covariant derivatives. The point of (12.44, 45) is that they make it evident that B is self-adjoint and indeed positive.

The relation between B and L will be a rather simple exercise in the fermion creation/annihilation operator calculus introduced by *Witten* [370], and which we have used in Chap. 11 and Sects. 12.1, 3:

Theorem 12.19 (The Weitzenböck Formula). Let $R_{(4)}$ be the operator

$$R_{(4)} = \sum R_{ijkl}(a^i)^* a^j (a^k)^* a^l \ . \tag{12.46}$$

Then

$$L = B + R_{(4)} \ . \tag{12.47}$$

Proof. We will exploit (12.14, 15) for d and (12.28) for $[V_i, V_j]$. Since $d = \sum_i (a^i)^* V_i$ [by (12.14)] and $\{AB, C\} = \{A, C\}B + A[B, C]$,

$$L = \{d, d^*\} = \sum_i \{(a^i)^*, d^*\} V_i + (a^i)^*[V_i, d^*] \ . \tag{12.48}$$

By using the adjoint of (12.14), $d^* = \sum_j V_j^* a^j$ and $\{A, BC\} = B\{A, C\} + [A, B]C$ we have that

$$\begin{aligned} \{(a^i)^*, d^*\} &= \sum_j V_j^* \{(a^i)^*, a^j\} + \sum_j [(a^i)^*, V_j^*] a^j \\ &= \sum_j V_j^* g^{ij} - \sum_{j,k} \Gamma_{jk}^i (a^k)^* a^j \end{aligned} \tag{12.49}$$

by (12.17). By using (12.15b), $d^* = -\sum a^j V_j$ and $[A, BC] = [A, B]C + B[A, C]$, we have that

$$\begin{aligned} [V_i, d^*] &= -\sum_j a^j [V_i, V_j] - \sum_j [V_i, a^j] V_j \\ &= \sum_{j,k,l} R_{ijkl} a^j (a^k)^* a^l + \sum \Gamma_{ik}^j a^k V_j \end{aligned} \tag{12.50}$$

by (12.28) and the adjoint of 1 (12.17) [using (12.16)]. Placing (12.49, 50) into (12.48), we obtain (12.46, 47). □

In the above, the Γ terms exactly cancel. By using normal coordinates, we could have avoided the explicit check that the unwanted terms cancel.

The fermion calculus is not only useful in the proof of the Weitzenböck formula, but because it povides a compact formula for $R_{(4)}$ (first noted by *Nelson* [263] and exploited by *Patodi* [273] in his proof of GBC). Using $a^j(a^k)^* = g^{jk} - (a^k)^* a^j$, one has

$$R_{(4)} = R_{(2)} + \tilde{R}_{(4)} \ , \tag{12.51a}$$

$$R_{(2)} = \sum S_{ij}(a^i)^* a^j \ , \tag{12.51b}$$

$$\tilde{R}_{(4)} = -\sum R_{ijkl}(a^i)^*(a^k)^* a^j a^l \ , \tag{12.51c}$$

where S_{ij} is the Ricci tensor ($S_{ij} = \sum_{l,k} R_{ilkj} g^{lk}$). The point is that (12.47) is usually written $L = B + R_{(2)} + \tilde{R}_{(4)}$ because R_4 does not have a compact form in terms of coordinate operations, while $R_{(2)}$ and $\tilde{R}_{(4)}$ do. As we will see, the splitting into $R_{(2)}$ and $\tilde{R}_{(4)}$ which have opposite signs if R has positivity properties can lead to problems. The splitting (12.51) is useful if we look at 1-forms since $\tilde{R}_4 \equiv 0$ on $\bigwedge^1$; thus, on 1-forms

$$L_1 = B_1 + R_{(2)} \ . \tag{12.52}$$

If we expand B using (12.10c), we obtain a formula for L acting in local coordinates which has terms with no a, terms with one a and a^* and terms with two a's and a^*'s. If we write the a, a^* terms with an a on the left, the term with no a's must be just $B_0 = L_0$ acting on the coordinate functions. By writing out the second and fourth order terms explicitly and noting Γ's and ∂g^{ij}'s occur, we obtain

Theorem 12.20. In terms of local coordinates, L acts on p-forms in a coordinate system center at m_0 by (12.43) where B_0 is Bochner Laplacian acting on coordinate components of the p-form, and where, in a coordinate system with $x^i(m_0) = 0$, $g^{ij}(0) = \delta_{ij}$, $\Gamma^i_{jk}(0) = 0$, $[\partial_k g_{ij}](0) = 0$:

$$R_{(4)}(0) = \sum_{ijkl} R_{ijkl}(m_0)(a^i)^* a^j (a^k)^* a^l \ ,$$

$$A_{(2)} = \sum_{ijk} N^k_{ij}(x) \partial_k (a^i)^* a^j + \sum_{i,j} P_{ij}(x)(a^i)^* a^j \ ,$$

$$A_{(4)} = \sum_{i,j,k,l} Q_{ijkl}(*)(a^i)^* a^j (a^k)^* a^l \ ,$$

with $N^k_{ij}(0) = 0$, $Q_{ijkl}(0) = 0$.

The remainder of this section will not be needed in our proof of GBC and may be skipped: We will first give some simple applications of the Weitzenböck formula and then discuss another, more usual, way of defining B. Finally, we will briefly describe the probabilistic view of B. Here, as a typical application of (12.52), is a result of *Lichnerowicz* [230]:

Theorem 12.21. Let M be a compact Riemannian manifold and let $0 = \lambda_0 < \lambda_1 < \cdots$ be the eigenvalues of $L_0 = B_0$ (on functions). Let $s(m)$ be the minimum eigenvalue of the Ricci tensor $S_{ij}(m)$ thought of as an operator on $T_m(M)$. Suppose

$$\inf_{m \in M} s(m) > 0 \ .$$

Then

$$\lambda_1 - \lambda_0 \geq \inf_{m \in M} s(m) \ .$$

Proof. By supersymmetry, λ_1 is also an eigenvalue of L_1 (in fact, if $L_0 f = \lambda_1 f$, $df \neq 0$ and $L_1(df) = \lambda_1 df$). Thus, if $e_1 \equiv \inf \operatorname{spec}(L_1)$, we see that

$$\lambda_1 \equiv \lambda_1 - \lambda_0 \geq e_1 \ . \tag{12.53}$$

But, if $u \in T_m^*(M)$, then

$$\begin{aligned}(u, R_{(2)}(m)u) &= \sum S_{ij}(m)(a^i u, a^j u) \\ &\geq s(m) \sum_i (a^i u, a^i u) = s(m)(u, u) \ ,\end{aligned}$$

since $\sum a_i^* a_i = 1$ on 1-forms. Thus, as an operator, $R_{(2)} \geq \inf_m s(m)$. Since $B_1 \geq 0$, (12.52) implies that

$$e_1 \geq \inf_{m \in M} s(m) \ , \tag{12.54}$$

so that (12.53) yields the theorem. □

Remark. It is a theorem of *Lichnerowicz* [230] that

$$\lambda_1 - \lambda_0 \geq c_n \inf_{m \in M} s(m) \ ,$$

where $c_n = n(n-1)^{-1}$ and n is the dimension of M. This result is not merely better than Theorem 12.21, the constant c_n is optimal in that equality holds for the n-sphere S^n with the constant curvature metric. In fact, *Obata* [265] has proven that equality in this last inequality only holds for this constant curvature case. See *Chavel* [62] for further discussion.

Equation (12.54) immediately implies that if

$$\inf_{m \in M} s(m) > 0 \ ,$$

then $e_1 > 0$, so L_1 has no zero eigenvalues. We thus have recovered a celebrated theorem of Bochner and Meyer:

Theorem 12.22 (Bochner [50], Meyer [242]). Let M be a compact Riemannian manifold, and let $s(m)$ be the minimum eigenvalue of the Ricci tensor $S_{ij}(m)$. If

$$\inf_{m \in M} s(m) > 0$$

("the manifold has strictly positive Ricci curvature"), then b_1, the first number, is zero.

Remark. (1) By (12.45), if $\langle u, B_1 u \rangle = 0$, $\nabla_X u = 0$ for all X so that one can show that $|u(m)|$ is constant on M. Thus, one can conclude that $b_1 = 0$ if

$$\inf_{m \in M} s(m) \geq 0$$

with $s(m) > 0$ at one point. If $s(m) \equiv 0$, one can conclude $b_1 \leq \dim(M)$, for $\nabla_X u = 0$ means that u is determined by its values at one point via parallel transport, so $\{u \,|\, Bu = 0\}$ has dimension at most $\dim(M)$. For an m-dimensional torus with its flat metric, one actually has $b_1 = m$.

(2) The above proof is a descendant of Bochner's proof. Meyer's proof is very different.

By using the full Weitzenböck formula, one can prove

Theorem 12.23 (*Meyer* [241]). Let M be a compact Riemannian manifold of dimension n. If the induced map $\mathrm{R}^{(2)}$ of (12.11) is positive definite at each $m \in M$, and strictly positive definite at one point m_0, then $b_1 = b_2 = \cdots = b_{n-1} = 0$.

Proof (related to *Gallot* and *Meyer* [128]). Pick normal coordinates at a general point m_1. In terms of the fermion calculus, let

$$N^{ij} = \tfrac{1}{2}[(a^i)^* a^j - (a^j)^* a^i] \ ;$$

so, by the antisymmetry of R_{ijkl} under $i \leftrightarrow j$ or $k \leftrightarrow l$,

$$R_{(4)} = \sum_{i,j,k,l} R_{ijkl}(m_1) N^{ij} N^{kl} \ .$$

It is easy to prove that if α is a $k \times k$ real matrix with $\sum \alpha_{ij} w_i w_j \geq b \sum |w_i|^2$ for all real $(w_1, \ldots, w_k)$, then $\sum_{i,j} \alpha_{ij} A_i^* A_j \geq b \sum_{i,j} A_i^* A_i$ for all k-tuples of operators, since N^{ij} is antisymmetric in i, j, it follows that if

$$\langle u, R^{(2)}(m_1) u \rangle \geq b(m_1) \langle u, n \rangle \ , \quad \text{then}$$

$$R_{(4)}(m_1) \geq b(m_1) \sum_{i,j} (N^{ij})^* (N^{ij}) \ .$$

But $(N^{ij})^* = -N^{ij}$ and $-(N^{ij})^2 = \tfrac{1}{4}[N_i(1 - N_j) + N_j(1 - N_i)]$, where $N_i =$

$(a^i)^* a^i$ obeys $N_i^2 = N_i$. Thus, since $N_i(1 - N_i) = 0$,

$$\sum_{i,j} N_{ij}^* N_{ij} = \tfrac{1}{2} \sum_{i,j} N_i(1 - N_j)$$

$$= \tfrac{1}{2}\left(\sum_i N_i\right)\left(n - \sum_i N_i\right) .$$

On p-forms, $\sum N_i = p$, so

$$R_{(4)}(m_1) \geq \tfrac{1}{2} b(m_1)(n - p)p \; .$$

Thus, $R_{(4)} \geq 0$ if $b \geq 0$, and so $Lu = 0$ implies $Bu = 0$ and $R_{(4)}u = 0$. But $Bu = 0$ implies $\nabla_i u = 0$, which implies that $|u|$ is constant. Suppose $p \neq 0, n$. Since $b(m_0) > 0$, it must be that $u(m_0) = 0$ so $u \equiv 0$, i.e. $b_p = 0$. $\square$

The difficulty in writing (12.51) is shown by the fact that, following Bochner's proof of Theorem 12.22 in 1946, the studies of the vanishing of the higher Betti numbers by *Bochner* and *Yano* [51] required both upper and lower bounds on $R^{(2)}$ with the upper and lower bounds allowed to differ only by a factor of 2. This was later improved by various authors, but the result without upper bounds was only proven in 1971.

Next, we want to describe another way of writing B. When we write $\nabla_X \nabla_Y$ we *always* mean the product of the operators ∇_X and ∇_Y on $\bigwedge^p$.

Definition 12.24. Given two vector fields X, Y, the *second covariant derivative* $\nabla^2_{(X,Y)}$ is defined by

$$\nabla^2_{(X,Y)} = \nabla_X \nabla_Y - \nabla_Z; \quad Z = \nabla_X Y \; .$$

The point of the ∇_Z term is that $\nabla^2_{(X,Y)}$ is tensorial in Y, i.e. $\nabla^2_{(fX,gY)} = fg\nabla^2_{(X,Y)}$ so that $\nabla^2_{(X,Y)}\, u$ at m only depends on $X(m)$ and $Y(m)$. If one thinks of $\nabla_{\cdot} u$ as an element of $\bigotimes^{p+1} T^*(M)$, then $\nabla^2_{(X,Y)}\, u$ is just $\nabla_X(\nabla_{\cdot} u)$ valued at $\cdot = Y$. If $\{X_i\}_{i=1}^n$ are coordinate vector fields, we denote $\nabla^2_{(X_i, X_j)}$ by $\nabla^2_{(i,j)}$. We warn the reader that many books denote $\nabla^2_{(i,j)}$ by $\nabla_i \nabla_j$, while with *our* meaning of $\nabla_i \nabla_j$ as the product of ∇_i and ∇_j we have

$$\nabla^2_{(i,j)} = \nabla_i \nabla_j - \sum_k \Gamma_{ij}^k \nabla_k \; .$$

One should note that $\nabla^2_{(X,Y)} - \nabla^2_{(Y,X)} = R(X, Y)$ for general X, Y, on account of the torsion free nature of ∇. The usual definition of the Bochner Laplacian is given by the following:

Theorem 12.25:

$$B = -\sum_{i,j} g^{ij} \nabla^2_{(i,j)} \; . \tag{12.55}$$

Proof. Both sides of (12.55) are coordinate independent, so we need only check it in normal coordinates. In such coordinates centered at m_0, $(\nabla_i u)(m_0) = (\partial_i u)(m_0)$, $(\nabla_i^* u)(m_0) = -(\partial_i u)(m_0)$, $(\nabla_i X_j)(m_0) = 0$ and $\partial_k(g^{ij}u)(m_0) = g^{ij}(\partial_k u)(m_0)$. Thus,

$$(Bu)(m_0) = -\sum_i [\partial_i(\nabla_i f)](m_0) = -\sum_{i,j} (g^{ij}\nabla^2_{(i,j)} f)(m_0) \ .$$

Notice that $(\partial_i \nabla_j f)(m_0) \neq (\partial_i \partial_j f)(m_0)$, since $(\partial_i \Gamma^j_{kl})(m_0)$ may not be zero. □

The reader can go back to Sect. 11.4 and check (by calculating in normal coordinates) that the operator A in the expansion of Witten's deformed Laplacian has the following invariant expression due to Witten:

$$A = \sum_{i,j} (\nabla^2_{(i,j)} f)[a_i^*, a_j] \ .$$

Finally, we want to mention that while L is the "natural" Laplacian for topologists, in many ways B is the "natural" Laplacian for probabilists (and to some extent analysts). Let M be a compact Riemann manifold. $B_0 = L_0$ acting on functions generates a positivity preserving semigroup (this follows from the Beurling-Deny criteria—see [295], pp. 209), $\exp(-tB_0)$ taking 1 to itself. As we shall see in the next two sections, it has a smooth integral kernel, $\exp(-tB_0)(x, y)$, so in the usual way, one can define Brownian motion on M: i.e. a family of probability measures, $\{E_x\}_{x \in M}$, on continuous paths starting at x so that (i) $E_x(f(b(s))) = [\exp(-sB_0)f](x)$ and (ii) $\{E_x\}$ is a Markov process in the usual sense (see [53, 108, 109] for background in probability, and [93, 238] for discussions of Brownian motion on manifolds).

While we have not discussed it, given any *smooth* path γ on M, there is a natural map $P^t_\gamma\colon \bigwedge^p T^*_{\gamma(0)} \to \bigwedge^p T^*_{\gamma(t)}$ called parallel translation, discussed in most differential geometry books (e.g. [52, 348, 349]). By using the theory of stochastic differential equations, one can define P^t_b also for almost all Brownian paths $b(s)$, started at each x. One can show that

$$(e^{-tB}u)(x) = E_x((P^t_b)^{-1} u(b(t))) \tag{12.56}$$

for any u in $\bigwedge^p$; note that $(P^t_b)^{-1} u(b(t))$ lies in $\bigwedge^p T_x$, so the expectation is over a vector-valued function. Equation (12.56) is discussed in [93, 238] and demonstrates that B is the natural Laplacian for Brownian motion. The corresponding formula for $\exp(-tH)$ can be obtained from the Weitzenböck formula; it will involve a "time ordered product" of $P^{\delta t}_b$'s and $\exp[-(\delta t)R_4]$ terms.

As the name "parallel translation" suggests, $|P^t_\gamma \eta| = |\eta|$ (the norms are with respect to the inner products at $\gamma(t)$ and $\gamma(0)$ respectively) and similarly $(P^t_\gamma)^{-1}$ is norm preserving. Thus, (12.56) implies that

$$|(e^{-tB}u)(x)| \leq E_x(|(P^t_b)^{-1} u(b(t))|) \leq E_x(|u(b(t))|) = (e^{-tB_0}|u|)(x) \ .$$

That is, if $u \in \bigwedge^p$

$$|(e^{-tB}u)(x)| \le (e^{-tB_0}|u|)(x) \tag{12.57}$$

(12.57) has a "direct" proof exploiting Kato's inequality [163, 322, 327] rather than path integrals; indeed, it seems to have first been noted explicitly by *Hess, Schrader* and *Uhlenbrock* [164] in this context. It implies that $\exp(-tB)$ is a contraction on each L^q space, while $\exp(-tL)$ may not be if curvatures are not positive; in this sense $\exp(-tB)$ is also more natural analytically.

12.5 Elliptic Regularity

Thus far in this chapter, we have had lots of fun with geometry and a little algebra, but there has not been any analysis ($\equiv$ estimates). We will remedy this lamentable defect in this and the next section. In this section, we will prove estimates which, roughly speaking, say that forms in $D(L^k)$ have $2k$ derivatives in L^2 and $2k-(n/2)-\varepsilon$ classical derivatives. The estimates will also allow us to prove essential self-adjointness of L on $\bigwedge^p$, the C^∞ forms, and compactness of $\exp(-tL)$. These results were used in Sect. 11.3. They will also provide the input for our machinery in the next section dealing with the asymptotics of integral kernels. We note that the basic results below are usually proven (see e.g. *Taylor* [353] or *Gilkey* [133]) using the pseudodifferential operator ($\psi D0$) calculus. We will use our "bare hands" instead—while this is perhaps a trifle more elementary, we strongly recommend the reader learn the powerful ($\psi D0$) machinery. We first consider operators on R^n and will later localize to get information on compact manifolds M. To try to keep the exposition self-contained, we will begin with a brief review of the theory of Sobolev spaces on R^n.

Definition 12.26. Let $s \ge 0$. The *Sobolev space* $\mathrm{H}_s(R^\nu)$ is the set of function $f \in L^2(R^\nu)$ whose Fourier transform $\hat{f}$ obeys

$$\|f\|_s^2 \equiv \int |\hat{f}(k)|^2(1+k^2)^s d^\nu k < \infty \ .$$

For integral s (which is all we wish to consider), it is not hard to show that $f \in \mathrm{H}_s$ if and only if, for all multi-indices α with $|\alpha| \le s$, the distributional derivative $D^\alpha f(\equiv \partial^{|\alpha|}f/\partial^{\alpha_1}x_1 \ldots \partial^{\alpha_\nu}x_\nu)$ lie in L^2, and that $\sum_{|\alpha|\le s}\|D^\alpha f\|_{L^2}$ and $\|f\|_s$ are equivalent norms. In particular, this implies:

Proposition 12.27. If $f, \partial_j f \in \mathrm{H}_s$, then $f \in \mathrm{H}_{s+1}$ and $\|f\|_{s+1}^2 \le \|f\|_s^2 + \|\partial_j f\|_s^2$.

The equivalence of norms and Leibniz' rule $[D^\alpha(fg) = \sum_{\beta\le\alpha}\binom{\alpha}{\beta}(D^\beta f) \times (D^{\alpha-\beta}g)$ for f C^∞ and g a distribution$]$ immediately implies:

Proposition 12.28. If $f \in C^\infty$ with $\sup_x \|D^\alpha f\|_\infty = e_\alpha < \infty$ for all α, then the map $g \mapsto fg$ takes H_s to H_s for all s with a norm bounded by c_s (depending on $\sum_{|\alpha|\le s} e_\alpha$) for a suitable s-dependent constant c_s.

The key that makes Sobolev spaces so useful for studying classical smoothness is:

Theorem 12.29. If $f \in \mathrm{H}_s(R^\nu)$ and $s > (\nu/2) + l$, then f is C^l (in classical sense) with derivatives $D^\alpha f$, $|\alpha| \leq l$ uniformly bounded.

Proof. We begin with the case $l = 0$. Write $s = t + r$ with $t > \nu/2$ and $0 < r < 1$. Since

$$|e^{ik\cdot x} - e^{ik\cdot y}| \leq C_r |k|^r |x - y|^r \ , \tag{12.58}$$

we have that $|f(x) - f(y)| \leq C_r |x - y|^r (2\pi)^{-\nu/2} \int |k|^r |\hat{f}(k)| \, d^\nu k$ and continuity of f results if we note that by the Schwarz inequality

$$\left[\int |k|^r |\hat{f}(k)| \, d^\nu k\right]^2 \leq \int (1 + |k|^2)^r (1 + |k|^2)^t |f(k)|^2 \, d^\nu k$$

$$\times \int (1 + |k|^2)^{-t} \, d^\nu k$$

and that $\int (1 + |k|^2)^{-t} \, d^\nu k < \infty$ since $t > \nu/2$. The general case results by replacing (12.58) with $(0 < r < 1)$

$$\left| e^{ik\cdot(x-y)} - \sum_{|\alpha| \leq l} e^{ikx} (ik)^\alpha y^\alpha (\alpha!)^{-1} \right| \leq C_{l,r} |k|^{l+r} |y|^r \ ,$$

which follows from the Taylor expansion of $\exp(ik \cdot y)$. $\square$

This result (or rather its proof) can be restated by saying that if $s > \nu/2$, the linear functional $\delta_x(f) = f(x)$ is a bounded linear functional on H_s. Let $\mathrm{H}_{-s} \equiv (\mathrm{H}_s)^*$ so $\delta_x \in \mathrm{H}_{-s}$. The stronger smoothness result can be restated by saying that the H_{-s} valued function, $x \mapsto \delta_x$, is l times differentiable if $s > l + (\nu/2)$. We also define

$$\mathrm{H}_\infty = \bigcap_s \mathrm{H}_s, \quad \mathrm{H}_{-\infty} = \bigcup_s \mathrm{H}_s; \quad -\infty < s < \infty \ .$$

It is useful to extend the inclusions $\mathrm{H}_s \subset \mathrm{H}_t$ for $s \geq t > 0$ by thinking of $L \in \mathrm{H}_t^*$ as lying in H_s^* by restriction. Thus, if $f \in \mathrm{H}_s$ and $g \in \mathrm{H}_t$ with $s + t \geq 0$, we can define $\langle f, g \rangle$ in a way that extends the inner product when $f, g \in L^2$ (H_s has a natural Hilbert space structure; if f, $g \in \mathrm{H}_s$ with $s > 0$, one should not confuse $\langle f, g \rangle$ with this natural Hilbert space structure, for the H_s inner product is $\langle f, (1 - \Delta)^s g \rangle_{L^2}$ not $\langle f, g \rangle_{L^2}$). Sobolev spaces are useful for discussing integral kernels.

Definition 12.30. We say that an operator $A: L^2(R^\nu) \to L^2(R^\nu)$ has an integral kernel $A(x, y)$ if $A(x, y)$ is a jointly measurable function with

$$\int |A(x, y)|^2 \, dy < \infty \tag{12.59}$$

for a.e. x and with

$$(Af)(x) = \int A(x, y) f(y)\, dy \tag{12.60}$$

for all $f \in L^2$ and a.e. x in R^ν.

Theorem 12.31. If A: $L^2 \to L^2$ extends to a map of H_{-s} to H_s for some $s > \nu/2$, then A has an integral kernel which is jointly continuous in L^2. Indeed $A(x, y) = (\delta_x, A\delta_y)$. If it is a map of H_{-s} to H_s for $s > (\nu/2) + l$, $A(x, y)$ is C^l, and if it maps $\mathrm{H}_{-\infty}$ to H_∞, $A(x, y)$ is C^∞.

Proof. Since $\delta_y \in \mathrm{H}_{-s}$, $A\delta_y$ lies in H_s and thus, by Theorem 12.29, $A\delta_y$ is a continuous function $A(x, y)$. Moreover, by duality A^* also maps H_{-s} to H_s and $A^*(x, y) = \overline{A(y, x)}$ so $\int |A(x, y)|^2\, dy = \int |A^*(y, x)|^2\, dy = \|A^*\delta_x\|_L{}^2$ and (12.59) holds. If suffices to check (12.60) for $f \in C_0^\infty(R^\nu)$. For such an f, $f = \int f(x)\delta_x$ as a strong integral in H_{-s} so (12.60) is immediate. The additional smoothness follows from the smoothness of $x \mapsto \delta_x$ as map into H_{-s} if $s > l + (\nu/2)$. □

Now we want to consider a $\nu \times \nu$ real symmetric matrix valued function $a^{ij}(x)$ on R^ν obeying

(i) $a^{ij}(x)$ is C^∞ in x with $\sup_x |D^\alpha(a^{ij})(x)| < \infty$

for all α, i, j.

(ii) $\sum_{i,j=1}^{\nu} u_i u_j a^{ij}(x) \geq a_0 \sum_{i,j=1}^{\nu} |u_i|^2$,

all $u_i \in R^\nu$ and $a_0 > 0$ independent of x. The map

$$H = -\sum \partial_i a^{ij} \partial_j$$

defines a map of $C_0^\infty(R^\nu)$ to itself. We call H *uniformly elliptic strictly second order (uesso)* if a obeys (i), (ii). The basic elliptic estimates we need are the following:

Theorem 12.32. Let H be a uesso operator on R^ν. Then for each $s \geq 0$, there exist constants c_s and d_s so that for all $\varphi \in C_0^\infty(R^\nu)$

$$(\varphi, (H + 1)^s \varphi) \leq c_s \|\varphi\|_s^2 , \tag{12.61}$$

$$\|\varphi\|_s^2 \leq d_s(\varphi, (H + 1)^s \varphi) . \tag{12.62}$$

Moreover, c_s, d_s depend only on $\|D^\alpha a^{ij}\|_\infty$ for $|\alpha| \leq 2s$, and on the number a_0 in condition (ii).

Proof. By integrating by parts and using condition (i), it is easy to see that

$$(\varphi, (H + 1)^s \varphi) \leq c_{1,s} \sum_{|\alpha| \leq s} \|D^\alpha \varphi\|_{L^2}$$

from which (12.61) follows. To prove (12.62), let $H_0 = -\Delta$ and take $s = 1$. Then, by condition (ii), for any φ and x

$$\sum_{i,j=1}^{\nu} a_{ij}(x)(\partial_i\varphi)(x)(\partial_j\varphi)(x) \geq a_0 \sum_{i=1}^{\nu} (\partial_i\varphi)(x)^2 .$$

Integrating over x, we obtain $(\varphi, H\varphi) \geq a_0(\varphi, H_0\varphi)$, so (12.62) holds for $s = 1$ with $d_1 = a_0^{-1}$.

Suppose we have (12.62) for $s - 1$. Let us prove it for s. By Proposition 12.27 and the induction hypothesis

$$\begin{aligned}\|\varphi\|_s^2 &= \|\varphi\|_{s-1}^2 + \sum_{j=1}^{\nu} \|\partial_j\varphi\|_{s-1}^2 \\ &\leq d_{s-1}\left[(\varphi, (H+1)^{s-1}\varphi) + \sum_j (\partial_j\varphi, (H+1)^{s-1}\partial_j\varphi)\right] . \end{aligned} \tag{12.63}$$

Now $(\varphi, H^s\varphi) \geq 0$ for all φ, s [for s even $(\varphi, H^s\varphi) = \|H^{s/2}\varphi\|^2$ and for s odd $(\varphi, H^s\varphi) = (H^{(s-1)/2}\varphi,\ HH^{(s-1)/2}\varphi)$ and H is positive by condition (ii)]. Thus, $(\varphi, (H+1)^s\varphi) \geq (\varphi, (H+1)^{s-1}\varphi)$ so the first term in (12.63) can be bounded by $d_{s-1}(\varphi, (H+1)^s\varphi)$. Moreover, we have the Schwartz inequality

$$|(\psi, (H+1)^t\eta)| \leq (\varphi, (H+1)^t\varphi)^{1/2}(\eta, (H+1)^t\eta)^{1/2} \tag{12.64}$$

for $t = 0, 1, 2, \ldots$ and all $\varphi, \eta \in C_0^\infty$. We can write

$$(\partial_j\varphi, (H+1)^{s-1}\partial_j\varphi) = a_j + b_j; \quad a_j = (\varphi, (H+1)^{s-1}\partial_j^2\varphi) ;$$
$$b_j = (\varphi, [\partial_j, (H+1)^{s-1}]\partial_j\varphi) .$$

By (12.64) for $t = s - 2$, since $s \geq 2$

$$\begin{aligned}|a_j| &\leq ((H+1)\varphi, (H+1)^{s-2}(H+1)\varphi)^{1/2}(\partial_j^2\varphi, (H+1)^{s-2}\partial_j^2\varphi)^{1/2} \\ &\leq \tfrac{1}{2}\varepsilon(\partial_j^2\varphi, (H+1)^{s-2}\partial_j^2\varphi) + \tfrac{1}{2}\varepsilon^{-1}(\varphi, (H+1)^s\varphi) \\ &\leq \tfrac{1}{2}c_{s-2}\varepsilon\|\partial_j^2\varphi\|_{s-2}^2 + \tfrac{1}{2}\varepsilon^{-1}(\varphi, (H+1)^2\varphi) \\ &\leq \delta\|\varphi\|_s^2 + C_\delta(\varphi, (H+1)^s\varphi) ,\end{aligned}$$

where we used $xy \leq \frac{1}{2}\varepsilon x^2 + \frac{1}{2}\varepsilon^{-1}y^2$ in Step 1 and (12.61) in Step 2. In the final inequality δ may be taken arbitrarily small.

Now b_j is an expectation of terms involving $2s - 1$ derivatives, so integrating by parts

$$\begin{aligned}|b_j| &\leq c\left(\sum_{|\alpha|\leq s-1} \|D^\alpha\varphi\|_{L^2}\right)^{1/2}\left(\sum_{|\alpha|\leq s} \|D^\alpha\varphi\|_{L^2}\right)^{1/2} \\ &\leq c_1\varepsilon\|\varphi\|_s^2 + c_1\varepsilon^{-1}\|\varphi\|_{s-1}^2 \\ &\leq c_1\varepsilon\|\varphi\|_s^2 + c_1\varepsilon^{-1}d_{s-1}(\varphi, (H+1)^{s-1}\varphi) \\ &\leq \delta\|\varphi\|_s^2 + D_\delta(\varphi, (H+1)^s\varphi) .\end{aligned}$$

Putting together the above estimates on the terms in (12.63), we obtain

$$\|\varphi\|_s^2 \leq 2\delta\nu\|\varphi\|_s^2 + (d_{s-1} + \nu D_\delta + \nu C_\delta)(\varphi, (H+1)^s\varphi) .$$

Choosing $\delta = 1/4\nu$, we obtain (12.62) for s. The proof shows that the constants only have the stated dependence. □

Since $(1 + H)$ is a map of C_0^∞ to itself, $1 + H$ and powers of $(1 + H)$ map distributions to themselves. It is quite natural to expect that if $(1 + H)^l T \in L^2$ for some l, then T lies in H_{2l}. This is certainly true of $H = H_0$ and we want to prove this in general since this will allow us to localize on M and deduce smoothness of eigenfunctions of $\bar{L}$ (note the closure) and essential self-adjointness of L on $C_0^\infty(M)$. The key is that powers of H are essentially self-adjoint on $C_0^\infty(R^\nu)$.

Theorem 12.33. Let H be a uesso operator. Then, for each l, H^l is essentially self-adjoint on $C_0^\infty(R^\nu)$ and $D(\overline{H^l}) = \mathrm{H}_{2l}$. Moreover, $Q(\overline{H^l}) = \mathrm{H}_l$.

Proof. Let $H(\theta) = (1-\theta)H_0 + \theta H = -\partial_i[\theta a^{ij} + (1-\theta)\delta^{ij}]\partial_j$. Thus, the operators $H(\theta)$ are uesso with constants, including a_0, uniformly bounded in θ, so

$$(\varphi, (H(\theta)+1)^s\varphi) \leq c_s\|\varphi\|_s^2 , \tag{12.65a}$$

$$\|\varphi\|_s^2 \leq d_s(\varphi, (H(\theta)+1)^s\varphi) \tag{12.65b}$$

for all $\varphi \in C_0^\infty$ and $0 \leq \theta \leq 1$. Moreover, it is evident that

$$\|[(H(\theta)+1)^l - (H(\theta')+1)^l]\varphi\|_{L^2} \leq e_l|\theta - \theta'|\,\|\varphi\|_{2l}$$

uniformly in $\varphi \in C_0^\infty$, $0 \leq \theta \leq 1$. Combining this with (12.65b), we see that if $|\theta - \theta'| \leq \alpha_l$ (for suitable $\alpha_l > 0$), then $[H(\theta)+1]^l - [H(\theta')+1]^l$ is an operator bounded perturbation (see Sect. 1.1) of $[H(\theta')+1]^l$ with relative bound smaller than 1. Starting from $\theta' = 0$ and applying the Kato-Rellich theorem (Theorem 1.4) a finite number of times we see that $(H+1)^l$ is essentially self-adjoint on C_0^∞ and $D(\overline{H^l}) = D((\bar{H})^l) = D(\bar{H}_0^l) = \mathrm{H}_{2l}$. Moreover, then $Q(\bar{H}^l) = Q(\bar{H}_0^l) = \mathrm{H}_l$. □

Corollary 12.34. Let H be a uesso operator. If $u \in L^2$ and the distributional operator H_{dist}^l obeys $H_{\mathrm{dist}}^l u \in L^2$ (i.e. there exists $f \in L^2$ so that for all $\varphi \in C_0^\infty$, $(H^l\varphi, u) = (\varphi, f)$), then $u \in D(\overline{H^l}) = \mathrm{H}_{2l}$. If $H_{\mathrm{dist}}^l u \in \mathrm{H}_1$, then $\varphi \in \mathrm{H}_{2l+1}$.

Proof. We prove the first statement; the other is similar. $(H^l)_{\mathrm{dist}}$ is by definition $[(H \restriction C_0^\infty)^l]^*$ which, by the theorem, is $(\bar{H})^l$, so the hypothesis $u \in D((H^l)^*)$ implies $u \in D(\bar{H}^l)$. □

Corollary 12.35. $u \in \bigcap_l D(\bar{H}^l)$ implies that u is C^∞.

Proof. $\bigcap_l D(\bar{H}^l) = \mathrm{H}_\infty$ contains only C^∞ function by Theorem 12.29. □

Henceforth we use H to denote the closure of $H \restriction C_0^\infty$.

Theorem 12.36. Let H be a uesso operator on R^ν. Then $\{\exp(-tH)\}_{t>0}$ has an integral kernel $\exp(-tH)(x, y)$, jointly C^∞ in x, y, t.

Proof. $(H+1)^{s/2}\exp(-tH)(H+1)^{s/2}$ is uniformly bounded from L^2 to L^2 for any s and C^∞ in t. Thus, by Theorem 12.33, $\exp(-tH)$ maps H_{-s} to H_s for any s and is C^∞ in t. The result now follows from Theorem 12.31. □

Remark. In fact, one can analytically continue in t to $\{t|\mathrm{Re}\, t > 0\}$.

Before going to manifolds, we need to extend the ideas of H's we discuss in three ways: (i) We want to allow vector valued functions of x, say, functions with values in the inner product space X. We will actually let $a^{ij}(x)$ be scalar, so this change is totally trivial, although we could allow vectorial dependence. (ii) We want to consider symmetric H's of the form

$$\tilde{H} = -\sum \partial_i a^{ij}(x)\partial_j + \sum b_i(x)\partial_i + c(x) = H + W \;, \tag{12.66}$$

where b_i and c are operators on the finite dimensional space X, and b_i, c are C^∞ with bounded derivatives. Since

$$\|(\tilde{H}^l - H^l)u\| \le c\|u\|_{2l-1} \;,$$

$\tilde{H}^l - H^l$ is H^l bounded with relative bound zero, so the first-order terms have no essential effect. (iii) We want to replaced $d^\nu x$ by $f(x)^2\, d^\nu x$ where f is smooth, identically 1 near infinity and bounded away from zero. Then $\tilde{H}$ on $L^2(R^\nu, f^2\, dx)$ is unitarily equivalent to $f\tilde{H}f^{-1}$ on $L^2(R^\nu, dx)$ and $f\tilde{H}f^{-1}$ still has the form (12.66), so this change is easy. By a *uniformly elliptic second order operator*, we mean a symmetric operator $\tilde{H}$ of the form (12.66) with all coefficients C^∞ with uniformly bounded derivatives and condition (ii) acting on $L^2(R^\nu; x; f^2\, d^\nu x)$. By the above remarks:

Metatheorem 12.37. All results for uniformly elliptic strictly second-order operators [and, in particular, Theorems 12.32–36 hold if the word "strictly" is dropped, so long as $H + 1$ is replaced by $(H + \mu)$ with μ a sufficiently large constant.

Now let M be a compact orientable Riemannian manifold of dimension n. Let $\{U_\alpha\}_{\alpha=1}^l$ be an open cover of M by coordinate neighborhoods and T_α: $U_\alpha \to R^n$ smooth coordinate maps. Let $g^{(\alpha)}$ be a metric on R^n which obey (i) $(g^{(\alpha)})_{ij} = \delta_{ij}$ near infinity (ii) $g_{ij}(m) = g_{ij}^{(\alpha)}(T_\alpha m)$ if $m \in U_\alpha$ (i.e. transfer the metric from U to $T_\alpha[U]$ and extend to a metric constant near infinity). By $\bigwedge_{(\alpha)}^p(R^n)$ we mean the set of smooth p-forms on R^n with the L^2-inner product given by $g^{(\alpha)}$ and $\bar{\bigwedge}_{(\alpha)}^p$ is its closure. Given $j \in C_0^\infty(U_\alpha)$, let

$$I_\alpha(j)\colon \bar{\bigwedge}^*(M) \to \bar{\bigwedge}_\alpha^*(R^n)$$

by

$$(I_\alpha(j)f)(x) = j(T_\alpha^{-1}x)f(T_\alpha^{-1}x) \quad x \in T_\alpha[U] \tag{12.67a}$$

$$= 0 \qquad\qquad x \notin T_\alpha[U] \;. \tag{12.67b}$$

Fix a partition of unity j_α, subordinate to U_α. We will define $\mathrm{H}_s(M)$ to be

$\{f \in \overline{\bigwedge}^*(M) | I_\alpha(j_\alpha)f \in \mathrm{H}_s(R^n)$ for all $\alpha\}$ with $\|f\|_s = \sum_\alpha \|I_\alpha(j_\alpha)f\|_s$. It is not hard to see that this definition of space is independent of the metric on M and the norms are changed to equivalent norms by a change of metric. It is not so obvious (but not even too hard by straightforward means) that H_s is independent of the choice of U_α and j_α; that is one consequence of:

Theorem 12.38. B_p and L_p are essentially self-adjoint on the smooth p-forms $\bigwedge^p$ and $Q(B_p^s) = Q(L_p^s) = \mathrm{H}_s(M) \cap \overline{\bigwedge}^p$. Moreover, B_p^s and L_p^s are essentially self-adjoint on $\bigwedge^p$ for $s = 1, 2, \ldots$.

We defer the proof of this result and first derive some consequences. Because of the Sobolev imbedding theorem (Theorem 12.29) on R^n, if $f \in \mathrm{H}_s(M)$ each function $(j_\alpha f)(m)$ is smooth on U_α and so since $\sum_\alpha j_\alpha = 1$, we have

Theorem 12.39. If $f \in \mathrm{H}_s(M)$ and $s > (v/2) + l$, then f is C^l in classical sense. If $A\colon \overline{\bigwedge}^* \to \overline{\bigwedge}^*$ extends to a map of H_{-s} to H_s for $s > v/2$, then A has a jointly continuous integral kernel $A(m, n)$ which is a linear map from $\bigwedge^*(T_n^*)$ to $\bigwedge^*(T_m^*)$. If it extends for all s, then $A(m, n)$ is C^∞.

As a corollary of the last two theorems, we obtain

Corollary 12.40. $\exp(-tL)$ has an integral kernel $\exp(-tL)(m, n)$ which is jointly C^∞ in m, n, t for $t > 0$.

Since $\exp(-tL)(m, n)$ is uniformly bounded, its Hilbert-Schmidt norm $\mathrm{tr}_{\bigwedge^* T^*_n}[\exp(-tL)(m, n) * \exp(-tL)(m, n)] < \infty$, so since the measure of M is finite, $\exp(-tL)$ is Hilbert-Schmidt for all t. Thus, $\exp(-tL) = [\exp(-tL/2)]^2$ is trace class for all t, and since its integral kernel is continuous, one can compute the trace by integrating the diagonal (see [330] for a discussion of trace class and Hilbert-Schmidt operators). We summarize with

Corollary 12.41. $\exp(-tL_p)$ is compact and trace class. It has a jointly continuous kernel $\exp(-tL)(x, y)\colon \bigwedge^p T_y^* \to \bigwedge^p T_x^*$ that is

$$\begin{aligned} (\mathrm{e}^{-tL_p}u)(x) &= \int \mathrm{e}^{-tL_p}(x, y)u(y)\,dy \\ \mathrm{Tr}(\mathrm{e}^{-tL_p}) &= \int \mathrm{tr}(\mathrm{e}^{-tL_p}(x, x))\,dx\ . \end{aligned} \tag{12.68}$$

In (12.68, 69), the measure is the natural one on M. Since $\exp(-tL_p)$ is compact, L_p has a complete set of eigenfunctions. These lie in $\bigcap_s D(L^s)$ and so in $\bigcap_s \mathrm{H}_s$ which is $\bigwedge^p$ by Theorem 12.39. Thus,

Corollary 12.42. L_p has a complete set of eigenfunctions which all lie in $\bigwedge^p$ (i.e. are smooth).

All that remains is the proof of Theorem 12.38. We first remark that it is enough to prove the result for B or for L alone, since the Weitzenböck formula says that $L^s - B^s$ is a relatively bounded perturbation of B^s (or L^s) of relative bound zero.

Next, we note that if $u \in C^\infty(M)$, it is very easy to see that $\|(B+1)^s u\|$ (so $\|(L+1)^s u\|$) are equivalent norms to the $H^s(M)$ norms since: (a) One can define operators B_α on $\bigwedge_\alpha^*(R^n)$ using the metric $g^{(\alpha)}$, so that if $k_\alpha \equiv 1$ on supp j_α, then $B_\alpha^s I_\alpha(j_\alpha)f = I_\alpha(k_\alpha)B^s(j_\alpha f)$. (b) By usng local coordinates, B_α is uniformly elliptic second order. (c) By the result on R^ν, any C^∞ function on M maps H_s to H_s. (d) We can find $\tilde{j}_\alpha$ so $\sum \tilde{j}_\alpha^2 = 1$ and $j_\alpha \tilde{j}_\alpha^{-1}$ are bounded C^∞ functions. Moreover, $\tilde{j}_\alpha = \sum_\beta j_\beta \tilde{j}_\alpha$ s by Step (c), $\sum_\alpha \|(B+1)^s \tilde{j}_\alpha f\|^2$ and $\sum_\alpha \|(B+1)\tilde{j}_\alpha f\|^2$ are equivalent norms.

The above remarks reduce the proof of Theorem 12.38 to the proof that B^s (or L^s) is essentially self-adjoint on $C_0^\infty(M)$. Thus, if one wants to use *Chernoff*'s result [63] (see also *Strichartz* [351]), the proof is done and the reader can ignore the technicalities that follow. Here we will prove essential self-adjointness directly. Since B maps C^∞ to itself and is formally self-adjoint, B extends to a map on distributions.

B defines a positive quadratic form on L^2 on $C^\infty(M)$ which is automatically closable. The resulting self-adjoint operator we also denote by B, so $Q(B)$ is the closure of $C^\infty(M)$ in the B-norm.

Lemma 12.42. Let $u \in L^2(M)$. Thus $u \in Q(B)$ if and only if the (distributional) derivatives of u in local coordinates lie in $L^2(M)$.

Proof. Let $\|u\|_B^2 \equiv (u, Bu) + (u, u)$. Since $\nabla_i(fu) = f\nabla_i u + (X_i f)u$, it is easy to see that if $u \in \bigwedge^p(M)$ and $f \in C^\infty$, then $\|fu\|_B \leq c\|u\|_B$ so $u \mapsto fu$ maps $Q(B)$ to itself. Writing out $\nabla_i u$ in terms of $\partial_i u$ we see then that on $\bigwedge^p(M)$, $\|\cdot\|_1^2$ is equivalent to

$$\left(\sum_\alpha \sum_{i=1}^{\nu} \|\partial_i^{(\alpha)} j_\alpha u\|^2\right) + \|u\|^2$$

with $\partial_1^{(\alpha)}$ coordinate derivative on U_α. From this it is easy to see that if $u \in Q(B)$, it has L^2 derivatives, and conversely if it has L^2 derivatives, by just convoluting the pieces in each in U_α we can approximate with elements in $\bigwedge^\infty(M)$ so $u \in Q(B)$. □

Lemma 12.43. (a) If $u \in H_{2s-1}$ and $B_{\text{dist}}^s u \in L^2$ and if $g \in C^\infty(M)$, then $B_{\text{dist}}^s(gu) \in L^2$.
(b) If $u \in H_{2s}$ and $B_{\text{dist}}^s u \in Q(B)$, and if $g \in C^\infty(M)$, then $B_{\text{dist}}^s(gu) \in Q(B)$.

Proof. We prove (a); the proof of (b) is similar given the last lemma. Formally $B_{\text{dist}}^s(gu) = gB_{\text{dist}}^s(u) +$ correction terms involving the first $(2s-1)$ derivative of u. Since $u \in H_{2s-1}$, it is easy to show that this formal equality actually defines an L^2 function equal to $B_{\text{dist}}^s(gu)$. □

Proof of Theorem 12.38. Given Lemma 12.41, we must show the pair of facts that B^s is essentially self-adjoint on $\bigwedge^p$, the smooth p-forms, and that $D(B^s) = H_{2s}$. We will prove this inductively. Suppose we have $u \in D([(B \restriction \bigwedge^p)^s]^*)$. By induction, $u \in D(B^{s-1})$ and $B^{s-1}u \in Q(B)$. By Lemma 12.43(b), $B_{\text{dist}}^{s-1}(j_\alpha u) \in D(B)$. But $(B_\alpha)^{s-1}(j_\alpha u \circ T_\alpha^{-1}) \equiv [B_{\text{dist}}^{s-1}(j_\alpha u)]\sigma T_\alpha^{-1}$ so $j_\alpha u \circ T_\alpha^{-1} \in H_{2s-1}$ by Corollary 12.34.

Thus, $u \in \mathrm{H}_{2s-1}(M)$. Repeating this argument, $u \in \mathrm{H}_{2s}$. Thus, $D(B^s) = D_s((B^s \restriction \bigwedge^p)^*) \subset \mathrm{H}_{2s}$. It is easy to smooth out any $u \in \mathrm{H}_s$ by localization and convolution to see that $\mathrm{H}_{2s} \subset D(\overline{(B \restriction \bigwedge^p)^s})$. □

12.6 A Canonical Order Calculus

The analytic techniques in the last section, while not the most usual ϕDO machinery, are relatively standard. In this section, we describe a technique for estimating heat kernels, which is due to one of us (B.S.) and presented here for the first time. As we shall see, it represents a kind of poor man's functional integral.

Definition 12.44. A family of operators $\{A_t\}_{t>0}$ on $L^2(R^n, d^n x)$ [or on $L^2(M, dx)$ (with M a compact n-dimensional Riemannian manifold) or on $\overline{\bigwedge}^p(R^n, d^n x)$ or on $\overline{\bigwedge}^p(M, dx)$] is said to have *canonical order* $m \in R$ if and only if (i) A_t maps $\mathrm{H}_{-\infty}$ to H_∞ for each $t > 0$. (ii) For any $k \geq l$ in $\{0, \pm 1, \pm 2, \dots\}$ there is a constant $c(k, l)$ so that for $0 < t < 1$

$$\|A_t u\|_k \leq c(k, l) t^{-a} \|u\|_l \tag{12.69}$$

with $a = \frac{1}{2}(k - l) - m$. The condition $k \geq l$ is crucial, since we do not want to imply that if $u \in \mathrm{H}_\infty$, then $\|A_t u\|_k$ goes to zero faster than a polynomial (it does, however, go to zero as t^m if $m > 0$). The definition is such that if A_t has canonical order m, it has canonical order m' for any $m' \leq m$. If A_t has canonical order m for all $m > 0$, we say it has *canonical order infinity*. The name and k, l dependence of a come from:

Proposition 12.45. If H is a uniformly elliptic second-order operator on R^n or a Laplacian (-Beltrami or Bochner) on $\bigwedge^*(M)$, then $\exp(-tH)$ has canonical order zero.

Proof. By elliptic regularity (Theorem 12.32 and 12.38), the H_s norm is equivalent to the $\|(H + \mu)^{s/2} \cdot \|$ norm, where μ is chosen so that $\mathrm{spec}(H) \subset [-\mu + 1, \infty)$. Thus,

$$\|\mathrm{e}^{-tH} u\|_k \leq C \, \|(H + \mu)^{+(k-l)/2} \mathrm{e}^{-tH}\|_{L^2} \|u\|_l \; .$$

By the functional calculus,

$$\|(H + \mu)^{+a} \mathrm{e}^{-tH}\|_{L^2} \leq \sup_{x \geq -\mu+1} \|(x + \mu)^{+a} \mathrm{e}^{-tx}\|$$

$$= \mathrm{e}^{t(\mu-1)} \sup_{y \geq 0} \|(y + 1)^{+a} \mathrm{e}^{-ty}\| \leq C t^{-a}$$

for $0 < t < 1$ since $a \geq 0$. □

The whole point of the canonical order calculus is the information it yields on integral kernels.

Proposition 12.46. If A_t has canonical order m, then A_t has an integral kernel $A_t(x, y)$ obeying

$$\lim_{t \downarrow 0} t^b \sup_{x,y} |A_t(x, y)| = 0 \tag{12.70}$$

for any $b > \frac{1}{2}n - m$.

Proof. Existence of the integral kernel follows from Theorem 12.31. Since $\delta_x \in \mathrm{H}_{-s}$ for any $s > n/2$ (Theorem 12.29), we have that

$$|A_t(x, y)| = |(\delta_x, A_t \delta_y)| \le \|A\|_{\mathrm{H}_s \to \mathrm{H}_{-s}} ,$$

so that (12.70) follows from (12.69). □

This theorem almost captures the exact leading behavior for the simple case $A_t = \exp(t\Delta)$, where $A_t(x, x) = (4\pi t)^{-n/2}$ so $b > (n/2) - m$ cannot even be replaced by $b \ge (n/2) - m$. Of course, this proposition alone does not rule out $t^{-n/2} \log(t^{-1})$ terms. Our strategy will be to use the proposition to separate out terms which are smaller than what we are interested in, and analyze the remainder by more explicit methods (essentially scaling).

We will need two abstract methods for obtaining operators with some canonical order from operators of other canonical orders. The first is very elementary:

Proposition 12.47. Let A_t have canonical order m, and suppose that B is a fixed operator so that, for some $b \ge 0$:

$$\|B\varphi\|_k \le \tilde{C}_k \|\varphi\|_{k+b} .$$

Then BA_t and $A_t B$ have canonical order $m - \frac{1}{2}b$.

Proof.

$$\|BA_t\varphi\|_k \le \tilde{C}_k \|A_t \varphi\|_{k+b} \le \tilde{C}_k c(k + b, l) t^a \|\varphi\|_l ,$$

where $a = \frac{1}{2}(k + b - l) - m = \frac{1}{2}(k - l) - (m - \frac{1}{2}b)$. In the above, we used $b \ge 0$ to be sure that $k + b \ge l$ if $k \ge l$. The proof for $A_t B$ is similar. □

Corollary 12.48. On R^n, if H is a uniformly elliptic second-order operator, then $\partial_j \exp(-tH)$ and $\exp(-tH)\partial_j$ have canonical order $-\frac{1}{2}$.

The other result is somewhat special looking if one is not familiar with DuHamel's expansion. For this reason, we recall:

Proposition 12.49. Let X, Y be bounded operators. Let

$$b_n = \int_{0<s_i<t} e^{-(t-s_1\ldots s_n)X} Y e^{-s_1 X} Y \ldots Y e^{-s_n X} \, ds_1 \ldots ds_n, \quad \sum_1^n s_i \le t$$

and let r_n be the same object with $\exp[-(t - s_1 \ldots s_n)X]$ replaced by $\exp[-(t - s_1 \cdots s_n)(X + Y)]$. Then, for any n

$$e^{-t(X+Y)} - e^{-tX} = \sum_{j=1}^{n} (-1)^j b_j + (-1)^{n+1} r_{n+1} \ .$$

Proof. We begin with the formula for $n = 0$ (DuHamel's formula):

$$e^{-t(X+Y)} - e^{-tX} = -\int_0^t e^{-(t-s)(X+Y)} Y e^{-sX} \, ds \ . \tag{12.71}$$

This follows by letting

$$B_t = e^{-t(X+Y)} e^{tX}$$

and noting that $dB_t/dt = -\exp[-t(X + Y)] Y \exp(tX)$ so that

$$B_1 - B_0 = -\int_0^t e^{-(t-s)(X+Y)} Y e^{(t-s)X} \, \mathrm{ds} \ .$$

Equation (12.71) follows by multiplying by $\exp(-tX)$. The general formula now holds inductively, since inserting (12.71) in the first factor of r_n implies that $r_n = b_n - r_{n+1}$. □

While we only proved this for bounded X, Y, we will later apply it early to the case where X and $X + Y$ are uniformly elliptic second order. We leave the extension to the reader with several remarks: (1) One should interpret the sum originally as maps from some H_s to H_{-s} (i.e. check matrix elements of φ in H_s with s large) and then notice that the operators make sense from H to H as bounded maps (if Y has second-order terms, we must be careful to avoid s_i^{-1} terms, but as there are $(n + 1)$ factors of $\exp(-sH)$ with $s_0 = t - s_1 \ldots s_n$ and at most $2n$ derivatives, we can use H_l for l fractional to arrange to bound by $s^{-a(n)}$, $a(n) = n/n + 1$). (2) It is useful to prove the expansion originally for t purely imaginary so $\exp[-t(X + Y)]\exp(tX)$ is well defined, and then analytically continue.

Given this expansion, the following is clearly of relevance.

Proposition 12.50. Let $A_t^{(0)}, \ldots, A_t^{(l)}$ be operators of canonical order $m_0, \ldots, m_l$ with $m_j > -1$. Then

$$B_t = \int_{\substack{s_i > 0 \\ s_1 + \cdots + s_l \le t}} A^{(0)}_{t-s_1-\cdots-s_l} A^{(1)}_{s_1} \ldots A^{(l)}_{s_l} \, ds_1 \ldots ds_l$$

is a convergent integral and B_t is an operator of canonical order $l + \sum_0^l m_j$.

Proof. Let $s_0 = t - s_1 - \cdots - s_l$. Write $B_t = \sum_{j=0}^{l} B_{t;j}$ where $B_{t;j}$ is the integral over the region R_j, where

$$\max_{0 \le k \le l} s_k = s_j \ .$$

Now in the region R_j, write

$$\|A_{s_0}^{(0)} \ldots A_{s_l}^{(l)} \varphi\|_k \le C\left(\prod_{i \ne j} s_i^{m_i}\right) s_j^{m_j - 1/2(k-p)} \|\varphi\|_p$$

by bounding A_{s_i} $(i \ne j)$ as a map from suitable H_q to H_q and A_{s_j} from H_p to H_k. Since $t \ge s_j \ge t/(n+1)$ on R_j and

$$\int_0^t \prod_{i \ne j} s_i^{m_i} \, ds_1 \ldots ds_l = ct^d; \quad d = l + \sum_{i \ne j} m_j$$

with $c < \infty$ since $m_i > -1$, we obtain the desired result. □

As our first application, we will control certain commutators of $\exp(-tH)$ and $f \in C^\infty$. For this, we need

Lemma 12.51. If A and B are bounded operators, then

$$[A, \mathrm{e}^{-tB}] = -\int_0^t \mathrm{e}^{-(t-s)B}[A, B]\mathrm{e}^{-sB} \, ds$$

Proof. Let

$$C_t = \mathrm{e}^{tB} A \mathrm{e}^{-tB} \quad \text{so} \ \frac{dC_s}{ds} = -\mathrm{e}^{sB}[A, B]\mathrm{e}^{-sB} \quad \text{and}$$

$$C_t - A = -\int_0^t \mathrm{e}^{sB}[A, B]\mathrm{e}^{-sB} \, ds \ .$$

Multiply by $\exp(-tB)$ and obtain the desired result. □

Again, we will use this for B, a uniformly elliptic second order operator, and A, a C^∞-function, leaving the proof for this case to the reader with the hint that one should prove it initially for t imaginary, and then analytically continue. The following will play a critical role:

Theorem 12.52. Let H be a uniformly elliptic second-order operator on R^n [resp. a Laplacian (– Beltrami or Bochner) on M]. Let f be a C^∞ function on R^n (resp. M) with all derivatives uniformly bounded. Then $[f, \exp(-tH)]$ has canonical order $\frac{1}{2}$.

Proof. We give the R^n proof; with localization, the proof for M is similar. By Lemma 12.51,

$$[f, \mathrm{e}^{-tH}] = -\int_0^t \mathrm{e}^{-(t-s)H}[f, H]\mathrm{e}^{-sH} \, ds \ .$$

But $[f, H] = \sum_j g_j \partial_j + h$ for C^∞ functions g_j and h. By Proposition 12.45 and 12.47, $\exp(-sH) g_j$ and $\exp(-sH)h$ have canonical order 0 and $\partial_j \exp(-sH)$ has canonical order $-\frac{1}{2}$. Thus, by Proposition 12.50, $[f, \exp(-tH)]$ has canonical order $1 - \frac{1}{2} = \frac{1}{2}$. □

As a typical application of this machinery, we want to discuss leading asymptotics of $\exp(-tH)(0,0)$. We exploit the fact that

$$(e^{t\Delta})(x,y) = (4\pi t)^{-n/2} \exp[-(x-y)^2/4t] \tag{12.72}$$

as can be seen by an elementary Fourier transform analysis. We prove the next two theorems on R^n. After the analysis in the next section, they will apply to M.

Theorem 12.53. Let $H = -\sum_{i,j} \partial_i a_{ij} \partial_j + \sum_i b_i \partial_i + c$ be a uniformly elliptic second-order operator on $L^2(\mathbb{R}^n)$ (with vector values allowed). Suppose that $a_{ij}(0) = \delta_{ij}$. Then

$$\lim_{t \downarrow 0} (4\pi t)^{n/2} e^{-tH}(0,0) = 1 \ .$$

Proof. Let $H_0 = -\Delta$ and let $\tilde{a}_{ij} = a_{ij} - \delta_{ij}$. By the DuHamel expansion

$$e^{-tH} - e^{-tH_0} = \int_0^t e^{-(t-s)H}(C_1 + C_2) e^{-sH} \, ds \ ,$$

where $C_1 = -\sum_i b_i \partial_i + c$ and $C_2 = \sum_{i,j} \partial_i \tilde{a}_{ij} \partial_j$. The C_1 term contributes an operator of canonical order $\frac{1}{2}$, so it makes a vanishing contribution to $(4\pi t)^{n/2}[\exp(-tH)(x,y) - \exp(+t\Delta)(x,y)]$ as $t \downarrow 0$. The C_2 term we write as $D_1 + D_2$.

$$D_1 = \sum_{i,j} \int_0^t (e^{-(t-s)H} \partial_i)(\partial_j e^{-sH}) \tilde{a}_{ij} \, ds$$

$$D_2 = \sum_{i,j} \int_0^t e^{-(t-s)H} \partial_i [\tilde{a}_{ij}, \partial_j e^{-sH}] \, ds \ .$$

By using Theorem 12.53, $[\tilde{a}_{ij}, \partial_j \exp(-sH)]$ has canonical order 0, while $\exp[-(t-s)H]\partial_i$ has canonical order $-\frac{1}{2}$, so D_2 has canonical order $\frac{1}{2}$. Thus, it too makes a vanishing contribution to $(4\pi t)^{n/2}[\exp(-tH)(x,y) - \exp(t\Delta)(x,y)]$. D_1 only has canonical order 0 and does make a contribution to this difference for *general* x, y. But since $\tilde{a}_{ij}(0) = 0$, $D_1(0,0)$ is identical to zero for all t. We have thus shown that

$$\lim_{t \downarrow 0} (4\pi t)^{n/2} [e^{-tH}(0,0) - e^{-tH_0}(0,0)] = 0 \ ,$$

so the theorem follows from (12.72). □

Corollary 12.54. Let H be a Bochner Laplacian or Laplace-Beltrami operator on

R^n with respect to a metric g_{ij} which is δ_{ij} near ∞. Let $\exp(-tH)(0,0)$ denote the integral kernel relative to the natural measure $dx (\equiv \sqrt{(\det g_{ij})}\, d^n x)$ on R^n thought of as a Riemannian manifold. Then

$$\lim_{t \downarrow 0} (4\pi t)^{n/2} e^{-tH}(0,0) = 1 .$$

Proof. Use normal coordinates at zero. Then H has the form given in the theorem. □

Remarks (1) Once we have the cutting and pasting results of the next section, we can replace $g_{ij} = \delta_{ij}$ near ∞ by uniform ellipticity.

(2) In Theorem 12.53, if $a_{ij}(0) \neq \delta_{ij}$, we just let $H_0 = -\partial_i a_{ij}(0)\partial_j$ and obtain $[\det a_{ij}(0)]^{1/2}(4\pi t)^{n/2} \exp(-tH)(0,0) \to 1$. If we then write the corollary in general coordinates, we obtain an extra $(\det g^{ij})^{-1/2}$ factor if the integral kernel is written relative to $d^n x$. This extra factor is then cancelled by writing the integral kernel relative to $dx = (\det g_{ij})^{1/2} d^n x$ since $(\det g_{ij})(\det g^{ij}) = 1$.

(3) We will later discuss higher order asymptotics; see Theorem 12.57.

If commutators with f were nice, one can hope the higher-order commutators with f are nicer. By Adf(B) we mean $[f, B]$. Then, we have that

Theorem 12.55. Let H and f be as in Theorem 12.52. Then for all j, $(\mathrm{Adf})^j[\exp(-tH)]$ has canonical order $j/2$.

Proof. Since $(\mathrm{Adf})^3(H) = 0$, induction starting from Lemma 12.51 shows that

$$(\mathrm{Adf})^j(e^{-tH})$$

$$= - \sum_{\substack{k+l+m=j \\ l=1 \text{ or } 2}} \frac{(j-1)!}{k!m!} \int_0^t [(\mathrm{Adf})^k(e^{-(t-s)H})][(\mathrm{Adf})^l(H)][(\mathrm{Adf})^m(e^{-sH})]\, ds .$$

From this formula and Theorem 12.50, one easily proves the desired result by induction in j: For given the result for $(\mathrm{Adf})^q[\exp(-tH)]$ with $q < j$ since $k, m < j$, we have that $(\mathrm{Adf})^k[\exp(-tH)]$ has canonical order $k/2$. $(\mathrm{Adf})^l(H)$ has terms of degree 1 in ∂_j if $l = 1$ and is a function of $l = 2$, so $[(\mathrm{Adf})^l(H)][(\mathrm{Adf})^m \exp(-sH)]$ has order $m/2 - (2 - l)/2$ by Proposition 12.47. Thus, the integral has canonical order $1 + k/2 + m/2 - 1 + l/2 = j/2$, proving the result for j. □

Corollary 12.56. Let H be as in Theorem 12.5, and let f, g obey the hypotheses on f in that theorem. Suppose that supp g is compact and supp $f \cap$ supp $g = \phi$. Then $g \exp(-tH) f$ has canonical order infinity.

Proof. Find h in C_0^∞ with $h \equiv 1$ on supp g and $h \equiv 0$ on supp f. Then $g(\mathrm{Adh})^j(\exp[-tH])f = g \exp(-tH)f$ for all j. □

The last few results show that the canonical order calculus is a kind of poor man's functional integral. $(\mathrm{Adf})^j \exp(-tH)$ has integral kernel $[f(x) -$

$f(y)]^j[\exp(-tH)](x,y)$ so that Theorem 12.56 is a version of the fact that, in time t, the paths can travel a distance which is $O(t^{1/2})$.

In the remainder of this section, we will discuss systematic higher-order asymptotic expansions on $[\exp(-tH)](0,0)$ and write out the first correction to $(4\pi t)^{-n/2}$ in terms of the curvature when H is the Laplacian on functions ($L_0 = B_0$). These things will *not* be used again, and can be skipped.

Theorem 12.57. Let $H = -\sum_{i,j} \partial_i \partial_{ij} \partial_j + \sum b_i \partial_i + c$ be a uniformly elliptic second-order operator on $L^2(R^n)$. Then

$$(4\pi t)^{n/2} \mathrm{e}^{-tH}(0,0) \sim \mathrm{e}_0 + t\mathrm{e}_1 + \cdots + t^j \mathrm{e}_j + \cdots$$

(in the sense that $|\mathrm{LHS} - \sum_{j=0}^{l} \mathrm{e}_j t^j| = O(t^{l+1})$ for each l) where e_j depends only on the Taylor coefficient of a up to order $2j$, the Taylor coefficient b up to order $2j-1$, and of c up to order $2j-2$.

Proof. The idea of the proof is quite simple. We first show that to go up to order t^l we need only use the first $2l$ terms of the DuHamel expansion; then we show we can replace the coefficients of the differential operator by Taylor polynomials, and finally, by scaling we evaluate the order in t of any given set of Taylor coefficients. Let $H_0 = -\Delta$.

By scaling, we can suppose that $a_{ij}(0) = \delta_{ij}$. We set $\tilde{a}_{ij} = a_{ij} - \delta_{ij}$; $Y_2 = -\sum_{i,j} \partial_i \tilde{a}_{ij} \partial_j$, $Y_1 = \sum_i b_i \partial_i$ and $Y_0 = c$ and $Y = Y_0 + Y_1 + Y_2$. We define

$$\tilde{r}_s^{(m)} = \int_{s_i \geq 0, \sum s_i \leq s} Y \exp[-(s - s_1 - s_2 \ldots s_{m-1})H_0] Y \exp(-s_1 H_0)$$

$$\ldots Y \exp(-s_{m-1} H_0)\, ds_1 \ldots ds_{m-1}$$

$$b_t^{(m)} = \int_0^t \mathrm{e}^{-(t-s)H_0} \tilde{r}_s^{(m)}\, ds$$

$$r_t^{(m)} = \int_0^t \mathrm{e}^{-(t-s)H} \tilde{r}_s^{(m)}\, ds$$

so that the DuHamel expansion says that

$$\mathrm{e}^{-tH} - \mathrm{e}^{-tH_0} = \sum_{j=1}^{m} (-1)^j b_t^{(j)} + (-1)^{m+1} r_t^{(m+1)} .$$

We seek operators $a_t^{(m)}$, $c_t^{(m)}$, $m \geq 0$ so that

(a) $a_t^{(m)} \delta_0 = \sum_{j=1}^{m} (-1)^j b_t^{(j)} \delta_0$
(b) $\tilde{r}_s^{(m+1)} \delta_0 = c_s^{(m)} \delta_0$
(c) $c_s^{(m)}$ has canonical order $(m-1)/2$ and for any $f^{(1)}, \ldots, f^{(j)}$ in C^∞, $(\mathrm{Ad} f^{(1)}) \cdots (\mathrm{Ad} f^{(j)})(c_s^{(m)})$ has canonical order $(j+m-1)/2$. If we find such operators, then $r_t^{(m)} \delta_0 = \int_0^t ds \exp[-(t-s)H] c_s^{(m-1)} \delta_0$ and $\int_0^t ds \exp[-(t-s)H] c_s^{(m-1)}$ has canonical order $m/2$, i.e.

$$e^{-tH}(0,0) = e^{-tH_0}(0,0) + \sum_{j=1}^{m} (-1)^j b_t^{(j)}(0,0) + O(t^{m/2-n/2-\varepsilon}) ,$$

and we need only obtain asymptotic series for $b_t^{(j)}(0,0)$. We define a, c inductively by

$$a_t^{(0)} = 0; \quad c_s^{(0)} = -\sum_{i,j} \partial_i[\tilde{a}_{ij}, \partial_j e^{-sH}] + (Y_1 + Y_0)e^{-sH}$$

$$a_t^{(m+1)} = a_t^{(m)} - \int_0^t e^{-(t-s)H_0} c_s^{(m)}$$

$$c_s^{(m+1)} = \int_0^t \left\{ -\sum_{i,j} \partial_i[\tilde{a}_{ij}, \partial_j e^{-(s-u)H_0} c_u^{(m)}] + (Y_0 + Y_1)e^{-(s-u)H_0} c_u^{(m)} \right\} du . \quad (12.73)$$

We leave the inductive verification of (a) – (c) to the reader.

Now suppose that in (12.73), $\tilde{a}_{ij}$ is replaced by $x_k \tilde{b}$, where $\tilde{b}$ also vanishes at zero. Then one can write

$$x_k \tilde{b} \partial_j e^{-(s-u)H_0} c_u^{(m)} \delta_0 = [x_k, [\tilde{b}, [\partial_j e^{-(s-u)H_0} c_u^{(m)}]]] \delta_0$$

and obtain an extra power of $\frac{1}{2}$ in canonical order. In that way, by using Taylor's theorem with remainder, we see that to go to order t^l we can replace $\tilde{a}_{ij}$, b, c by finite Taylor approximants. These commutator formulae show that the s integrals all converge. By using the explicit form of the integral kernel for $\exp(t\Delta)$, one sees that a term with j integrals over s, and j_1 integrands of $\partial_i x^{a_\alpha} \partial_j$, j_2 integrands of the form $x^{b_\alpha}\partial_i$ and j_3 of the form $x^{c_\alpha} (j_1 + j_2 + j_3 = j)$ is of the form $ct^{j+1/2(a+b+c)}$ where

$$a = \sum_{\alpha=1}^{j_1} (a_\alpha - 2),\ b = \sum_{\alpha=1}^{j_2} (b_\alpha - 1),\ c = \sum_{\alpha=1}^{j_3} c_\alpha .$$

If $a + b + c$ is odd, the integral defining c is zero, since there is a symmetry taking x to $-x$, ∂_i to $-\partial_i$ and leaving $\exp(tH)$ invariant. Thus, only integral orders enter in the expansion. □

The above proof makes the expansion seem fairly complicated, and in higher order it is; however, we want to illustrate that lower orders are fairly easy to compute.

Theorem 12.58. Let L be the Laplacian *on functions* for a metric g_{ij} on R^n. Then

$$e^{-tL}(0,0) = (4\pi t)^{-n/2}[1 + \tfrac{1}{6}Rt + O(t^2)] ,$$

where R is the scalar curvature at 0.

Proof. Pass to normal coordinates. Let $G_{ij;kl} = \partial^2 g_{ij}/\partial x^k \partial x^l$. So [using the facts that $\det g \equiv 1$ and that $g^{ij} = \delta^{ij} - \frac{1}{2}\sum_{k,l}(G_{ij;kl}x^k x^l + O(|x|^3)$—the minus sign coming from the fact that g^{ij} is the inverse of g_{ij}]

$$L = -\sum_{i,j} \partial_i g^{ij} \partial_j$$

$$= -\Delta + \tfrac{1}{2} \sum_{i,j,k,l} G_{ij;kl} \partial_i x^k x^l \partial_j + 0(\partial_i x^3 \partial_y)$$

and so (with $H_0 = -\Delta$)

$$e^{-tL}(0,0) = (4\pi t)^{-n/2} - \tfrac{1}{2} \sum_{i,j,k,l} G_{ij;kl} \int_0^t [e^{-(t-s)H_0} \partial_i x^k x^l \partial_j e^{-sH_0}](0,0)\, ds$$
$$+ O(t^{-n/2} t^2) \ . \tag{12.74}$$

Now given any operator A, let

$$\langle A \rangle_s = (4\pi t)^{n/2} (e^{-(t-s)H_0} \delta_0, A e^{-sH_0} \delta_0) \ .$$

The factor of $(4\pi t)^{n/2}$ is chosen so $\langle 1 \rangle_s = 1$.
$(4\pi t)^{n/2} \exp(-sH_0)(x,0) \exp[-(t-s)H_0](x,0)\, d^n x$ is a Gaussian measure and a simple calculation shows that the exponential term is

$$\exp[-tx^2/4s(t-s)]$$

so that when A is a function of x, it is easy to find $\langle A \rangle_s$; explicitly

$$\langle x^i x^j \rangle_s = \delta_{ij} 2s(t-s) t^{-1}$$
$$\langle x^i x^j x^k x^l \rangle_s = \rho_{ijkl} 4s^2 (t-s)^2 t^{-2} \ , \tag{12.75}$$

where $\rho_{ijkl} = \delta_{ij}\delta_{kl} + \delta_{ik}\delta_{jl} + \delta_{il}\delta_{jk}$. Taking derivatives of $\exp(-sH_0)(x,0)$ and $\exp[-(t-s)H_0](x,0)$, and integrating by parts we see that

$$\langle \partial_i x^k x^l \partial_j \rangle = -(2s)^{-1} [2(t-s)]^{-1} \langle x^i x^k x^l x^j \rangle$$
$$= -s(t-s) t^{-2} \rho_{ijkl} \ .$$

By (12.9a, b),

$$\sum_{i,j,k,l} \rho_{ijkl} G_{ij;kl} = 2R \ .$$

Thus,

$$\text{2nd term in (12.74)} = (4\pi t)^{-n/2} R \int_0^t s(t-s) t^{-2}\, ds$$
$$= (4\pi t)^{-n/2} \tfrac{1}{6} R t \ ,$$

proving the theorem. □

Remark. Equation (12.75) says that $\langle\cdot\rangle_s$ is essentially the Brownian bridge (see [331]); the above calculation is again a kind of slightly richer man's path integral.

Once one has the cutting and pasting results of the next section, the above calculation extends to an arbitrary manifold, and

Corollary 12.59. Let M be a compact orientable Riemann manifold, and let L_0 be the Laplacian on functions. Let $\tau = \int_M dx$, the volume of M, and $\kappa = \int_M R\,dx$ the total curvature. Then

$$\mathrm{Tr}(e^{-tL}) = (4\pi t)^{-n/2}[\tau + \tfrac{1}{6}t\kappa + O(t^2)] \ .$$

It is a useful exercise to check the above constants to compute the asymptotics for a two-dimensional unit sphere with its usual metric. *A* not entirely trivial analysis shows that

$$\mathrm{Tr}(e^{-tL_0}) = \sum_{l=0}^{\infty} (2l+1)e^{-tl(l+1)} = t^{-1} + \tfrac{1}{3} + O(t) \tag{12.76}$$

$\tau = 4\pi$, and since R has the constant value 2 ($R = 2R_{1221}$ and $R_{1221} = 1$), $\kappa = (4\pi)2$.

We close this section with an explanation of the "invariant theory" proof of Corollary 12.59; one first remarks, by one's procedure for proving asymptotics of the heat kernel, that the coefficient $c(x)$ of t in $(4\pi)^{-n/2}[1 + c(x)t + O(t^2)] = \exp(-tL_0)(x,x)$ is linear in the first and second derivatives of g_{ij} at 0, with coefficients only depending on dimension (by cutting and pasting). Since c is independent of coordinate systems, we can try to compute it in normal coordinates. There first derivatives of g vanish, so c must be built from $G_{ij;kl}$. By rotation invariance, only the contractions $\sum G_{ij;ij}$ and $\sum G_{ii;jj}$ can enter. But the second is zero by (12.19a). Thus, the only invariant one can build is R, i.e. $c(x) = a_n + b_n R(x)$. That leaves one the job of computing a_n and b_n. Since these are universal constants, one can try to compute in special cases. We begin by noting a product rule: If $M = M_1 \times M_2$ with the product metric, then $\exp(-tL)(0,0)$ is a product, and $R = R_1 + R_2$. This yields a relation on a_{n_1}, a_{n_2} and $a_{n_1+n_2}$, and similarly for the b's. Taking $M = M_1^j$, we find that $a_{nj} = ja_n$ and $b_n = b_{nj}$. Reversing the roles of n and j, we see that $b_n = b_j$ all n, j and $a_n/n = a_j/j$. Finally, one computes for a 2-sphere of radius ρ and standard metric and finds $R = 2/\rho^2$ and $c(x) = 1/3\rho^2$ so $a_2 = 0$, $b_2 = 1/6$.

To an analyst, the above proof is somewhat unsatisfactory. Normally one does a general computation and calculates a special case to check arithmetic. It is one thing to find a clever way of doing a calculation, and quite another to avoid it altogether! In the above case where the direct calculation is not so bad, this approach is not reasonable. *If* one had to do the cancellations in $\sum (t)^p \mathrm{Tr}[\exp(-tL_p)]$ by hand, the invariant theory approach would be reasonable, and indeed, this approach to the general index theorem is exactly that used by *Gilkey* [132–134] and *Atiyah, Bott* and *Patodi* [18]. However, with the

Berezin-Patodi formula, one can compute $\operatorname{str}[\exp(-tL)(x,x)]$ directly, and it seems clearer to do this computation than to avoid it.

While the cancellations will be made in general with the Berezin-Patodi formula, in principle one can do it by hand. For $n = 2$, we have just seen that

$$\operatorname{tr}[e^{-tL_0}(0,0)] = \operatorname{tr}[e^{-tL_2}(0,0)] \sim (4\pi t)^{-1}(1 + \tfrac{1}{6}tR) \ .$$

The reader should do the asymptotics for $\exp(-tL_1)$ and find that

$$\operatorname{tr}[e^{-tL_1}(0,0)] \sim (4\pi t)^{-1}(2 - \tfrac{2}{3}tR)$$

and so find that (following *McKean* and *Singer* [240])

$$\operatorname{str}[e^{-tL}(0,0)] \sim (4\pi)^{-1}R$$

and thereby prove the Gauss-Bonnet theorem.

12.7 Cutting and Pasting

Our goal in this section is to prove the following result:

Theorem 12.60. Let M be an orientable Riemann manifold with metric g. Let $U_\alpha \subset M$ be a coordinate neighborhood of m and $T\colon U_\alpha \to R^\nu$ a coordinate map with $T(m) = 0$. Let $\tilde{g}$ be a metric on R^ν so that $\tilde{g} \circ T = g$ (in the sense that if T_* is the induced map of $T(U_\alpha)$ to $T(R^\nu)$, then $\tilde{g}(T_* X, T_* Y) = g(X, Y)$ for X, Y vector fields over U_α). Let L be the Laplace-Beltrami operator on $\overline{\bigwedge^*(M)}$ and $\tilde{L}$ the Laplace-Beltrami operator on $\overline{\bigwedge^*(R^\nu)}$. Then $\|\exp(-tL)(m,m) - \exp(-t\tilde{L})(0,0)\| = O(t^k)$ for all k.

Before proving this, we note that with path integral (or other) methods, one can actually prove the difference is $O(\exp(-c/t))$ for suitable $c > 0$; we only need that it is $O(t^\varepsilon)$ for some $\varepsilon > 0$. This result says we need only consider heat kernels on R^ν; e.g. Corollary 12.54, Theorem 12.57 and Theorem 12.58 immediately extend to M from R^ν. We also note that the same proof relates $\exp(-tB)$ and $\exp(-t\tilde{B})$.

Proof of Theorem 12.60. We use the maps $I_\alpha(f)\colon \overline{\bigwedge^*(M)} \to \overline{\bigwedge^*(R^\nu)}$ defined in (12.67). Let $f, h \in C^\infty(M)$ with $f \equiv 1$ near m, $h \equiv 1$ in a neighborhood of $\operatorname{supp} f$ and $\operatorname{supp} h \subset U_\alpha$. We begin by looking at $\exp(-t\tilde{L})I_\alpha(h) - I_\alpha(h)\exp(-tL)$. Let $C_s = \exp(-(t-s)\tilde{L})I_\alpha(h)\exp(-sL)$. So we are interested in $C_0 - C_1 = -\int_0^1 (d/ds)C_s$, i.e.

$$e^{-t\tilde{L}}I_\alpha(h) - I_\alpha(h)e^{-tL} = \int_0^1 e^{-(t-s)\tilde{L}}\{I_\alpha(h)L - \tilde{L}I_\alpha(h)\}e^{-sL}\,ds \ .$$

By the relation of g and $\tilde{g}$, $I_\alpha(h)L - \tilde{L}I_\alpha(h)$ is a first-order differential operator with coefficients depending only on derivatives of h. In particular, since these derivatives have supports disjoint from f,

$$B_s \equiv \{I_\alpha(h)L - \tilde{L}I_\alpha(h)\}e^{-sL}f$$

has canonical order ∞ by Corollary 12.56. Thus, by Proposition 12.50, $\exp(-t\tilde{L})I_\alpha(h)f - I_\alpha(h)\exp(-tL)f$ has canonical order ∞. Multiply by $I_\alpha(f)*$ on the left and use $I_\alpha(h)f = I_\alpha(f)$, $I_\alpha(f) * I_\alpha(h) = f$ and find that

$$I_\alpha(f) * e^{-t\tilde{L}}I_\alpha(f) - fe^{-tL}f$$

has canonical order ∞. Thus, the integral kernel at (m, m) is $O(t^k)$ for all k. □

The reader familiar with scattering theory will note a Cook-type argument is used above.

12.8 Completion of the Proof of the Gauss-Bonnet-Chern Theorem

We can now put together all the elements and prove Theorem 12.14. We begin with Theorem 12.20, which allowed us to write L acting globally on $\overline{\bigwedge^*(R^\nu)}$ in normal coordinates centered at 0 by

$$L = B_0 + R_{(4)}(0) + A_{(2)} + A_{(4)} \ ,$$

where $A_{(2)}$ and $A_{(4)}$ obey the vanishing conditions given by Theorem 12.20. Let $L^\# = B_0 + R_{(4)}(0)$ and let $(C)_n$ denote the part of an operator C on $\bigwedge_0^*$ with n a^*'s and n a's when expanded via $C = \sum C_{I,J}a_I^* a_J$. Then we claim that

$$(e^{-tL^\#}(0,0))_n - (e^{-tL}(0,0))_n = 0(t^{1/2-\varepsilon}) \tag{12.77}$$

[it then automatically follows that it is $O(t)$]. Expand both $\exp(-tL^\#)$ and $\exp(-tL)$ as perturbations of $\exp(-tB_0)$ using a DuHamel expansion. To get n a's in a term of order q, we need $l[R_4(0) + A_{(4)}]$ terms and $jA_{(2)}$ terms where $l + j = q$, $2l + j \geq n$ (with more than n a's and a^*'s we will get out $a^*_{\{1,\ldots,n\}}a^*_{\{1,\ldots,n\}}$ terms by using the commutation relations). Since A_2 has canonical order $-\frac{1}{2}$ and $R_4(0) + A_{(4)}$ has order 0, we have that the order of this term in the DuHamel expansion is (by Proposition 12.50), $q - (j/2) = l + (j/2) \geq (n/2)$. Thus, except for terms with $2l + j$ exactly equal to n, we have terms of order $t^{1/2-\varepsilon}$ or greater. The terms which differ for L and $L^\#$ have at least one $A_{(2)}$ term or one $A_{(4)}$. The part of the $A_{(2)}$ term with no derivative is already of order $+\frac{1}{2}$ more than we computed above. The others have a coefficient vanishing at 0, so by the commutator trick we have used many times already, we can find an operator which has order $\frac{1}{2}$ higher which yields the same answer applied to δ_0. Thus, (12.77) holds.

Notice that B_0 acts only on functions and $R_{(4)}(0)$ only on the vectorial component so B_0 and $R_{(4)}(0)$ commute and

$$\mathrm{str}(\mathrm{e}^{-tL^{\#}}(0,0)) = \mathrm{e}^{-tB_0}(0,0)\mathrm{str}(\mathrm{e}^{-tR(4)(0)}) \ .$$

Now, $\exp(-tB_0)(0,0) = (4\pi t)^{-k}[1 + \mathrm{O}(t)]$ and the first term in the expansion of $\exp[-tR_{(4)}(0)]$ involving n a's has order k, so

$$\mathrm{str}(\mathrm{e}^{-tR(4)(0)}) = \frac{(-t)^k}{k!}\mathrm{str}[R_{(4)}(0)^k] + \mathrm{O}(t^{k+1}) \ .$$

Thus,

$$\mathrm{str}[\mathrm{e}^{-tL^{\#}}(0,0)] = (-1)^k(4\pi)^{-k}(k!)^{-1}\,\mathrm{str}[R_{(4)}(0)^k] + \mathrm{O}(t) \ .$$

Thus, we are reduced to computing $\mathrm{str}[R_{(4)}(0)^k]$ and so the proof is completed by the second of the lemmas below. □

Lemma 12.61. $\mathrm{str}[(a_{i_1}^* a_{j_1} a_{k_1}^* a_{l_1})(a_{i_2}^* \dots a_{l_k})]$ is zero unless $(i_1, k_1, i_2, \dots, k_l)$ and $(j_1, l_1, *_2, \dots, l_k)$ are permutations of $\{1, \dots, n\}$ and $\mathrm{str}[a_{\pi(1)}^* a_{\sigma(1)} a_{\pi(2)}^* a_{\sigma(2)} \dots a_{\pi(n)}^* a_{\sigma(n)}] = (-1)^\sigma(-1)^\pi$.

Proof. The first assertion is an immediate consequence of the Berezin-Patodi formula (Theorem 12.9). As for the second, the Berezin-Patodi formula says that the supertrace of terms coming from $\{a_{\pi(j)}^*, a_{\sigma(l)}\}$ is zero, so since there are $(n-1) + (n-2) + \dots = n(n-1)/2$ anticommutations

$$\begin{aligned}&\mathrm{str}[a_{\pi(1)}^* a_{\sigma(1)} \dots a_{\pi(n)}^* a_{\sigma(n)}] \\ &\quad = (-1)^{n(n-1)/2}\,\mathrm{str}[a_{\pi(1)}^* \dots a_{\pi(n)}^* a_{\sigma(1)} \dots a_{\sigma(n)}] \ .\end{aligned}$$

Next, we note that

$$\begin{aligned} a_{\sigma(1)} \dots a_{\sigma(n)} &= (-1)^\sigma a_1 \dots a_n \\ a_{\pi(1)}^* \dots a_{\pi(n)}^* &= (-1)^\pi a_1^* \dots a_n^* = (-1)^\pi(-1)^{n(n-1)/2} a_n^* \dots a_1^* \\ &= (-1)^\pi(-1)^{n(n-1)/2}(a_1 \dots a_n)^* \ . \end{aligned}$$

Since $(-1)^n = 1$, Theorem 12.9 yields the required result. □

Lemma 12.62.

$$\mathrm{str}[R_{(4)}(0)^k] = 2^{-k}\sum_{\pi,\sigma}(-1)^\sigma R_{\pi(1)\pi(2)\sigma(1)\sigma(2)} \cdots R_{\pi(n-1)\pi(n)\sigma(n-1)\sigma(n)} \ .$$

Proof. By the last lemma and definition of $R_{(4)}$

$$\begin{aligned}\mathrm{str}[R_{(4)}(0)^k] = &\sum_{\pi,\sigma}(-1)^\pi(-1)^\sigma R_{\pi(1)\sigma(1)\pi(2)\sigma(2)} \\ &\times R_{\pi(3)\sigma(3)\pi(4)\sigma(4)} \cdots R_{\pi(n-1)\sigma(n-1)\pi(n)\sigma(n)} \ .\end{aligned}$$

Fix σ and $\pi(3), \ldots, \pi(n)$ and consider the sum over the two choices of $\pi(1)$, $\pi(2)$. We then have, with $j = \sigma(1)$, $l = \sigma(2)$,

$$R_{ijkl} - R_{kjil} = R_{ijkl} + R_{jkil} = -R_{kijl}$$
$$= \tfrac{1}{2}(R_{ikjl} - R_{kijl}) \ ,$$

where we have used the various symmetry properties of R. Thus,

$$\operatorname{str}[R_{(4)}(0)^k]$$
$$= \sum_{\pi,\sigma} (-1)^\pi (-1)^\sigma [\tfrac{1}{2} R_{\pi(1)\pi(2)\sigma(1)\sigma(2)}] R_{\pi(3)\sigma(3)\pi(4)\sigma(4)} \cdots R_{\pi(n-1)\sigma(n-1)\pi(n-1)\pi(n)} \ .$$

Repeating the above calculation for each factor of R from $R_{\pi\sigma\pi\sigma}$ to $R_{\pi\pi\sigma\sigma}$, we obtain the claimed result. □

12.9 Mehler's Formula

These final two sections are intended as an introduction to more general index theorems than the GBC theorem. The key input one needs to go beyond GBC is a formula known as Mehler's formula, originally found by Mehler in connection with generating functions for Hermite polynomials, but in modern language it is the formula for the integral kernel of $\exp(-tH)$ where

$$H = -\frac{d^2}{dx^2} + x^2 - 1 \tag{12.78}$$

on $L^2(\mathbb{R}, dx)$ is the harmonic oscillator Hamiltonian.

The discovery by Hirzebruch of various special cases of what would become the general Atiyah-Singer index theorem introduced in various contexts some hyperbolic functions, or more properly, finite truncations of their Taylor series. When *Patodi* [274] extended his proof of GBC to a more complicated case of the index theorem—the Riemann-Roch-Hirzebruch theorem, he obtained the hyperbolic functions "by hand", and it appears that the difficulty in that case dissuaded him from trying to extend his approach to the more general index theorem (in addition, it was not known until later [18] that one could obtain the general index theorem from the index theorem for twisted Dirac complexes—in fact, it is only for these twisted Dirac complexes that the direct heat kernel proof has been made to work).

It was *Getzler* [129] who realized that one could easily get the hyperbolic functions from Mehler's formula. To some extent, this was implicit in the earlier work on supersymmetric proofs in that their proofs obtain hyperbolic functions via a calculation of a "boson determinant" in a path integral, and it is known that such determinants are also the key to one proof of Mehler's formula. In any event, this way of thinking of Mehler's formula emphasizes why it can be viewed as the third supersymmetric element of our proof: It is a kind of boson analog of the Berezin-Patodi formula. For this reason, we will give a proof of Mehler's formula, essentially due to *Simon* and *Hoegh-Krohn* [344], which emphasizes the fermion/boson analog. There are three shorter alternate proofs which we sketch afterwards. We begin by stating Mehler's formula:

Theorem 12.63. Let H be the operator (12.78). Then $\exp(-tH)$ has the integral kernel

$$Q_t(x,y) = \pi^{-1/2}(1 - e^{-4t})^{-1/2}\exp[-F_t(x,y)] \ , \qquad (12.79a)$$

$$F_t(x,y) := (1 - e^{-4t})^{-1}[\tfrac{1}{2}(1 + e^{-4t})(x^2 + y^2) - 2e^{-2t}xy] \ . \qquad (12.79b)$$

In particular, Theorem 12.63 says that

$$Q_t(0,0) = e^t[(2\pi)\sinh(2t)]^{-1/2} \ , \qquad (12.80)$$

which will be the source of hyperbolic functions. The proof we will give first exploits the boson analogs of the fermion creation/annihilation operators a^* and a. The reader already familiar with the creation operator analysis of the harmonic oscillator and not worried about domain subtleties may skip to the end of the proof of Proposition 12.65.

We define

$$b = \frac{1}{\sqrt{2}}\left(x + \frac{d}{dx}\right) , \qquad (12.81a)$$

$$b^\dagger = \frac{1}{\sqrt{2}}\left(x - \frac{d}{dx}\right) . \qquad (12.81b)$$

Unlike a and a^* which are bounded (in fact, essentially defined on a finite dimensional space), b and $b^\dagger$ are unbounded, so we need to specify domains. Initially, at least, we will define them as operators with domain $S(\mathbb{R})$, the Schwarz space. It is trivial to see that b and $b^\dagger$ map S to itself and that b and $b^\dagger$ are related by $(\varphi, b\psi) = (b^\dagger\varphi, \psi)$ for all $\varphi, \psi \in S$. Shortly, we will see that their closures are each other's adjoint. As operators on S, b and $b^\dagger$ obey

$$[b, b^\dagger] = 1 \ . \qquad (12.82)$$

It is precisely the similarity to the fermion relation $\{a, a^*\} = 1$ that leads one to think of b as the boson analog of a.

Proposition 12.64. (a) The operator H of (12.78) is essentially self-adjoint on S. Henceforth, H denotes the closure of (12.78) on S.

(b) $D(\bar{b}) = D(\overline{b^\dagger}) = Q(H)$

(c) $(\bar{b})^* = \overline{(b^\dagger)}$, $(b^\dagger)^* = \bar{b}$.

Proof. (a) This has many distinct proofs, of which we mention three! First, one can appeal to general theorems concerning positive potentials like Theorem 1.14. Second, one can exploit the known solutions of $Hu = Eu$. To be able to avoid parabolic cylinder functions, we take $E = 0$, so that the differential equation $Hu = 0$ has the solutions

$$\Omega_0(x) = \pi^{-1/4} e^{-x^2/2}$$
$$\psi_0(x) = x e^{x^2/2} \ . \tag{12.83}$$

The factor of $\pi^{-1/4}$ is put in (12.83) so that $\int_{-\infty}^{\infty} \Omega_0(x)^2\, dx = 1$. Thus, any L^2 distributional solution of $Hu = 0$ has $u = c\Omega_0$ and it follows that $\overline{[\operatorname{Ran} H \restriction S]} = \{u \mid u = c\Omega_0\}^{\perp}$. Thus, given $f \in D(H^*)$, we can find u in $D(\bar{H})$ and c so $H^*f = Hu + c\Omega_0$. Taking inner products with Ω_0 implies $c = 0$. Thus, $H^*(f - u) = 0$ so $f = u + c\Omega_0$, and thus $f \in D(\bar{H})$, i.e. $\bar{H}$ is self-adjoint. The third proof of (a) uses the N-representation of S ([292], Appendix to Sect. V.3) and is left as an exercise to the reader.

(b) We begin with the formula, easy to check on S,

$$H = 2b^\dagger b \ . \tag{12.84}$$

Thus,

$$2\|b\varphi\|^2 = (\varphi, H\varphi) \ ,$$

and by (12.82),

$$2\|b^\dagger\varphi\|^2 = (\varphi, (H + 2)\varphi) \ . \tag{12.85}$$

From this we conclude that on S the $Q(H)$-form norm is equivalent to the graph norms of $b^\dagger$ or of b. Since S is a core for H, it is a form core for H, and so the closure results follow.

(c) The proof is a variant of the second proof of (a). If $(b^\dagger)^*u = 0$, then, in distributional sense $u' = -xu$, so u' is locally L^2 and so u is continuous. Thus, u is C^1 in classical sense and thus $u = c\Omega_0$. Similarly, if $b^*u = 0$, one sees that $u = c \exp(x^2/2)$ and thus, since this u is not in L^2, we see that $b^*u = 0$ has no solution.

Next, note that by (12.85), $\|b^\dagger\varphi\| \geq \|\varphi\|$ and so $\operatorname{Ran} \overline{b^\dagger}$ is closed. Since $\operatorname{Ker}(b^\dagger)^* = \{c\Omega_0\}$, we see that $\operatorname{Ran} \overline{b^\dagger} = \{c\Omega_0\}^\perp$. Thus, given $\varphi \in D(b^*)$, we can find $f \in D(\overline{b^\dagger})$ so $b^*\varphi = \overline{b^\dagger} f + c\Omega_0$. Taking inner products with Ω_0, we see that $c = 0$ and then that $b^*(\varphi - f) = 0$. Thus, $\varphi = f$ lies in $D(\overline{(b^\dagger)})$. Thus, $\overline{b^\dagger} = b^*$. Taking adjoints $(b^\dagger)^* = \bar{b}$. □

Henceforth, we use b for the closure of the operator on S, and since $\overline{b^\dagger}$ is just b^*, we start using b^* rather than $b^\dagger$.

Proposition 12.65. Define

$$\Omega_n = (n!)^{-1/2} (b^*)^n \Omega_0 \ .$$

Then

(a) $\{\Omega_n\}_{n=0}^{\infty}$ are an orthonormal basis for $L^2(R, dx)$,
(b) $H\Omega_n = (2n)\Omega_n$,
(c) $\text{Spec}(H) = \{0, 2, 4, \ldots\}$.

Proof. On S, it is easy to see, on account of (12.82) and (12.84), that $[H, b^*] = 2b^*$ or $Hb^* = b^*(H + 2)$. Since $H\Omega_0 = 0$, one immediately obtains (b). This shows distinct Ω_j's are orthogonal. To check normalization, we compute, using (12.85), that

$$2\|b^*\Omega_{n-1}\|^2 = (\Omega_{n-1}, (H + 2)\Omega_{n-1}) = 2n\|\Omega_{n-1}\|^2 ;$$

so by induction $\Omega_n = n^{-1/2}b^*\Omega_{n-1}$ is normalized.

Next we check (c). We will exploit supersymmetry ideas! Form the Hilbert space $L^2 \oplus L^2$ and let

$$Q = \begin{pmatrix} 0 & b^* \\ b & 0 \end{pmatrix}$$

on the domain $Q(H) \oplus Q(H)$. By Proposition 12.64, Q is self-adjoint. Thus, Q commutes with spectral projections of

$$2Q^2 = \begin{pmatrix} H & 0 \\ 0 & H + 2 \end{pmatrix} .$$

It follows that

$$bE_{(\alpha,\beta)}(H) = E_{(\alpha,\beta)}(H + 2)b .$$

Since $\text{Ker}\, b = E_{\{0\}}(H)$, we see that if $E_{(\alpha,\beta)}(H + 2) = 0$ and $0 \notin (\alpha, \beta)$, then $E_{(\alpha,\beta)}(H) = 0$. Taking $(\alpha, \beta) = (0, 2)$ and using $H \geq 0$, we conclude that $E_{(0,2)}(H) = 0$. Then, by induction, we see that $E_{(2n, 2n+2)}(H) = 0$. This proves (c).

It remains to prove completeness or equivalently that each point $2n$ in spec(H) is an eigenvalue of multiplicity 1. For $n = 0$, we note that $H\varphi = 0$ implies $(\varphi, H\varphi) \equiv \|b\varphi\|^2 = 0$, which implies that $b\varphi = 0$. This O.D.E. we have already solved, and since its only solution is $\varphi = c\Omega_0$, zero has multiplicity 1. For general n, one proceeds by induction. If $H\varphi = 2n\varphi$, then the commutation relations show that $H(b\varphi) = (2n - 2)(b\varphi)$, so by induction, $b\varphi = c\Omega_{n-1}$. Thus, $\varphi = (2n)^{-1}H\varphi = (2n)^{-1}b^*(b\varphi) = c'b^*\Omega_{n-1} = c''\Omega_n$. □

Next, we define operators $:x^n:$ called the "Wick-ordered product" by writing $x = (\sqrt{2})^{-1}(b + b^*)$ expanding $(b + b^*)^n$ and rewriting any monomial $b^\# \ldots b^\#$ with jb's and $(n - j)b^\#$'s as $(b^*)^{n-j}b^j$, i.e. we define

$$:x^n: = 2^{-n/2} \sum_{j=0}^{n} \binom{n}{j} (b^*)^{n-j} b^j . \tag{12.86}$$

Proposition 12.66. (a) $:x^n:$ is the operator of multiplication by a function which we also denote by $:x^n:$,

(b) $:x^n:\Omega_0$ is an unnormalized eigenfunction of H with eigenvalue $2n$. $\|:x^n:\Omega_0\| = 2^{-n/2}(n!)^{1/2}$,

(c) $:x^n: = x:x^{n-1}: - \frac{1}{2}\frac{d}{dx}:x^{n-1}:$,

(d) $:x^n: = x^n +$ polynomial in x of degree $n-2$,

(e) $:x^n:$ are the orthogonal polynomials for the measure $\exp(-x^2)$.

Proof. (a) and (c): By the definition, we see that

$$:x^n: = 2^{1/2}(b^*:x^{n-1}: + :x^{n-1}:b)$$
$$= x:x^{n-1}: + [:x^{n-1}:, 2^{-1/2}b] .$$

By induction, this shows that $:x^n:$ is a multiplication operator, since $[f, b] = -2^{-1/2}f'$. Moreover, we immediately see the relation (c).

(d) This follows by induction from (c),

(b) and (e): By (12.86),

$$:x^n:\Omega_0 = 2^{-n/2}(b^*)^n\Omega_0 = 2^{-n/2}\sqrt{n!}\,\Omega_n.$$

Thus, (b) is proven, and the orthogonality of the Ω_n implies that, for $n \neq m$

$$\int :x^n::x^m:(\Omega_0^2)\,dx = 0 . \quad \square$$

Statement (e) tells us that the Wick-ordered polynomials $:x^n:$ are precisely the classical Hermite polynomials. From the relation (c), one easily writes down the first few $:x^n:$ explicitly:

$$:x^0: = 1, :x: = x, :x^2: = x^2 - \tfrac{1}{2}, :x^3: = x^3 - x, :x^4: = x^4 - \tfrac{5}{2}x^2 + \tfrac{1}{2}.$$

Here are some additional useful properties of the $:x^n:$

Proposition 12.67. (a) $(\Omega_0, :x^n:\Omega_0) = 0$,

(b) $\frac{d}{dx}:x^n: = n:x^{n-1}:$,

(c) $:x^n: = x:x^{n-1}: - \tfrac{1}{2}(n-1):x^{n-2}:$.

Proof. Statement (a) is just a special case of (e) of Proposition 12.66 and (c) follows from (c) of Proposition 12.66 and (b) of the present proposition. That means we need only prove (b). We do this by induction; $n = 1$ is obvious. By (c) of the last propositions and induction:

$$\frac{d}{dx}:x^n: = \frac{d}{dx}[x:x^{n-1}: - \tfrac{1}{2}(n-1):x^{n-2}:]$$
$$= :x^{n-1}: + (n-1)\left[x:x^{n-2}: - \frac{1}{2}\frac{d}{dx}:x^{n-2}:\right]$$
$$= n:x^{n-1}: . \quad \square$$

We note that (a), (b) inductively determine $:x^n:$.

Now we define the Wick-ordered exponential (a generating function for the Hermite functions) by

$$:\exp(tx): = \sum_{n=0}^{\infty} (n!)^{-1} t^n :x^n: \ . \tag{12.87}$$

Proposition 12.68. (a) For any t in C, (12.87) converges uniformly in x on compact subsets of $(-\infty, \infty)$.

(b) The convergence takes place locally in the C^1-topology.

(c) $:\exp(tx): = \exp(tx - \frac{1}{4}t^2)$.

Proof. We will show that $\sum_{n=0}^{\infty} (n!)^{-1} t^n :x^n: \Omega_0$ converges uniformly on all of R. A Sobolev estimate implies

$$\begin{aligned}\|f\|_\infty &\le c(\|f'\|_2 + \|f\|_2) \le c(\|H^{1/2}f\|_2 + \|f\|_2) \\ &\le c(\|Hf\|_2 + \|f\|_2) \ .\end{aligned}$$

Thus,

$$\|:x^n:\Omega_0\|_\infty \le c(2n+1)2^{-n/2}\sqrt{n!} \ ,$$

from which the required uniform convergence follows.

(b) By the same Sobolev estimate, one can show that

$$\|f'\|_\infty \le c(\|Hf\|_2 + \|f\|_2) \ .$$

(c) By (b), we can interchange the infinite sum and d/dx.
Since $(d/dx):x^n: = n:x^{n-1}:$ we see that $f(t,x) \equiv :\exp(tx):$ obeys

$$\frac{d}{dx} f(t,x) = tf(t,x) \ . \tag{12.88}$$

Since

$$:x^{n+1}: = \left(x - \frac{1}{2}\frac{d}{dx}\right):x^n:,$$

we see that

$$\begin{aligned}\frac{d}{dt} f(t,x) &= \left(x - \frac{1}{2}\frac{d}{dx}\right) f(t,x) \\ &= (x - \tfrac{1}{2}t) f(t,x)\end{aligned}$$

by (12.88). Since $f(0,x) = 1$, we obtain the desired result by integrating this differential equation. □

The last element we need in the proof of Theorem 12.63 is

Proposition 12.69. Let $\phi_s \equiv :\exp sx:\Omega_0$. Then

$$e^{-tH}\phi_s = \phi_{e^{-2t}s} .$$

Proof. The same proof used for uniform convergence (only easier) shows that (12.87) applied to Ω_0 converges in L^2. Since

$$e^{-tH}:x^n:\Omega_0 = e^{-2tn}:x^n:\Omega_0 ,$$

we see that

$$\begin{aligned} e^{-tH}(:\exp sx:\Omega_0) &= \sum_{n=0}^{\infty} (n!)^{-1}(se^{-2t})^n :x^n:\Omega_0 \\ &= :\exp(se^{-2t})x:\Omega_0 . \quad \square \end{aligned}$$

First Proof of Theorem 12.63. Since H has eigenvalues $2n$, $\exp(-tH)$ is Hilbert Schmidt, and thus it has an integral kernel, $Q_t(x, y)$. Let $G_t(x, y) = Q_t(x, y)\Omega_0(x)^{-1}\Omega_0(y)$ so that

$$[e^{-tH}(\eta\Omega_0)](x) = \Omega_0(x) \int G_t(x, y)\eta(y)\, dy .$$

By Proposition 12.68(c), Proposition 12.69 can be rephrased as saying that

$$\int G_t(x, y)e^{isy}\, dy = \exp[i(e^{-2t}s)x]\exp[-\tfrac{1}{4}s^2(1 - e^{-4t})] .$$

Thus, by taking inverse Fourier transforms (one can explicitly compute Fourier transforms of Gaussians), one obtains an explicit formula for G_t and so for Q_t. This formula yields (12.79). $\square$

We will *sketch* three alternate proofs of Theorem 12.63!

Second Proof of Theorem 12.63. This is a variant of a proof of *Doob* [89]; where he uses Stochastic processes, we exploit the Trotter product formula; see *Simon* [331] for more details. The integral kernels of $\exp[t(d^2/dx^2)]$ and $\exp(-tx^2)$ both are Gaussian. Since a partial integral of Gaussians is Gaussian, $\{\exp[tn^{-1}(d/dx^2)]\exp(-tx^2n^{-1})\}^n$ has a Gaussian integral kernel. Since a limit of Gaussians is Gaussian, the Trotter product formula implies that $Q_t(x, y)$ is Gaussian. Let $A_t(x, y) = \Omega_0(x)\Omega_0(y)Q_t(x, y)$ which is also Gaussian. Clearly

$$\int A_t(x, y)\, dx\, dy = (\Omega_0, e^{-tH}\Omega_0) = 1 ,$$

so $A_t(x, y)\, dx\, dy$ is a Gaussian probability measure. We can determine A_t by computing the 2×2 matrix α with

$$\alpha_{11} = \int x^2 A_t(x, y)\, dx\, dy; \quad \alpha_{22} = \int y^2 A_t(x, y)\, dx\, dy$$

$$\alpha_{21} = \alpha_{12} = \int xyA_t(x, y)\, dxy$$

and inverting it (see e.g. [331]). But

$$\alpha_{11} = \alpha_{22} = (x^2\Omega_0, e^{-tH}\Omega_0) = (x^2\Omega_0, \Omega_0) = \tfrac{1}{2}$$

$$\alpha_{12} = \alpha_{21} = (x\Omega_0, e^{-tH}x\Omega_0) = e^{-2t}(x^2\Omega_0, \Omega_0) = \tfrac{1}{2}e^{-2t} \ ,$$

since $x\Omega_0$ is an eigenfunction of H. Inverting α will introduce terms involving $\det\alpha = [1 - \exp(-4t)]/4$ which explains where the prefactors come from. Arithmetic yields (12.79). □

Third Proof of Theorem 12.63. This is the algebraist's proof. It depends on the fact that the Lie algebra generated by x^2 and p^2 is the Lie algebra $sl(2, R)$ associated to the group of 2×2 real unimodular matrices. This is no coincidence. $SL(2, R)$ is also the group of symplectic matrices, and so of linear canonical transformations in classical mechanics; the quantum theories associated with p^2, x^2, etc. are quantum linear canonical transformations. For each *linear* dynamics, quantum expectation values obey the classical equations of motion. For more on this point of view, see *Hagedorn, Loss* and *Slawny* [148].

To be explicit, define the operators $p = -id/dx$ and

$$A = \frac{i}{2}(xp + px) \quad B = ix^2 \quad C = ip^2 \ ,$$

so we have the commutation relations

$$[A, B] = 2B \quad [A, C] = -2C \quad [B, C] = -4A \ .$$

The two by two matrices

$$\tilde{A} = \begin{pmatrix} 1 & 0 \\ 0 & -1 \end{pmatrix} \quad \tilde{B} = \begin{pmatrix} 0 & 2 \\ 0 & 0 \end{pmatrix} \quad \tilde{C} = \begin{pmatrix} 0 & 0 \\ -2 & 0 \end{pmatrix} .$$

obey the same commutation relations. It is simple matrix theory to write $\exp[(t(\tilde{B} - \tilde{C})]$ in terms of $\exp[s(\tilde{A})]$, $\exp(u\tilde{B})$ and $\exp(v\tilde{C})$ and so since only algebra is involved, $\exp[t(B + C)]$ in terms of $\exp(sA)$, $\exp(uB)$ and $\exp(vC)$. Since these operators have explicit integral kernels, we obtain a formula for $Q_t(x, y)$ for t pure imaginary and small. By analytic continuation, one obtains (12.79). □

Last Proof of Theorem 12.63. This is the shortest, but probably least satisfactory proof. Q_t is explicitly given in (12.79). It is straightforward to check that $Q_t \to \delta(x - y)$ as $t \downarrow 0$ and the Q_t obeys the required differential equation. This can be used to show it must be the correct integral kernel. □

For the application in the next section, we will need to generalize Theorem 12.63 in several ways. The first step is to replace H by $-\Delta + \sum A_{ij}x^i x^j$. Then,

$$e^{-tH}(0, 0) = (4\pi t)^{-n/2}\{\det[(2t\sqrt{A})^{-1}\sinh(2t\sqrt{A})]\}^{-1/2} \ . \qquad (12.89)$$

For, by Theorem 12.63,

$$\exp\left[-t\left(-\frac{d^2}{dx^2}+x^2\right)\right](0,0)=[2\pi\sinh(2t)]^{-1/2} .$$

A simple scaling argument then shows that

$$\exp\left[-t\left(-\frac{d^2}{dx^2}+\omega^2x^2\right)\right](0,0)=[2\pi\omega^{-1}\sinh(2\omega t)]^{-1/2} .$$

By separation of variables

$$\exp\left[-t\left(-\Delta+\sum_{i=1}^{n}\omega_i x_i^2\right)\right](0,0)=(4\pi t)^{-n/2}\sum_{i=1}^{n}[(2t\omega_i)^{-1}\sinh(2t\omega_i)]^{-1/2} .$$

By diagonalizing A, this yields (12.89).

We need to be a little more precise about the analytic structure of $\det[(2t\sqrt{A})^{-1}\sinh(2t\sqrt{A})]$. We first note that since

$$x^{-1}\sinh x = 1+\tfrac{1}{6}x^2+\tfrac{1}{120}x^4\cdots$$

is an entire function of x^2, we have that

$$(2t\sqrt{A})^{-1}\sinh(2t\sqrt{A})=1+\tfrac{2}{3}t^2A+\tfrac{2}{15}t^4A^2\cdots \tag{12.90}$$

is an analytic function of A. Since

$$\det(1+y)=1+\mathrm{Tr}(y)-\tfrac{1}{2}[\mathrm{Tr}(y^2)-\mathrm{Tr}(y)^2]+\cdots \tag{12.91}$$

is analytic in y near $y=0$, we see that

Proposition 12.70. $\{\det[(2t\sqrt{A})^{-1}\sinh(2t\sqrt{A})]\}^{-1/2}$ is analytic about $t=0$ and it equals $1+t^2f_1(A)+t^4f_2(A)+\cdots+t^{2m}f_m(A)+\cdots$, where f_m is a polynomial in the matrix elements of A homogeneous of degree m.

For example, using (12.90, 91) and

$$(1+y)^{-1/2}=1-\tfrac{1}{2}y+\tfrac{3}{8}y^2+\cdots,$$

we see that

$$f_1(A)=-\tfrac{1}{3}\sum_i A_{ii} ,$$

$$f_2(A)=\tfrac{2}{45}\sum_{i,j}A_{ij}A_{ji}+\tfrac{1}{18}\left(\sum_i A_{ii}\right)^2 .$$

Thus, $\{\det[\ldots]\}$ is shorthand for some rather complicated functions of the A_{ij}.

Now we are interested in a situation where, in fact, the full Mehler formula may not hold. Rather than consider scalar valued functions on R^n, we consider vector valued functions (with values in V, a finite dimensional vector space) on R^n and the operator $-\Delta + \sum A_{ij}x^i x^j$ where the A_{ij} are operators in $\mathscr{L}(V)$ not necessarily commuting. What we require is that, in a sense we will make precise, Mehler's formula gives the correct leading behavior for $\exp(-tH)(0,0)$. To even explain what one means by this, we must restrict to a rather specific situation which is close to the situation relevant for the next section.

Thus, let $n = 2k$ and let V be a complex vector space of dimension 2^k. On V we let $\{\gamma_\mu\}_{\mu=1}^n$ be the Dirac matrices described in Sect. 12.2. We call this structure a spin space. As we proved there, $\mathscr{L}(V)$ is spanned by the set of $\gamma_A \equiv \gamma_{\mu_1} \cdots \gamma_{\mu_p}$ if $A = \{\mu_1, \dots, \mu_p\} \subset \{1, \dots, n\}$ with $\mu_1 < \cdots < \mu_p$. Let $\mathscr{L}_p(V)$ denote the span of those γ_A with $\#(A) = p$ so $\mathscr{L}(V) = \oplus_p \mathscr{L}_p(V)$. Given any operator $B \in \mathscr{L}(V)$, we write $B_p \in \mathscr{L}_p(V)$ for its components in this direct sum decomposition. We also define

$$\mathscr{L}_{\leq p}(V) \equiv \bigoplus_{l=0}^{p} \mathscr{L}_l(V) \ .$$

We want to consider the case where the A_{ij} lie in $\mathscr{L}_{\leq 4}(V)$. There are technical problems which arise if one considers $-\Delta + \sum A_{ij}x^i x^j$ because the x^i are unbounded. These must be irrelevant to the applications we have in mind where $x^i x^j$ occurs as a Taylor approximation to a bounded function, so we modify x^i to make it bounded.

Theorem 12.71. Fix $n = 2k$ and let V be the spin space of dimension 2^k. Let $\{A_{ij}\}_{1 \leq i,j \leq n}$ be a family of operators in $\mathscr{L}_{\leq 4}(V)$ with $A_{ij} = A_{ji}$. Let $g_i(x)$ be bounded C^∞ functions with bounded derivatives so that $g_i(x) = x^i$ in a neighborhood of $x = 0$. Let H be the operator $H = -\Delta + \sum_{i,j} A_{ij} g_i(x) g_j(x)$. Then, for $0 \leq p \leq n$

$$\lim_{t \downarrow 0} (4\pi)^{n/2} t^{1/2(n-p)} [\mathrm{e}^{-tH}(0,0)]_p = 0 \quad \text{if } p \not\equiv 0 \ (\mathrm{mod}\ 4) \ , \tag{12.92a}$$

$$= [f_{p/4}(A)]_p \quad \text{if } p \equiv 0 \ (\mathrm{mod}\ 4) \ , \tag{12.92b}$$

where $f_m(\cdot)$ are the polynomials in A_{ij} defined by Proposition 12.70.

Remarks. (1) The quantity $f_{p/4}$ is a polynomial of degree $p/4$. Since $\mathscr{L}_p(V) \times \mathscr{L}_q(V) \subset \mathscr{L}_{\leq (p+q)}$, $f_{p/4}(A)$ is a sum of objects in $\mathscr{L}_l$ with $l \leq p$. By $[f_{p/4}(A)]_p$, we mean its $\mathscr{L}_p$ component.

(2) If $B \in \mathscr{L}_p$ and $C \in \mathscr{L}_q$ and p is even, then $BC - CB$ lies in $\mathscr{L}_{\leq (p+q-2)}$ so $[f_{p/4}(A)]_p$ is independent of any questions of ordering of factors due to the noncommutativity of the A_{ij}. It is essentially for this reason that the noncommutative nature of the A_{ij} does not spoil the asymptotic result here.

Proof. Since $\sum A_{ij} g_i(x) g_j(x) := W$ is a bounded function, we can expand the semigroup via a DuHamel expansion with remainder and use the canonical order calculus to control the remainder:

$$e^{-tH}(0,0) = e^{t\Delta}(0,0) + \int_0^t e^{(t-s)\Delta}(0,x)W(x)e^{s\Delta}(x,0) + \cdots .$$

We see that there is a term of order $t^{-n/2}$ in $\mathscr{L}_0$. A term with m factors of W, by the commutator arguments, has an *extra* factor of order t^m from the W's in addition to the factor from the integral. Hence, there is a term of order $t^{-n/2}t^2$ in $\mathscr{L}_{\leq 4}$, of order $t^{-n/2}t^4$ in $\mathscr{L}_{\leq 8}$. From this, the result for $p \not\equiv 0 \bmod 4$ is immediate. To get the $p \equiv 0 \pmod 4$ result, we need only note that on the order of $\mathscr{L}_p$, multiplication is commutative and the terms will look exactly like they do in the case when A_{ij} are numbers where we know the limit is $f_{p/4}(A)$. □

There is a final complication which will lead us to the situation we require in the next section. Rather than want $-\Delta + x^2$, we want $(-\mathrm{i}\nabla - Ax)^2$ for a rather special class of A that make the cross terms irrelevant. As a warm-up, we consider the case of a constant magnetic field in three dimensions. In the azimuthal gauge the Hamiltonian has the form

$$H = (-\mathrm{i}\nabla - \tfrac{1}{2}\boldsymbol{B} \times \boldsymbol{x})^2 . \tag{12.93}$$

If $\boldsymbol{B} = (0,0,B)$ and $\rho = \sqrt{x^2 + y^2}$, then (12.93) becomes

$$H = -\Delta + \tfrac{1}{4}B^2\rho^2 - BL_z ,$$

where L_z generates rotations about the z axis. Notice that L_z commutes with both $-\Delta$ and ρ^2, so

$$e^{-tH} = e^{-t\tilde{H}}e^{+tBL_z}$$

with $\tilde{H} = -\Delta + \frac{1}{4}B^2\rho^2$. But $\exp(+tBL_z)\delta_0 = \delta_0$, so

$$e^{-tH}(0,0) = e^{-t\tilde{H}}(0,0) ;$$

that is, the cross terms are irrelevant as claimed.

We can therefore state the final theorem of this section which will be needed in the next:

Theorem 12.72. Fix $n = 2k$ and let V be the spin space of dimension 2^k. Let $\{C_{ij}\}_{1\leq i,j\leq n}$ be a family of operators in $\mathscr{L}_{\leq 2}(V)$ with $C_{ij} = -C_{ji}$, $C_{ij}^* = -C_{ij}$. Let $g_i(x)$ be bounded C^∞ functions with bounded derivatives so that $g_i(x) = x^i$ in a neighborhood of $x = 0$. Let H be the operator

$$H = -\sum_i \left[\partial_i - \sum_j C_{ij}g_j(x) \right]^2 .$$

Let $A_{ij} = \sum_k C_{ki}^* C_{kj}$. Then $\exp(-tH)$ obeys (12.92).

Proof. Let $\tilde{H} := -\Delta + \sum A_{ij}g_i(x)g_j(x) := -\Delta + W$ and let

$$B = \sum_{i,j} \partial_i [C_{ij} g_j(x)] + [C_{ij} g_j(x)] \partial_i \quad \text{and}$$

$$\tilde{B} = \sum_{i,j} \partial_i [C_{ij} x^j] + [x^j C_{ij}] \partial_i \ .$$

Notice that $\tilde{B}$ commutes with $-\Delta$. A simple calculation using the antisymmetry of the C's shows that *if* the C's commute, then $\tilde{B}$ commutes with W. In general it does not, but the same computations and $[\mathscr{L}_{\leq 2}, \mathscr{L}_{\leq 4}] \subseteq \mathscr{L}_{\leq 4}$ shows that

$$[\tilde{B}, W] \subseteq \mathscr{L}_{\leq 4} \ . \tag{12.94}$$

Now expand $\exp(-tH)$ in a DuHamel expansion with $B + W$ as the perturbation. We need only show that any term with a B does not contribute to the leading order of $[\exp(-tH)(0,0)]_p$. The kind of simple commutation argument we have used so far shows that the B terms contribute no more than the W terms: Explicitly, each two γ's in B introduce one order of perturbation theory (a factor of t), one ∂_j (a factor of $t^{-1/2}$) and one factor of x^j (a factor of $t^{1/2}$ in a naive commutator) so $p\gamma$'s yield $p/2$ factors of t.

We will obtain extra factors to make the terms vanish to leading order as follows: First, we can replace B by $\tilde{B}$. The errors are of order x^m for any m, and so by the commutator calculus, they contribute $O(t^m)$ for all m to the diagonal integral kernel. Since $\tilde{B}\delta_0 = 0$, we can, as usual, write the commutator of $\tilde{B}$ with a product of semigroups $\exp(+s_i\Delta)$, factors of $\tilde{B}$ and factors of W. $\tilde{B}$, by construction, commutes with $\exp(s_i\Delta)$ and $\tilde{B}$. In the commutator of $\tilde{B}$ and W, we preserve the total order of factors of x_i and ∂_i (namely $\tilde{B}$ has one x_i, one ∂_i for order $t^{1/2}t^{-1/2}$ and W has x_2^2 for order $(t^{1/2})^2$; the commutator has x_i^2 for order $(t^{1/2})^2$). However, by (12.94), we also decrease the number of γ's by 2 so that the term no longer contributes to the leading order. □

12.10 Introduction to the Index Theorem for Dirac Operators

In this final section, we will make some remarks on the proof of the index theorem for Dirac and twisted Dirac operators. The latter result is particularly significant because it was proven by Atiyah, Bott and Patodi [18] that it implies the general Atiyah-Singer index theorem. Indeed, when *Getzler* [135] and *Bismut* [49] claim to prove the general index theorem, they really mean that they have proven the special case of twisted Dirac operators. Since the argument in [18] is not analytic, this means that we are still not in possession of direct analytic proof of the general index theorem. It would be extremely interesting to find such a direct proof, perhaps using some variant of the ideas of this chapter.

Since we do not wish to make an already lengthy chapter even longer, we will not try to describe the notion of spin structures, their natural connections and the precise definition of a twisted Dirac operator. Rather, we will be very sketchy, emphasizing how to use the ideas of the last section to replace a scaling argument in Getzler's paper. The reader may consult the papers of Getzler and Bismut and the references therein for further details.

As usual, M will be a compact, Riemannian manifold of dimension $n = 2k$. Under suitable topological restriction on M, one can build over M a natural vector bundle of spinors of dimension 2^k. It is very important that this bundle is complex, that is, the fibers are complex vector spaces and not merely real ones; indeed, we have given the dimension as a complex space. There exist operators $\gamma_\mu(x)$ acting on the fiber of the bundle at the point x where μ runs from 1 to n. These operators generalize the Dirac operators described in Sect. 12.2. In particular, they obey the anticommutation relations

$$\{\gamma_\mu(x), \gamma_\nu(x)\} = g_{\mu\nu} \; . \tag{12.95}$$

One can define a covariant derivative ∇_μ acting on spinor valued functions. As for the case of form-valued functions and the covariant derivative defined in Sect. 12.1, ∇_μ has a term involving derivatives of coordinate functions and a term involving something like a Riemann-Christoffel symbol; explicitly,

$$\nabla_\mu = \partial_\mu - \sum_{\alpha\beta} \tilde{\Gamma}_\mu^{\alpha\beta} \gamma_\alpha \gamma_\beta \; , \tag{12.96}$$

where $\tilde{\Gamma}$ is a relative of the Riemann-Christoffel symbol (actually with an extra factor of $\frac{1}{2}$ since "a spinor is only half a vector").

On the bundle of spinors there is a basic operator which we will denote by η. It is the analog of what, in the usual four-dimensional Dirac theory, is γ_5. The operator η is just the product of the n γ's divided by $\sqrt{\det g}$. Thus, $\eta^2 = (-1)^{n(n-1)/2}$. If the γ's are suitably defined and the spinor bundle can be defined, then η is independent of coordinate system. It plays the role that $(-1)^p$ plays in the GBC theory presented above. The operator η anticommutes with each of the γ's and so with the Dirac operator

$$\not{D} = \sum_\mu \gamma^\mu \nabla_\mu \tag{12.97}$$

The covariant derivative is defined in such a way that $i\not{D}$ is self-adjoint.

If k is even, i.e. n is divisible by 4, then η has eigenvalues ± 1 and the space of spinors naturally decomposes into two subspaces on which η is 1, and on which η is -1. Since η anticommutes with $\not{D}$, the usual supersymmetry argument is applicable and all nonzero eigenvalues of $\not{L} \equiv -\not{D}^2$ are doubled. The index of $\not{D}$ is defined to be the difference of the dimensions of the kernel of Ł on even spinors and on odd spinors. The doubling of the eigenvalues therefore yields the familiar heat kernel formula

$$\operatorname{index}(\not{D}) = \operatorname{tr}(\eta e^{-t\not{L}}) \; . \tag{12.98}$$

In following the path we used in the GBC case, one must begin with the analog of the Weitzenböck formula. The commutation properties of the γ's easily provide a formula for $Ł$ as a sum of a Bochner-type Laplacian and a curvature term. The curvature term, not surprisingly, is a multiple of

$$\sum R_{\mu\nu\alpha\beta}\gamma^{\mu}\gamma^{\nu}\gamma^{\alpha}\gamma^{\beta} , \tag{12.99}$$

but the anticommutation properties of the γ's and the antisymmetry properties of the curvature tensor conspire to gobble up the γ's. That is, this curvature has no γ terms—rather, it acts as a scalar on spinors; in fact, the scalar is just a multiple of the scalar curvature. Since the direct curvature term has no γ's, it will contribute nothing to the supertrace of the heat kernel when one makes a DuHamel expansion of this kernel and so, for purposes of computing the index, one can safely ignore it.

The fact that this direct curvature is irrelevant represents the first major difference from the GBC situation. There is a second, even more important difference. The covariant derivative in each case represents a term of zeroth order in a's or γ's with one derivative and a term with no derivatives but two a's (one a and one a^*) or γ's. However, to contribute to the supertrace one needs $2n$ a's (n a's and n a^*'s) *but only* n γ's. If one repeats the analysis which, in the GBC case, showed that the Riemann-Christoffel symbol did not matter so long as it vanished to first order at the point of interest (normal coordinates), one finds that because fewer γ's are needed, one must keep the first-order terms in the Riemann-Christoffel $\tilde{\Gamma}$ symbol, but can drop the higher-order terms. Since these first-order terms are essentially the curvature, the canonical order calculus allows one to replace $Ł$ by

$$\tilde{Ł} = -\sum_{\mu} g^{\mu\nu}(\partial_{\mu} - \sum r_{\mu\kappa\alpha\beta}x^{\kappa}\gamma^{\alpha}\gamma^{\beta})(\partial_{\nu} - \sum r_{\nu\kappa\alpha\beta}x^{\kappa}\gamma^{\alpha}\gamma^{\beta}) , \tag{12.100}$$

where r is a close relative of the curvature. The reader will now recognize an operator precisely of the form studied in Theorem 12.72. Mehler's formula allows one to figure out a precise expression for the index of the Dirac operator. The object obtained by keeping the leading order of the various factors of numbers of γ's is essentially what is called the $\hat{A}$-genus.

A twisted Dirac operator is associated with a bundle which is locally just a tensor product of the spinor bundle and an auxiliary bundle. When one makes the detailed analysis, the curvature of the auxiliary bundle contributes to the direct curvature term so we obtain a term much like the $\exp(-tR_4)$ in the GBC analysis. This combination is essentially the so-called complete Chern class of the auxiliary bundle. What results is a formula for the index of twisted Dirac operators which combines the $\hat{A}$-genus of the manifold and the Chern class of the auxiliary bundle.

Bibliography

1. Agmon, S.: Lower bounds for solutions of Schrödinger equations. J. Anal. Math. **23**, 1–25 (1970)
2. Agmon, S.: Spectral properties of Schrödinger operators and scattering theory. Ann. Norm. Sup. Pisa C1 Sci II **2**, 151–218 (1975)
3. Agmon, S.: *Lectures on Exponential Decay of Solutions of Second-Order Elliptic Equations* (Princeton University Press, Princeton 1982)
4. Agmon, S.: "On Positive Solutions of Elliptic Equations with Periodic Coefficients in R^N, Spectral Results and Extensions to Elliptic Operators on Riemannian Manifolds, in *Differential Equations*, ed. by I. Knowles, R. Lewis (North Holland, Amsterdam 1984)
5. Aguilar, J., Combes, J.M.: A class of analytic perturbations for one-body Schrödinger Hamiltonians. Commun. Math. Phys. **22**, 269–279 (1971)
6. Aharonov, Y., Casher, A.: Ground state of spin-1/2 charged particle in a two-dimensional magnetic field. Phys. Rev. **A19**, 2461–2462 (1979)
7. Aizenman, M., Simon, B.: Brownian motion and Harnack's inequality for Schrödinger operators. Commun. Pure Appl. Math. **35**, 209–273 (1982)
8. Albeverio, S., Ferreira, L., Streit, L. *Resonances-Models and Phenomena*, Lecture Notes in Physics, Vol. 211, (Springer, Berlin, Heidelberg 1984)
9. Allegretto, W.: On the equivalence of two types of oscillation for elliptic operators. Pac. J. Math. **55**, 319–328 (1974)
10. Allegretto, W.: Spectral estimates and oscillation of singular differential operators. Proc. Am. Math. Soc. **73**, 51 (1979)
11. Allegretto, W.: Positive solutions and spectral properties of second order elliptic operators. Pac. J. Math. **92**, 15–25 (1981)
12. Alvarez-Gaumez, L.: Supersymmetry and the Atiyah-Singer index theorem: Commun. Math. Phys. **80**, 161–173 (1983)
13. Amrein, W.O.: *Nonrelativistic Quantum Dynamics*, Mathematical Physics Studies 2 (Reidel, Dordrecht 1981)
14. Amrein, W., Georgescu, V.: On the characterization of bound states and scattering states in quantum mechanics. Helv. Phys. Acta **46**, 635–658 (1973)
15. Andre, G., Aubry, S.: Analyticity breaking and Anderson localization in incommensurate lattices. Ann. Isr. Phys. Soc. **3**, 133 (1980)
16. Antonets, M.A., Zhislin, G.M., Shereshevskii, J.A.: On the discrete spectrum of the Hamiltonian of an N-particle quantum system. Theor. Math. Phys. **16**, 800–808 (1972)
17. Atiyah, M., Bott, R.: A Lefschetz fixed point formula for elliptic complexes, I. Ann. Math. **86**, 374–407 (1967)
18. Atiyah, M., Bott, R., Patodi, V.K.: On the heat equation and the index theorem. Invent. Math. **19**, 279–330 (1973)
19. Avron, J.: Bender-Wu formulas for the Zeeman effect in hydrogen. Ann. Phys. **131**, 73 (1981)
20. Avron, J. et al.: Bender-Wu formula, the SO (4, 2) dynamical group and the Zeeman effect in hydrogen, Phys. Rev. Lett. **43**, 691–693 (1979)
21. Avron, J.E., Herbst, I.: Spectral and scattering theory of Schrodinger operators related to the Stark effect. Commun. Math. Phys. **52**, 239–254 (1977)
22. Avron, J., Herbst, I., Simon, B.: Schrödinger operators with magnetic fields, I. General interactions. Duke Math. J. **45**, 847–883 (1978)
23. Avron, J., Herbst, I., Simon, B.: Schrödinger operators with magnetic fields, II. Separation of the center mass in homogeneous magnetic fields Ann. Phys. **114**, 431–451 (1978)

24. Avron, J., Herbst, I., Simon, N.: Strongly bound states of hydrogen in intense magnetic field. Phys. Rev. **A20**, 2287–2296 (1979)
25. Avron, J., Herbst, I., Simon, B.: Schrödinger operators with magnetic fields, III. Atoms in magnetic fields. Commun. Math. Phys. **79**, 529–572 (1981)
26. Avron, J., Seiler, R.: Paramagnetism for nonrelativistic electrons and Euclidean massless Dirac particles. Phys. Rev. Lett. **42**, 931–934 (1979)
27. Avron, J., Simon, B.: A counterexample to the paramagnetic conjecture. Phys. Lett. **A75**, 41 (1979)
28. Avron, J., Simon, B.: Transient and recurrent spectrum. J. Funct. Anal. **43**, 1–31 (1981)
29. Avron, J., Simon, B.: Almost periodic Schrödinger operators, I. Limit periodic potentials. Commun. Math. Phys. **82**, 101–120 (1982)
30. Avron, J., Simon, B.: Singular continuous spectrum for a class of almost periodic Jacobi matrices. Bull. AMS **6**, 81–85 (1982)
31. Avron, J., Simon, B.: Almost periodic Schrödinger operators, II. The integrated density of states. Duke Math. J. **50**, 369–391 (1983)
32. Avron, J., Tomares, Y.: unpublished
33. Babbitt, D., Balslev, E.: A characterization of dilation-analytic potentials and vectors. J. Funct. Anal. **18**, 1–14 (1975)
34. Babbitt, D., Balslev, E.: Local distortion technique and unitarity of the S-matrix for the 2-body problem. J. Math. Anal. Appl. **54**, 316–347 (1976)
35. Balslev, E.: Absence of positive eigenvalues of Schrödinger operators. Arch. Ration. Mech. Anal. **59**, 343–357 (1973)
36. Balslev, E.: Analytic scattering theory of two-body Schrödinger operators. J. Funct. Anal. **29**, 375–396 (1978)
37. Balslev, E.: Resonances in three-body scattering theory. Adv. Appl. Math. **5**, 260–285 (1984)
38. Balslev, E.: Local spectral deformation techniques for Schrödinger operators. J. Funct. Anal. **58**, 79–105 (1984)
39. Balslev, E., Combes, J.M.: Spectral properties of many-body Schrödinger operators with dilation-analytic interactions. Commun. Math. Phys. **22**, 280–294 (1971)
40. Bellissard, J., Lima, R., Scoppola, E.: Localization in ν-dimensional incommensurate structures. Commun. Math. Phys. **88**, 465–477 (1983)
41. Bellissard, J., Lima, R., Testard, D.: A metal-insulator transition for the almost Mathieu model. Commun. Math. Phys. **88**, 207–234 (1983)
42. Bellissard, J., Simon, B.: Cantor spectrum for the almost Mathieu equation. J. Funct. Anal. **48**, 408–419 (1982)
43. Ben-Artzi, M.: An application of asymptotic techniques to certain problems of spectral and scattering theory of Stark-like Hamiltonians. Trans. AMS **278**, 817–839 (1983)
44. Benderskii, M., Pasteur, L.: On the spectrum of the one-dimensional Schrödinger equation with a random potentia. Math. USSR Sb. **11**, 245 (1970)
45. Bentosela, F. et al.: Schrödinger operators with an electric field and random or deterministic potentials. Commun. Math. Phys. **88**, 387–397 (1983)
46. Berezanskii, J.: Expansion in eigenfunctions of selfadjoint operators. Transl. Math. Monogr. **17**, (1968) Am. Math. Soc. (1968)
47. Berezin, F.A.: *The Method of Second Quantization* (Academic, New York 1966)
48. Berthier, A.M.: *Spectral Theory and Wave Operators for the Schrödinger Equation*, Pitman Research Notes in Mathematics (Pitman, London 1982)
49. Bismut, J.M.: The Atiyah-Singer theorems: A probabilistic approach, I. The index theorem. J. Funct. Anal. **57**, 56–99 (1984)
50. Bochner, S.: Vector fields and Ricci curvature. Bull. Am. Math. Soc. **52**, 776–797 (1946)
51. Bochner, S., Yano, K.: Tensor fields in non-symmetric connections. Ann. Math. **56**, 504 (1952)
52. Boothby, W.: *An Introduction to Differentiable Manifolds and Riemannian Geometry* (Academic, New York 1975)
53. Breiman, L.: *Probability* (Addison-Wesley, London 1968)

54. Brezis, H., Kato, T.: Remarks on the Schrödinger operator with singular complex potentials. J. Math. Pures Appl. **58**, 137–151 (1979)
55. Brossard, J.: Noyeau de Poisson pour l'Operateur de Schrödinger. *Duke Math. J.* **52**, 199–210 (1985)
56. Brossard, J.: Perturbations aleatoires de potentiels periodique (preprint)
57. Callias, C.: Axial anomalies and index theorems on open spaces. Commun. Math. Phys. **62**, 213 (1978)
58. Carmona, R.: Regularity properties of Schrödinger and Dirichlet semigroups. J. Funct. Anal. **17**, 227–237 (1974)
59. Carmona, R.: Exponential localization in one dimensional disordered systems. Duke Math. J. **49**, 191 (1982)
60. Carmona, R.: One-dimensional Schrödinger operators with random or deterministic potentials: New spectral types. J. Funct. Anal. **51**, 229 (1983)
61. Carmona, R.: "Random Schrödinger Operators," in *École d'Été de Probabilités de Saint-Flour XIV-1984*, ed. by P.L. Henneguin, Lecture Notes in Mathematics, Vol. 1180 (Springer, Berlin, Heidelberg, New York 1986)
62. Chavel, I.: *Eigenvalues in Riemannian Geometry* (Academic, New York 1984)
63. Chernoff, P.: Essential self-adjointness of powers of generators of hyperbolic equations. J. Funct. Anal. **12**, 401–414 (1973)
64. Chulaevsky, V.: On perturbations of a Schrödinger operator with periodic potential. Russ. Math. Surv. **36**, 143 (1981)
65. Combes, J.M.: Seminar on Spectral and Scattering Theory, ed. by S. Kuroda, RIMS Pub. **242**, 22–38 (1975)
66. Combes, J.M., Duclos, P., Seiler, R.: Krein's formula and one-dimensional multiple well. J. Funct. Anal. **52**, 257–301 (1983)
67. Constantinescu, F., Fröhlich, J., Spencer, T.: Analyticity of the density of states and replica method for random Schrödinger operators on a lattice. J. Stat. Phys. **34**, 571–596 (1984)
68. Craig, W.: Pure point spectrum for discrete almost periodic Schrödinger operators. Commun. Math. Phys. **88**, 113–131 (1983)
69. Craig, W., Simon, B.: Log Hölder continuity of the integrated density of states for stochastic Jacobi matrices. Commun. Math. Phys. **90**, 207–218 (1983)
70. Craig, W., Simon, B.: Subharmonicity of the Lyaponov index. Duke Math. J. **50**, 551–560 (1983)
71. Cycon, H.L.: On the stability of selfadjointness of Schrödinger operators under positive perturbations. Proc. R. Soc. Edinburgh **86A**, 165–173 (1980)
72. Cycon, H.L.: Resonances defined by modified dilations. Helv. Phys. Acta **58**, 969–981 (1985)
73. Cycon, H., Leinfelder, H., Simon, B.: unpublished
74. Cycon, H.L., Perry, P.A.: Local time-decay of high energy scattering states for the Schrödinger equation. Math. Z. **188**, 125–142 (1984)
75. Davies, E.B.: On Enss' approach to scattering theory. Duke Math. J. **47**, 171–185 (1980)
76. Davies, E.B.: The twisting trick for double well Hamiltonians. Commun. Math. Phys. **85**, 471–479 (1982)
77. Deift, P.: Applications of a commutation formula. Duke Math. J. **45**, 267–310 (1978)
78. Deift, P., Simon, B.: A time-dependent approach to the completeness of multiparticle quantum systems. Commun. Pure Appl. Math. **30**, 573–583 (1977)
79. Deift, P., Simon, B.: Almost periodic Schrödinger operators, III. The absolutely continuous spectrum in one dimension. Commun. Math. Phys. **90**, 389–411 (1983)
80. Delyon, F.: Apparition of purely singular continuous spectrum in a class of random Schrödinger operators. J. Stat. Phys. **40**, 621 (1985)
81. Delyon, F., Kunz, H., Souillard, B.: One dimensional wave equations in disordered media. J. Phys. A. **16**, 25 (1983)
82. Delyon, F., Levy, Y., Souillard, B.: Anderson localization for multidimensional systems at large disorder or large energy. Commun. Math. Phys. **100**, 463 (1985)

83. Delyon, F., Levy, Y., Souillard, B.: An approach "a la Borland" to Anderson localization in multidimensional disordered systems, Phys. Rev. Lett. **55**, 618 (1985)
84. Delyon, F., Simon, B. Souillard, B.: From power law localized to extended states in a disordered system. Ann. Inst. Henri Poincare **42**, 283 (1985)
85. Delyon, F., Souillard, B.: The rotation number for finite difference operators and its properties. Commun. Math. Phys. **89**, 415 (1983)
86. Delyon, F., Souillard, B.: Remark on the continuity of the density of states of ergodic finite difference operators. Commun. Math. Phys. **94**, 289 (1984)
87. Dinaburg, E., Sinai, Ya.: The one-dimensional Schrödinger equation with a quasi-periodic potential. Funct. Anal. Appl. **9**, 279–289 (1975)
88. Dodds, P., Fremlin, D.H.: Compact operators in Banach lattices. Isr. J. Math. **34**, 287–320 (1979)
89. Doob, J.: *Stochastic Processes* (Wiley, New York 1953)
90. Eastham, M.S.P., Kalf, H.: *Schrödinger-type Operators with Continuous Spectra*, Pitman Research Notes in Mathematics (Pitman, London 1982)
91. Efimov, V.N.: Energy levels arising from resonant two-body forces in a three-body system. Phys. Lett. **B33**, 563–654 (1970)
92. Efimov, V.N.: Weakly-bound states of three resonantly-interacting particles. Sov. J. Nucl. Phys. **12**, 589–595 (1971)
93. Elworthy, K.D.: *Stochastic Differential Equations on Manifolds*, London Math. Soc. Lect. Notes No. 70 (Cambridge University Press, Cambridge 1982)
94. Enss, V.: A note on Hunziker's theorem. Commun. Math. Phys. **52**, 233–238 (1977)
95. Enss, V.: Asymptotic completeness for quantum-mechanical potential scattering, I. Short-range potentials. Commun. Math. Phys. **61**, 285–291 (1978)
96. Enss, V.: Asymptotic completeness for quantum-mechanical potential scattering, II. Singular and long-range potentials. Ann. Phys. **119**, 117–132 (1979)
97. Enss, V.: Two-cluster scattering of N charged particles. Commun. Math. Phys. **65**, 151–165 (1979)
98. Enss, V.: Asymptotic observables on scattering states. Commun. Math. Phys. **89**, 245–268 (1983)
99. Enss, V.: "Completeness of Three-Body Quantum Scattering", in *Dynamics and Processes* ed. by P. Blanchard, L. Streit, Lecture Notes in Mathematics, Vol. 1031 (Springer, Berlin, Heidelberg, New York 1983) pp. 62–88
100. Enss, V.: "Geometrical Methods in Spectral and Scattering Theory of Schrödinger Operators, in *Rigorous Atomic and Molecular Physics* ed. by G. Velo, A.S. Wightman (Plenum, New York 1981) pp. 7–69
101. Enss, V.: Propagation properties of quantum scattering states. J. Funct. Anal. **52**, 219–251 (1983)
102. Enss, V.: "Scattering and Spectral Theory for Three Particle Systems," in *Differential Equations* ed. by I. Knowles, R. Lewis (North Holland, Amsterdam 1984) pp. 173–204
103. Enss, V.: Topics in scattering theory for multiparticle quantum mechanics, a progress report Physica **124A**, 269–292 (1984)
104. Enss, V.: "Topics in Scattering Theory for Multiparticle Quantum Mechanics" in *Mathematical Physics VII*, ed. by W. Britten, K. Gustafson, W. Wyss (North Holland, Amsterdam 1984) pp. 269–292
105. Enss, V.: "Quantum Scattering Theory of Two- and Three-Body Systems with Potentials of Short and Long Range", in *Schrödinger Operators*, ed. by S. Graffi, Lecture Notes in Mathematics, Vol. 1159 (Springer, Berlin, Heidelberg, New York 1985)
106. Faddeev, L.: *Mathematical Aspects of the Three Body Problem in Quantum Scattering Theory* (Steklov Institute 1963)
107. Faris, W.G.: *Self-Adjoint Operators*, Lecture Notes in Mathematics, Vol. 433 (Springer, Berlin, Heidelberg, New York 1975)
108. Feller, W.: *An Introduction to Probability Theory and Its Applications*, Vol. I (Wiley, New York 1957)
109. Feller, W.: *An Introduction to Probability Theory and Its Applications*, Vol. II (Wiley, New York 1966)

110. Figotin, A., Pastur, L.: An exactly solvable model of a multidimensional incommensurate structure. Commun. Math. Phys. **95**, 401–425 (1984)
111. Fishman, S., Grempel, D., Prange, R.: Chaos, quantum recurrences, and Anderson localization. Phys. Rev. Lett. **49**, 509 (1982)
112. Fishman, S., Grempel, D., Prange, R.: Localization in a *d*-dimensional incommensurate structure. Phys. Rev. **B29**, 4272–4276 (1984)
113. Friedan, D., Windy, P.: Supersymmetric derivation of the Atiyah-Singer index theorem and the Chiral anomaly. Nucl. Phys. **B253**, 395–416 (1984)
114. Froese, R., Herbst, I.: Exponential bounds and absence of positive eigenvalues for N-body Schrödinger operators. Commun. Math. Phys. **87**, 429–447 (1982)
115. Froese, R., Herbst, I.: A new proof of the Mourre estimate. Duke Math. J. **49**, 4 (1982)
116. Froese, R. et al.: L^2-exponential lower bounds to solutions of the Schrödinger equation. Commun. Math. Phys. **87**, 265–286 (1982)
117. Fröhlich, J.: Applications of commutator theorems to the integration of representations of Lie algebras and commutation relations. Commun. Math. Phys. **54**, 135–150 (1977)
118. Fröhlich, J. et al.: Anderson localization for large disorder or low energy. Commun. Math. Phys. **101**, 21 (1985)
119. Fröhlich, J., Spencer, T.: Absence of diffusion in the Anderson tight binding model for large disorder or low energy. Commun. Math. Phys. **88**, 151–189 (1983)
120. Fujiwara, D.: A construction of the fundamental solution for the Schrödinger equation. J. Anal. Math. **35**, 41–96 (1979)
121. Fujiwara, D.: On a nature of convergence of some Feynman path integrals, I. Proc. Jpn. Acad. Ser. A Math. Sci. **55**, 195–200 (1979)
122. Fujiwara, D.: Remarks on convergence of Feynman path integrals. Duke Math. J. **47**, 559–600 (1980)
123. Fukushima, M.: On the spectral distribution of a disordered system and the range of a random walk. Osaka J. Math. **11**, 73–85 (1974)
124. Fukushima, M., Nagai, H., Nakao, S.: On an asymptotic property of spectra of a random difference operator. Proc. Jpn. Acad. **51**, 100–102 (1975)
125. Furstenberg, H.: Non-commuting random products. Trans. Am. Math. Soc. **108**, 377–428 (1963)
126. Furstenberg, H., Kesten, H.: Products of random matrices. Ann. Math. Stat. **31**, 457–469 (1960)
127. Gagliardo, E.: Ulteriori proprieta di alcune classi di funzioni au piu variabili. Ric. Mat. Napoli **8**, 24–51 (1959)
128. Gallot, S., Meyer, D.: Sur la premiere valeur proper de *p*-spectra pour les varietes a operateur de courbure positive. C. R. Acad. Sci. (Paris) **276A**, 1619–1621 (1973)
129. Getzler, E.: Pseudo-differential operators on supermanifolds and the Atiyah-Singer index theorem. Commun. Math. Phys. **82**, 163–178 (1983)
130. Getzler, E.: A short proof of the Atiyah-Singer index theorems. Topology **25**, 111–117 (1986)
131. Gilbarg, D., Trudinger, N.: *Elliptic Partial Differential Equations of Second Order* (Springer, Berlin, Heidelberg, New York 1977)
132. Gilkey, P.: Curvature and the eigenvalues of the Laplacian. Adv. Math. **10**, 344–382 (1973)
133. Gilkey, P.: *The Index Theorem and the Heat Equation* (Publish or Perish, Berkeley 1974)
134. Gilkey, P.: "Lefschetz Fixed Point Formulas and the Heat Equation", in *Partial Differential Equations and Geometry* ed. by C. Byrnes (Dekker, New York 1979) pp. 97–147
135. Ginibre, J.: La methode "dependent du temps" dans le probleme de la completude asymptotique. Publ. IRMA, Strasbourg, RCP **25 29**, 1–66 (1981)
136. Ginibre, J., Moulin, M.: Hilbert space approach to the quantum mechanical three-body problem. Ann. Inst. Henri Poincare **A21**, 97–145 (1974)
137. Goldsheid, I.: Talk at 1984 Tashkent Conf. on Information Theory
138. Goldsheid, I., Molchanov, S., Pastur, L.: A pure point spectrum of the stochastic one-dimensional Schrödinger equation. Funct. Anal. Appl. **11**, 1–10 (1977)
139. Gordon, A.: Usp. Mat. Nauk. **31**, 257 (1976) [in Russian].
140. Graffi, S., Grecchi, V.: Resonances in Stark effect and perturbation theory. Commun. Math. Phys. **62**, 83–96 (1978)

141. Graffi, S., Yajima, K.: Exterior scaling and the AC-Stark effect in a Coulomb field. Commun. Math. Phys. **89**, 277–301 (1983)
142. Grempel, D.R., Fishman, S., Prange, R.E.: Localization in an incommensurate potential: An exactly solvable model. Phys. Rev. Lett. **49**, 833 (1982)
143. Grossman, A.: "Momentum-like Constants of Motion," in *Statistical Mechanics and Field Theory* ed. by R.N. Sen, C. Weil (Halsted, Jerusalem 1972)
144. Grumm, H.: Quantum mechanics in a magnetic field. Acta Phys. Austriaca **53**, 113 (1981)
145. Guerra, F., Rosen, L., Simon, B.: The vacuum energy for $P(\varphi)_2$: Infinite volume limit and coupling constant dependence. Commun. Math. Phys. **29**, 233–247 (1973)
146. Le Guillou, J., Zinn-Justin, J.: The hydrogen atom in strong magnetic fields: Summation of the weak field series expansion. Ann. Phys. **147**, 57–84 (1983)
147. Hagedorn, G.A.: A link between scattering resonances and dilation-analytic resonances in few-body quantum mechanics. Commun. Math. Phys. **65**, 181–201 (1979)
148. Hagedorn, G., Loss, M., Slawny, J.: Non-stochasticity of time-dependent quadratic Hamiltonians and the spectra of canonical transformations. J. Phys. A. **19**, 521–531 (1986)
149. Harrell, E.: On the rate of asymptotic eigenvalue degeneracy. Commun. Math. Phys. **60**, 73–95 (1978)
150. Harrell, E., Simon, B.: The mathematical theory of resonances which have exponentially small widths. Duke Math. J. **47**, 845–902 (1980)
151. Hartman, P.: *Ordinary differential equations* (Wiley & Sons, New York 1964)
152. Helffer, B., Sjostrand, J.: Multiple wells in the semi-classical limit, I. Commun. Partial Diff. Equ. **9**, 337–408 (1984)
153. Herbert, D., Jones, R.: Localized states in disordered systems. J. Phys. **C4**, 1145–1161 (1971)
154. Herbst, I.: Unitary equivalences of Stark-Hamiltonians. Math. Z. **155**, 55–70 (1977)
155. Herbst, I.: Dilation analyticity in constant electric field, I. The two-body problem. Commun. Math. Phys. **64**, 279–298 (1979)
156. Herbst, I.: Exponential decay in the Stark effect. Commun. Math. Phys. **75**, 197–205 (1980)
157. Herbst, I.: Schrödinger Operators with External Homogeneous Electric and Magnetic Fields", in *Rigorous Atomic and Molecular Physics*, Proc. Fourth Int. School of Math. Phys., Erice, ed. by G. Velo, A.S. Wightman (Plenum, New York 1981)
158. Herbst, I., Simon, B.: Some remarkable examples in eigenvalue perturbation theory. Phys. Lett. **B78**, 304–306 (1978)
159. Herbst, I., Simon, B.: Stark effect revisited. Phys. Rev. Lett. **47**, 67–69 (1978)
160. Herbst, I., Simon, B.: Dilation analyticity in constant electric field, II. The N-body problem, Borel summability. Commun. Math. Phys. **80**, 181–216 (1981)
161. Herbst, I., Sloan, A.: Perturbations of translation invariant positivity preserving semigroups in $L^2(R^N)$. Trans. Am. Math. Soc. **236** 325–360 (1978)
162. Herman, M.: Une methode pour minorer les exposants de Lyapunov et quelques exemples montrant le caractere local d'un theoreme d'Arnold et de Moser sur le tore en dimension 2. Commun. Math. Helv. **58**, 453–502 (1983)
163. Hess, H., Schrader, R., Uhlenbrock, D.A.: Domination of semigroups and generalization of Kato's inequality. Duke Math. J. **44**, 893–904 (1977)
164. Hess, H., Schrader, R., Uhlenbrock, D.A.: Kato's inequality and the spectral distribution of Laplacians on compact Riemannian manifolds. J. Diff. Geo. **15**, 27–37 (1980)
165. Hirschfelder, J., Johnson, B., Yang, K.H.: Interaction of atoms, molecules, and ions with constant electric and magnetic fields. Rev. Mod. Phys. **55**, 109–153 (1983)
166. Ho, Y.K.: The method of complex coordinate rotation and its application to atomic collision processes. Phys. Rep. **99**, 1–68 (1983)
167. Holden, H., Martinelli, F.: On the absence of diffusion for a Schrödinger operator on $L^2(R^\nu)$ with a random potential. Commun. Math. Phys. **93**, 197 (1984)
168. Howland, J.S.: Stationary scattering theory for time-dependent Hamiltonians. Math. Ann. **207**, 315–335 (1974)

169. Howland, J.S.: Scattering theory for Hamiltonians periodic in time. Indiana. Math. J. **28**, 471–494 (1979)
170. Howland, J.S.: "Two Problems with Time Dependent Hamiltonians", in *Mathematical Methods and Applications of Scattering Theory*, ed. by J.A. Santo, Lecture Notes in Physics, Vol. 730 (Springer, Berlin, Heidelberg, New York 1980)
171. Hunziker, W.: On the spectra of Schrödinger multiparticle Hamiltonians. Helv. Phys. Acta **39**, 451–462 (1966)
172. Hunziker, W.: Schrödinger Operators with Electric or Magnetic Fields, in *Mathematical Problems in Theoretical Physics*, Lecture Notes in Physics, Vol. 116, ed. by K. Osterwalder (Springer, Berlin, Heidelberg, New York 1979) pp. 25–44
173. Hunziker, W.: Private communication
174. Hunziker, W.: Distortion analyticity and molecular resonance curves. (in preparation)
175. Ikebe, T.: Eigenfunction expansions associated with the Schrödinger operators and their application to scattering theory. Arch. Ration. Mech. Anal. **5**, 1–34 (1960)
176. Ishii, K.: Localization of eigenstates and transport phenomena in one-dimensional disordered systems. Suppl. Prog. Theor. Phys. **53**, 77 (1973)
177. Ismagilov, R.: Conditions for the semiboundedness and discreteness of the spectrum for one-dimensional differential equations. Sov. Math. Dokl. **2**, 1137–1140 (1961)
178. Isozaki, H., Kitada, H.: Micro-local resolvent estimates for two-body Schrödinger operators. J. Funct. Anal. **57** 270–300 (1984)
179. Isozaki, H., Kitada, H.: Modified wave operators with time-independent modifiers. J. Fac. Sci. Univ. Tokyo, Sec. IA **32**, 77–104 (1985)
180. Isozaki, H., Kitada, H.: A remark on the micro-local resolvent estimates for two-body Schrödinger operators. RIMS, Kyoto Univ., (1986)
181. Iwatsuka, A.: Examples of absolutely continuous Schrödinger operators in magnetic fields, Publ. RIMS, Kyoto Univ. **21**, 385–401 (1985)
182. Jauch, J.M.: *Foundations of quantum mechanics* (Addison-Wesley, London 1973)
183. Jensen, A.: Local distortion technique, resonances and poles of the S-matrix. J. Math. Anal. Appl. **59**, 505–513 (1977)
184. Jensen, A., Mourre, E., Perry, P.A.: Multiple commutator estimates and resolvent smoothness in quantum scattering theory. Ann. Inst. Henri Poincare, Physique Theorique **41**, 207–225 (1984)
185. Jensen, A., Perry, P.A.: Commutator methods and Besov space estimates for Schrödinger operators. J. Opt. Theory **14**, 181–188 (1985)
186. Johnson, R., Moser, J.: The rotation number for almost periodic potentials. Commun. Math. Phys. **84**, 403 (1982)
187. Jörgens, K., Weidmann, J.: *Spectral Properties of Hamiltonian Operators*, Lecture Notes in Mathematics, Vol. 373 (Springer, Berlin, Heidelberg, New York 1973)
188. Kac, M.: Can you hear the shape of a drum? Am. Math. Mon. **73**, 1–23 (1966)
189. Kalf, H.: The quantum mechanical virial theorem and the absence of positive energy bound states of Schrödinger operators. Isr. J. Math. **20**, 57–69 (1975)
190. Kato, T.: Growth properties of solutions of the reduced wave equation with a variable coefficient. Commun. Pure Appl. Math. **12**, 403–425 (1959)
191. Kato, T.: Smooth operators and commutators. Stud. Math. **31**, 535–546 (1968)
192. Kato, T.: Linear evolution equations of "hyperbolic" type. J. Fac. Sci. Univ. Tokyo **17**, 241–258 (1970)
193. Kato, T.: Schrödinger operators with singular potentials. Isr. J. Math. **13**, 135–148 (1972)
194. Kato, T.: Two space scattering theory with applications to many body problems. J. Fac. Sci. Univ. Tokyo **24**, 503–514 (1977)
195. Kato, T.: Remarks on Schrödinger operators with vector potentials. Int. Equ. Operator Theory **1**, 103–113 (1978)
196. Kato, T.: *Perturbation Theory for Linear Operators, 2nd ed.* (Springer, Berlin, Heidelberg, New York 1980)

197. Kato, T.: "Remarks on the Essential Self-Adjointness and Related Problems for Differential Operators, in *Spectral Theory of Differential Operators*, ed. by J.W. Knowles, R.T. Lewis (North Holland, Amsterdam 1981)
198. Kato, T., Masuda, K.: Trotter's product formula for nonlinear semigroups generated by subdifferentials of convex functionals. J. Math. Soc. Jpn. **30**, 169–178 (1978)
199. Katznelson, Y.: *An Introduction To Harmonic Analysis* (Dover, New York 1976)
200. Kingman, J.: Subadditive ergodic theory. Ann. Prob. **1**, 883–909 (1973)
201. Kirsch, W.: On a class of random Schrödinger operators. Adv. Appl. Math. **6**, 177–187 (1985)
202. Kirsch, W.: "Random Schrödinger operators and the Density of States, in *Stochastic Aspects of Classical and Quantum Systems*, ed. by S. Albeverio, P. Combe, M. Sirugue-Collin, Lecture Notes in Mathematics, Vol. 1109 (Springer, Berlin, Heidelberg, New York 1985) pp. 68–102
203. Kirsch, W., Kotani, S., Simon, B.: Absence of absolutely continuous spectrum for some one-dimensional random but deterministic Schrödinger operators. Ann. Inst. Henri Poincare **42**, 383 (1985)
204. Kirsch, W., Martinelli, F.: On the spectrum of Schrödinger operators with a random potential. Commun. Math. Phys. **85**, 329 (1982)
205. Kirsch, W., Martinelli, F.: On the ergodic properties of the spectrum of general random operators. J. Reine Angew. Math. **334**, 141–156 (1982)
206. Kirsch, W., Martinelli, F.: On the density of states of Schrödinger operators with a random potential. J. Phys. **A15**, 2139–2156 (1982)
207. Kirsch, W., Martinelli, F.: Large deviations and Lifshitz singularity of the integrated density of states of random Hamiltonians. Commun. Math. Phys. **89**, 27–40 (1983)
208. Kirsch, W., Simon, B.: Lifshitz tails for periodic plus random potentials. J. Stat. Phys. **42**, 799–808 (1986)
209. Kitada, H.: On a construction of the fundamental solution for Schrödinger equations. J. Fac. Sci. Univ. Tokyo IA **27**, 193–226 (1980)
210. Kitada, H.: Time-decay of the high energy part of the solution for a Schrödinger equation, J. Fac. Sci. Univ. Tokyo Sect. IA 109–146 (1984)
211. Kitada, H., Kumanoago, H.: A family of Fourier integral operators and the fundamental solution for a Schrödinger equation. Osaka J. Math. 291–360 (1981)
212. Kitada, H., Yajima, K.: A scattering theory for time dependent long-range potentials. Duke Math. J. **49**, 341–376 (1982)
213. Kitada, H., Yajima, K.: Remarks on our paper "A scattering theory for time-dependent long-range potentials". Duke Math. J. **50**, 1005–1016 (1983)
214. Klaus, M.: On $-d^2/dx^2 + V$ where V has infinitely many "bumps". Ann. Inst. Henri Poincare **38**, 7–13 (1983)
215. Kotani, S.: "Ljaponov Indices Determine Absolutely Continuous Spectra of Stationary Random One-dimensional Schrödinger Operators", in *Stochastic Analysis*, ed. by K. Ito (North Holland, Amsterdam 1984) pp. 225–248
216. Kotani, S.: Support theorems for random Schrödinger operators. Commun. Math. Phys. **97**, 443–452 (1985)
217. Kotani, S.: Lyaponov exponents and spectra for one-dimensional random Schrödinger operators, to appear in *Proc. AMS Conference on Random Matrices and Their Applications*
218. Kotani, S.: Lyaponov exponent and point spectrum for one-dimensional Schrödinger operators (in preparation)
219. Kovalenko, V., Semenov, Yu.: Some problems on expansions in generalized eigenfunctions of the Schrödinger operator with strongly singular potentials. Russ. Math. Surv. **33**, 119–157 (1978)
220. Kramers, H.A.: Nonrelativistic quantum electric dynamics and correspondence principle. Rapp. 8e Cons. Solvay 241–268 (1948)
221. Kunz, H., Souillard, B.: Sur le spectre des operateurs aux differences finies aleatoires. Commun. Math. Phys. **78**, 201–246 (1980)
222. Kunz, H., Souillard, B.: The localization transition on the Bethe lattice. J. Phys. (Paris) Lett. **44**, (1983) L411; paper in prep.

223. Kuroda, S.: Scattering theory for differential operators, I. II. J. Math. Soc. Jpn. **25**, 75–104, 222–234 (1973)
224. Landau, L.D., Lifschitz, E.M.: *Quantum Mechanics–Nonrelativistic Theory*, 2nd ed. (Pergamon, New York 1965)
225. Lavine, R.: Absolute continuity of Hamiltonian operators with repulsive potentials. Proc. Am. Math. Soc. **22**, 55–60 (1969)
226. Lavine, R.: Absolute continuity of positive spectrum for Schrödinger operators with long-range potentials. J. Funct. Anal. **12**, 30–54 (1973)
227. Leinfelder, H.: Habilitations summary, Universität Bavreuth (1981)
228. Leinfelder, H.: Gauge invariance of Schrödinger operators and related spectral properties. J. Opt. Theory, 163–179 (1983)
229. Leinfelder, H., Simader, C.: Schrödinger operators with singular magnetic vector potentials. Math. Z. **176**, 1–19 (1981)
230. Lichnerowicz, A.: *Geometry of Transformation Groups* (Dunod, Paris 1958)
231. Lieb, E.: Atomic and molecular ionization. Phys. Rev. Lett. **52**, 315 (1984)
232. Lieb, E.: Bound on the maximum negative ionization of atoms and molecules. Phys. Rev. **A29**, 3018–3028 (1984)
233. Lieb, E. et al.: Asymptotic neutrality of large Z ions. Phys. Rev. Lett. **52** 994 (1984)
234. Lifshitz, I.: Energy spectrum structure and quantum states of disordered condensed systems. Sov. Phys. Usp. **7**, 549 (1965)
235. Lloyd, P.: Exactly solvable model of electronic states in a three-dimensional disordered Hamiltonian: Non-existence of localized states. J. Phys. **C2**, 1717–1725 (1969)
236. Loss, M., Sigal, I.M.: The three body problem with threshold singularities. ETH preprint
237. Lowdin, P. (ed.): Sanibel workshop on complex scaling. Int. J. Quantum Chem. **14** (1978)
238. Malliavin, P.: *Geometrie Differentiables Stochastique* (University of Montreal Press, Montreal 1978)
239. Martinelli, F., Scoppola, E.: A remark on the absence of absolutely continuous spectrum for the Anderson model at large disorder or low energy. Commun. Math. Phys. **97**, 465 (1985)
240. McKean, H., Singer, I.M.: Curvature and eigenvalues of the Laplacian. J. Diff. Geo. **1**, 43–69 (1967)
241. Meyer, D.: Sur les varietes Riemanniennes a operateur de courbure positive. C. R. Acad. Sci. Paris **272A**, 482–485 (1971)
242. Meyer, S.: Riemannian manifolds with positive mean curvature. Duke Math. J. **8**, 401–404 (1941)
243. Miller, K.: "*Bound States of Quantum Mechanical Particles in Magnetic Fields*"; Ph.D. Thesis, Princeton University (1982)
244. Miller, K., Simon, B.: Quantum magnetic Hamiltonians with remarkable spectral properties. Phys. Rev. Lett. **44**, 1706–1707 (1980)
245. Milnor, J.: *Morse Theory*, Ann. Math. Stud. No. 51 (Princeton University Press, Princeton 1963)
246. Minakshisundaram, S., Pieijel, A.: Some properties of the eigenfunctions of the Laplace operator on Riemannian manifolds. Can. J. Math. **1**, 242–256 (1949)
247. Minami, N.: An extension of Kotani's theorem to random generalized Sturm-Liouville operators. Commun. Math. Phys. **103**, 387–402 (1986)
248. Molchanov, S.: The structure of eigenfunctions of one-dimensional unordered structures. Math. USSR Izv. **12**, 69 (1978)
249. Molchanov, S.: The local structure of the spectrum of the one-dimensional Schrödinger operator. Commun. Math. Phys. **78**, 429 (1981)
250. Morgan, J.D.: Schrödinger operators whose potentials have separated singularities. J. Opt. Theory **1**, 109–115 (1979)
251. Morgan, J.D., Simon, B.: On the asymptotics of Born-Oppenheimer curves for large nuclear separation. Int. J. Quantum Chem. **17**, 1143–1166 (1980)
252. Moser, J.: An example of a Schrödinger equation with an almost periodic potential and nowhere dense spectrum. Commun. Math. Helv. **56**, 198 (1981)

253. Moser, J., Pöschel, J.: An extension of a result by Dinaburg and Sinai on quasi-periodic potentials. Commun. Math. Helv. **59**, 39–85 (1984)
254. Moss, W., Piepenbrink, J.: Positive solutions of elliptic equations. Pac. J. Math. **75**, 219–226 (1978)
255. Mourre, E.: Link between the geometrical and the spectral transformation approaches in scattering theory. Commun. Math. Phys. **68**, 91–94 (1979)
256. Mourre, E.: Absence of singular continuous spectrum for certain selfadjoint operators. Commun. Math. Phys. **78**, 391–408 (1981)
257. Mourre, E.: Operateurs conjugues et proprietes de propagations Commun. Math. Phys. **91**, 279 (1983)
258. Muthuramalingam, Pl.: A note on time dependent scattering theory for $P_1^2 - P_2^2 + (1 + |Q|)^{-1-\varepsilon}$ and $P_1 P_2 + (1 + |Q|)^{-1-\varepsilon}$ on $L^2(R^2)$. Math. Z **188**, 339–348 (1985)
259. Muthuramalingam, Pl.: A time dependent scattering theory for a class of simply characteristic operators with short range local potentials. J. London Math. Soc. **32**, 259–264 (1985)
260. Nachbin, L.: *The Haar Integral* (Van Nostrand, Princeton 1964)
261. Naimark, M.A.: *Linear Differential Operators* (Akademie, Berlin 1963)
262. Nakao, S.: On the spectral distribution for the Schrödinger operator with a random potential. Jpn J. Math. **3**, 111 (1977)
263. Nelson, E.: *Tensor Analysis* (Princeton University, Princeton 1967)
264. Nirenberg, L.: An extended interpolation inequality. Sc. Normale Superiore Pisa, Ser. 3, 20, 733–737 (1966)
265. Obata, M.: Certain conditions for a Riemannian manifold to be isometric with a sphere. J. Math. Soc. Jpn. **14**, 333–340 (1962)
266. Osborn, T., Fujiwara, D.: Time evolution kernels: uniform asymptotic expansions. J. Math. Phys. **24**, 1093 (1983)
267. Osceledec, V.: A multiplicative ergodic theorem. Lyapunov characteristic numbers for dynamical systems. Trans. Moscow Math. Soc. **19**, 197–231 (1968)
268. Ovchinnikov, Yu. N., Sigal, I.M.: Number of bound states of three-body systems and Efimov's effect. Ann. Phys. **123**, 274–295 (1979)
269. Pastur, L.: Spectra of random selfadjoint operators. Usp. Mat. Nauk. **28**, 3 (1973)
270. Pastur, L.: Behaviour of some Wiener integrals as $t \to \infty$ and the density of states of Schrödinger equation with a random potential. *Theor. Math. Phys.* **32**, 88–95 (1977)
271. Pastur, L.: Spectral properties of disordered systems in one-body approximation. Commun. Math. Phys. **75**, 179 (1980)
272. Pastur, L., Figotin, A.: Localization in an incommensurate potential: An exactly solvable multidimensional model. JETP Lett. **37**, 686–688 (1983)
273. Patodi, V.K.: Curvature and eigenforms of the Laplace operator. J. Diff. Geo. **5**, 251–283 (1971)
274. Patodi, V.: An analytic proof of the Riemann-Roch-Hirzebruch theorem for Kaehler manifolds. J. Diff. Geo. **5**, 251–283 (1971)
275. Pearson, D.: Singular continuous measures in scattering theory. Commun. Math. Phys. **60**, 13 (1978)
276. Pearson, D.: "Pathological Spectral Properties", in *Mathematical Problems in Theoretical Physics* (Springer, Berlin, Heidelberg 1980) pp. 49–51
277. Perry, P.: Mellin transforms and scattering theory, I. Short-range potentials. Duke Math. J. **47**, 187–193 (1980)
278. Perry, P.: Propagation of states in dilation analytic potentials and asymptotic completeness. Commun. Math. Phys. **81**, 243–259 (1981)
279. Perry, P.: *Scattering Theory by the Enss Method* (Harwood Academic London 1983)
280. Perry, P.: Exponential bounds and semi-finiteness of point spectrum for N-body Schrödinger operators. Commun. Math. Phys. **92**, 481–483 (1984)
281. Perry, P., Sigal, I.M. Simon, B.: Spectral analysis of N-body Schrödinger operators. Ann. Math. **114**, 519–567 (1981)
282. Persson, A.: Bounds for the discrete part of the spectrum of a semi-bounded Schrödinger operator. Math. Scand. **8**, 143–153 (1960)

283. Piepenbrink, J.: Nonoscillatory elliptic equations. J. Diff. Eq. **15**, 541–550 (1974)
284. Piepenbrink, J.: A conjecture of Glazman. J. Diff. Eq. **24**, 173–177 (1977)
285. Pitt, L.D.: A compactness condition for linear operators on function spaces. J. Opt. Theory **1**, 49–54 (1979)
286. Pöschel, J.: Examples of discrete Schrödinger operators with pure point spectrum. Commun. Math. Phys. **88**, 447–463 (1983)
287. Povzner, A.: On eigenfunction expansions in terms of scattering solutions. Dokl. Akad. Nank **104**, (1955)
288. Prange, R., Grempel, D., Fishman, S.: A solvable model of quantum motion in an incommensurate potential. Phys. Rev. B **29**, 6500–6512 (1984)
289. Putnam, C.R.: *Commutation Properties of Hilbert Space Operators and Related Topics* (Springer, Berlin, Heidelberg 1967)
290. Radin, C., Simon, B.: Invariant domains for the time-dependent Schrödinger equation. J. Diff. Eq. **29**, 289–296 (1978)
291. Raghunathan, M.: A proof of Oseledec's multiplicative ergodic theorem. Isr. J. Math. **32**, 356–362 (1979)
292. Reed, M., Simon, B.: *Methods of Modern Mathematical Physics, I. Functional Analysis* (*rev. ed.*) (Academic, London 1980)
293. Reed, M., Simon, B.: *Methods of Modern Mathematical Physics, II. Fourier Analysis, Self-Adjointness* (Academic, London 1975)
294. Reed, M., Simon, B.: *Methods of Modern Mathematical Physics, III. Scattering Theory* (Academic, London, 1979)
295. Reed, M., Simon, B.: *Methods of Modern Mathematical Physics, IV. Analysis of Operators* (Academic, London 1978)
296. Reinhardt, W.P.: Dilation analyticity and the radius of convergence of the $1/Z$-perturbation expansion: Comment on a conjecture of Stillinger. Phys. Rev. **A15**, 802–805 (1977)
297. Reinhardt, W.P.: Complex coordinates in the theory of atomic and molecular structure and dynamics. Ann. Rev. Phys. Chem. **33**, 223–255 (1982)
298. Romerio, M., Wreszinski, W.: On the Lifshitz singularity and the tailing in the density of states for random lattice systems. J. Stat. Phys. **21**, 169 (1979)
299. Ruelle, D.: On the asymptotic condition in quantum field theory. Helv. Phys. Acta **35**, 147–163 (1962)
300. Ruelle, D.: A remark on bound states in potential scattering theory. Nuovo Cimento **61A**, 655–662 (1969)
301. Ruelle, D.: Ergodic theory of differentiable dynamical systems. Publ. IHES **50**, 275 (1979)
302. Ruskai, M.B.: Absence of discrete spectrum in highly negative ions. Commun. Math. Phys. **82**, 457–469 (1982)
303. Ruskai, M.B.: Absence of discrete spectrum in highly negative ions, II. Commun. Math. Phys. **85**, 325–327 (1982)
304. Russmann, H.: On the one-dimensional Schrödinger equation with a quasi-periodic potential. Ann. N.Y. Acad. Sci. **357**, 90–107 (1980)
305. Saito, Y.: *Spectral Representation for Schrödinger Operators With Long-Range Potentials, Lecture Notes in Mathematics*, Vol. 727 (Springer, Berlin, Heidelberg, New York 1979)
306. Sarnak, P.: Spectral behavior of quasi periodic potentials. Commun. Math. Phys. **84**, 377–401 (1982)
307. Sch'nol, I.: On the behavior of the Schrödinger equation. Mat. Sb. **42**, 273–286 (1957) [in Russian]
308. Schechter, M.: *Spectra of Partial Differential Operators* (North Holland, Amsterdam 1971)
309. Sigal, I.M.: *Mathematical Foundations of Quantum Scattering Theory for Multiparticle Systems*, Mem. Am. Math. Soc., Vol. 209 (1978)
310. Sigal, I.M.: Geometric methods in the quantum many-body problem. Nonexistence of very negative ions. Commun. Math. Phys. **85**, 309–324 (1982)
311. Sigal, I.M.: Complex transformation method and resonances in one-body quantum systems. Ann. Inst. Henri Poincare **41**, 103–114 (1984)

312. Sigal, I.M.: Geometric parametrices and the many-body Birman-Schwinger principle. Duke Math. J. **50**, 517–537 (1983)
313. Sigal, I.M.: How many electrons can a nucleus bind?. Ann. Phys. **157**, 307–320 (1984)
314. Sigal, I., Soffer, A.: *N*-particle scattering problem: Asymptotic completeness for short range systems. Preprint
315. Sigalov, A.G., Sigal, I.M.: Description of the spectrum of the energy operator of quantum mechanical systems that is invariant with respect to permutations of identical particles. *Theor. Math. Phys.* **5**, 990–1005 (1970)
316. Silverman, J.: Generalized Euler transformation for summing strongly divergent Rayleigh-Schrödinger perturbation series: The Zeeman effect. Phys. Rev. **A28**, 498–501 (1983)
317. Simader, C.: Selbstadjungiertheitsprobleme und spektraltheoretische Fragen für Schrödingeroperatoren. *Recent Trends Math.* **50**, 270–276 (1982)
318. Simon, B.: On positive eigenvalues of one-body Schrödinger operators. Commun. Pure Appl. Math. **12**, 531–538 (1969)
319. Simon, B.: Quadratic form techniques and the Balslev-Combes theorem. Commun. Math. Phys. **27**, 1–9 (1972)
320. Simon, B.: Resonances in *N*-body quantum systems with dilation analytic potentials and the foundations of time-dependent perturbation theory. Ann. Math. **97**, 247–274 (1973)
321. Simon, B.: Absence of positive eigenvalues in a class of multiparticle quantum systems. Math. Ann. **207**, 133–138 (1974)
322. Simon, B.: An abstract Kato inequality for generators of positivity preserving semigroups. Ind. Math. J. **26**, 1067–1073 (1977)
323. Simon, B.: Geometric methods in multiparticle quantum systems. Commun. Math. Phys. **55**, 259–274 (1977)
324. Simon, B.: Resonances and complex scaling: A rigorous overview. J. Quant. Chem. **14**, 529–542 (1978)
325. Simon, B.: Classical boundary conditions as a tool in quantum physics. Adv. Math. **30**, 268–281 (1978)
326. Simon, B.: Phase space analysis of simple scattering systems: Extensions of some work of Enss. Duke Math. J. **46**, 119–168 (1979)
327. Simon, B.: Kato's inequality and the comparison of semigroups. J. Funct. Anal. **32**, 97–101 (1979)
328. Simon, B.: The definition of molecular resonance curves by the method of exterior complex scaling. Phys. Lett. **A71**, 211–214 (1979)
329. Simon, B.: Maximal and minimal Schrödinger forms. J. Opt. Theory **1**, 37–47 (1979)
330. Simon, B.: *Trace Ideals and Their Applications* (Cambridge University Press, Cambridge 1979)
331. Simon, B.: *Functional Integration and Quantum Physics* (Academic, London 1979)
332. Simon, B.: Spectrum and continuum eigenfunctions of Schrödinger operators. J. Funct. Anal. **42**, 66–83 (1981)
333. Simon, B.: Some Jacobi matrices with decaying potential and dense point spectrum. Commun. Math. Phys. **87**, 253–258 (1982)
334. Simon, B.: Schrödinger semigroups. Bull. Am. Math. Soc. **7**, 447–526 (1982)
335. Simon, B.: Almost periodic Schrödinger operators: A review. Adv. Appl. Math. **3**, 463–490 (1982)
336. Simon, B.: Kotani theory for one dimensional stochastic Jacobi matrices. Commun. Math. Phys. **89**, 227 (1983)
337. Simon, B.: Semiclassical analysis of low lying eigenvalues, I. Nondegenerate minima: Asymptotic expansions. Ann. Inst. Henri Poincare **38**, 295–307 (1983)
338. Simon, B.: The equality of the density of states in a wide class of tight binding Lorentzian models. Phys. Rev. **B27**, 3859–3860 (1983)
339. Simon, B.: Semiclassical analysis of low lying eigenvalues, II. Tunneling. Ann. Math. **120**, 89–118 (1984)
340. Simon, B.: Almost periodic Schrödinger operators, IV. The Maryland model. Ann. Phys. **159**, 157–183 (1985)

341. Simon, B.: Lifschitz tails for the Anderson model. J. Stat. Phys. **38**, 65–76 (1985)
342. Simon, B.: Regularity properties of the density of states: A review. *Proc. IMA Conf., 1985* (to appear)
343. Simon, B.: Localization in general one-dimensional random systems, I. The discrete case. Commun. Math. Phys. **102**, 327 (1985)
344. Simon, B., Hoegh-Krohn, R.: Hypercontractive semigroups and two-dimensional self-coupled Bose fields. J. Funct. Anal. **9**, 121–180 (1972)
345. Simon, B., Wolff, T.: Singular continuous spectrum under rank one perturbations and localization for random Hamiltonians, Comm. Pure Appl. Math. **39**, 75–90 (1986)
346. Spain, B.: *Tensor Calculus* (Oliver and Boyd, London 1960)
347. Spencer, T.: The Schrödinger equation with a random potential—a mathematical review, in *Critical Phenomena, Random Systems, Gauge Theories*, Les Houches, XLIII, ed. by K. Osterwalder and R. Stora
348. Spivak, M.: *A Comprehensive Introduction to Differential Geometry, Vol. I* (Publish or Perish, Berkeley 1970)
349. Spivak, M.: *A Comprehensive Introduction to Differential Geometry, Vol. V* (Publish or Perish, Berkeley 1979)
350. Stein, E.M.: *Singular Integrals and Differentiability Properties of Functions* (Princeton University, Princeton 1970)
351. Strichartz, R.: Analysis of the Laplacian on the complete Riemannian manifold. J. Funct. Anal. **52**, 48–79 (1983)
352. Stummel, F.: Singulare elliptische Differentialoperatoren in Hilbertschen Raumen. Math. Ann. **132**, 150–176 (1956)
353. Taylor, M.: *Pseudodifferential Operators* (Princeton University, Princeton 1981)
354. Thomas, L.: Asymptotic completeness in two and three particle quantum mechanical scattering. Ann. Phys. **90**, 127–165 (1975)
355. Thouless, D.J.: Electrons in disordered systems and the theory of localization. Phys. Rep. **13**, 93 (1974)
356. Tip, A.: Atoms in circularly polarized fields: The dilation-analytic approach. J. Phys. A Math. Gen. Phys. **16**, 3237–3259 (1983)
357. Titchmarsh E.C.: *Eigenfunction Expansions Associated with Second Order Ordinary Differential Equations, Part I* (Oxford University, Oxford 1962)
358. Treves, F.: *Topological Vector Spaces, Distributions and Kernels* (Academic, New York 1967)
359. Van Winter, C.: Theory of finite systems of particles, I. Mat. Fys. Skr. Danske Vid. Selsk. **1**, 1–60 (1964)
360. Vick, J.: *Homology Theory; An Introduction to Algebraic Topology* (Academic, New York 1973)
361. Von Neumann, J.: *Mathematische Grundlagen der Quantenmechanik* (Springer, Berlin 1932)
362. Von Neumann, J., Wigner, E.: Über merkwürdige diskrete Eigenwerte. Z. Phys. **30**, 465–467 (1929)
363. Vugalter, S.A., Zhislin, G.M.: Finiteness of the discrete spectrum of many-particle Hamiltonians in symmetry spaces. Theor. Math. Phys. **32**, 602–614 (1977)
364. Walter, J.: Absolute continuity of the essential spectrum of $-d^2/dt^2 + q(t)$ without monotony of q. Math. Z. **129**, 83–94 (1972)
365. Wegner, F.: Statistics of disorder chains. Z. Phys. **B22**, 273–277 (1975)
366. Wegner, F.: Bounds on the density of states in disordered systems. Z. Phys. **B44**, 9–15 (1981)
367. Weidmann, J.: The virial theorem and its application to the spectral theory of Schrödinger operators. Bull. Am. Math. Soc. **73**, (1967)
368. Windy, P.: Supersymmetric quantum mechanics and the Atiyah-Singer index theorem, CERN preprint TH3758
369. Witten, E.: Constraints on supersymmetry breaking. Nucl. Phys. **B202**, 253–316 (1982)
370. Witten, E.: Supersymmetry and Morse theory. J. Diff. Geo. **17**, 661–692 (1982)
371. Wüst, R.: Generalizations of Rellich's theorem on perturbation of essentially selfadjoint operators. Math. Z. **119**, 276–280 (1971)

372. Yafaev, D.R.: On the theory of the discrete spectrum of the three-particle Schrödinger operator. Math. USSR Sb. **23**, 535–559 (1974)
373. Yafaev, D.R.: On singular spectrum of a three body system. Mat. Sb. **106**, 622 (1978) [in Russian]
374. Yafaev, D.R.: Time dependent scattering theory for elliptic differential operators. Notes Sci. Semin. LOMI **115**, 285–300 (1982)
375. Yafaev, D.R.: On the asymptotic completeness for multidimensional time-dependent Schrödinger equation. Mat. Sb. **118**, 262–279 (1982)
376. Yajima, K.: Scattering theory for Schrödinger operators with potentials periodic in time. J. Math. Soc. Jpn. **29**, 729–743 (1977)
377. Yajima, K.: Spectral and scattering theory for Schrödinger operators with Stark-effect. J. Fac. Sci. Tokyo **26**, 377–390 (1979)
378. Yajima, K.: Spectral and scattering theory for Schrödinger operators with Stark-effect. II, J. Fac. Sci. Tokyo **28**, 1–15 (1981)
379. Yajima, K.: Resonances for the AC-Stark effect. Commun. Math. Phys. **78**, 331–352 (1982)
380. Zelditch, S.: Reconstruction of singularities for solutions of Schrödinger's equation. Commun. Math. Phys. **90**, 1–26 (1983)
381. Zhao, Z.: Uniform boundedness of conditional gauge and Schrödinger equations. Commun. Math. Phys. **93**, 19–31 (1984)
382. Zhislin, G.M.: Discussion of the spectrum of Schrödinger operator for systems of many particles. Tr. Mosk. Mat. Obs. **9**, 81–128 (1960)
383. Zhislin, G.M.: On the finiteness of the discrete spectrum of the energy operator of negative atomic and molecular ions. Theor. Math. Phys. **7**, 571–578 (1971)
384. Zhislin, G.M.: Finiteness of the discrete spectrum in the quantum N-particle problem. Theor. Math. Phys. **21**, 971–980 (1974)

List of Symbols

Subject Index